2008 China Interior Design Annual

2008中国室内设计年鉴（1）

《中国室内设计年鉴》编委会 编

辽宁科学技术出版社

扫盲过后的反思

叶铮 Ye zheng
上海泓叶室内装饰有限公司总经理
中国建筑学会室内设计分会理事
高级室内建筑师
中国饭店协会设计委员会常务理事
中国建筑装饰协会设计委员会委员
上海应用技术学院副教授

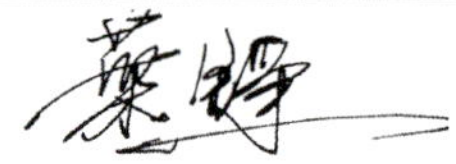

回顾这些年的历程，中国室内设计就如快速腾飞的中国经济，已到达了一个新的起点。站在这个时代的制高点上，我们不难发现，如今国内一批顶尖的室内设计师，能充满自信地面对来自于世界上任何一位优秀的设计师，并以自身的视角来解读审视着他们的成果。对于中国的室内设计师来讲，这同时也是一个能同过去、现在、未来进行对话的“起点”。

然而，这一起点对整体中国室内设计的发展而言，我仍视作一场专业“扫盲”运动的结束。整整一代人，尤其是那些出生在20世纪70年代前的当代中国室内设计师们，他（她）们花了近20年的时间，为今天的设计发展基本完成了规模空前的“扫盲”运动，这无论在技术上、理念上、手法上还是功能上、规模上、专业知识等问题的多样性与复杂性上，本土的室内设计师都迅速地经历了这场“扫盲”与启蒙。如此，为今后的行业发展在多方面奠定了专业基础。这一成果也无不反映在《2008中国室内设计年鉴》上。年鉴作为专业的年度检阅，为我们呈现了当下中国最新的设计水准。这是一次整体实力的集体亮相。

翻开《2008中国室内设计年鉴》，我们不难发现，那些更为出色的设计作品，都不同程度反映出设计师在各专题领域内的理解力与成熟度，继而将这种优势完整地注入空间设计中的各项要素，使之表现得更到位，作品更趋完整性与秩序感。这种能力，归根结底就是对空间各要素的把控能力，使形式语言与手法的运用更为自然流畅，空间韵味一气呵成。进而言之，装饰技法只有在空间组合关系的制约协调中，才能体现其审美价值。否则，过多无序又脱离空间逻辑的装饰，只能被视为道法缺失与气度平庸的表现，审美即品格也就能从中窥见一二。其实，历史上每一位在各种设计专题领域作出杰出成就的人，他（她）们在设计学的发展中往往能准确洞察并探究出新方向，从而开辟了学术思想上的新天地，而被世人推崇为一代宗师，他（她）们均是学术思想上承上启下，继往开来的领军者，是个人才华同集体学识的统一体。从这点而论，设计师更应将设计当学问做，这也是职业精神所在。

真正的探求之路，始于扫盲过后的集体反思。从08年鉴中不难看到来自各地设计师们风格迥异的最新作品，各类时尚的设计手法都能在此分别得以表现，也许这“丰富”的背后恰恰是“扫盲”过后必经之路的新起点。这说明在话语权缺乏的状况之下，更多的设计师在取悦时代的同时所做出无奈选择。生活中我们常常看到许多设计作品，说好了，是“概念”，说穿了，就是“噱头”。将设计降格为一场视觉魔术般的花招表演，其功效无非是一时之惊喜，而当惊奇过后，亦再没有什么可被留下的。就如同进入了无情的市场，继而被迅速地“产品替代”。究其缘由，就是对表象形式的刻意搬弄，甚至不解其意，将形式沦为卖点，而失去内在气息。如此下去，当舶来品相继被快速地仿制消费后，自身内在的贫乏随即陷入创作的迷茫而无所适从。有内在气质的形式产生于好的设计概念。概念是建立在时代总体文化背景的精神层面上的，同时对于室内建筑学亦是建立在设计空间自身发展规律之逻辑轨迹上的，它可表现为一种可被持续的发展模式，具有传承性和延续性。而“琐碎”的表面形式的刻意搬弄，只能是内空的反映——其本质是“空虚”。这种文化的“集体空虚”，最终使得自身人文气质的丧失。上述所指的“空虚”，不仅仅为精神层面上，亦包含专业技术层面的。我们常常清晰地认识到今后在这两大层面上的发展过程中存在的问题。一个艰难的任务摆在全体中国设计师和知识阶层面前。

首先是如何建树自身的文化精神，抑或称为文化气质。视觉概念仅仅是这种内在文化气质的外化形式。回顾中外历史，事实上任何一个强大的时代都有其强大的文化精神为背景，这样的事例不胜枚举。直观而言，这种气质，也许是浪漫的、优雅的、大气的、平和的、激荡的、英雄的、儒雅的、超然的、开明的、郁闭的、神秘的、粗放的、理性的、灵巧的、伟大的……，并且这种直观感受始终如一的能反射出那时、那地、那人的文化特质与精神内涵。而这些都以某种特定的形式概念得以呈现和流传。这是一种富有活力的视觉形式。即有内涵的形式。诚然，这需要一场新文化思想的运动。是整体社会人文精神的崛起，这样的崛起将为了诞生一个更有力的整体文化背景。而无论何种门类的文化艺术，在此背景中，都将使外在形式同内在人文精神得到完美的体现。这不仅仅是中国本土设计师的责任，更是全体知识分子的集体使命。空间设计师作为社会知识的一部分和环境文化的缔造者，责无旁贷。

如果说，这第一层面的问题是一个深厚宽泛的背景问题，那么第二层面的问题则纯属室内设计师自身的专业修养与学术课题。在近几年发展学术理论需求的一再呼声中，国内一些先锋设计师及教育家们不约而同地开始投入此番工作。研究成果旨在从个体才华的有限制约中，以更为理性的研究方式来超越局限，并无限持续。这种近乎科学研究的态度，首先要求室内设计师以一个学者的姿态来面对既理性又感性的室内空间领域，并在集体无意识的文化气质中，遵循设计自律性的原则不断研习与探索，将这门空间设计学不断的充实下去，使室内设计的研究成果真正得到积淀与创新。例如，对空间构成的认识；对抽象关系的发现；对思维方式的总结；对过程与表达的理解；对设计脉络的洞察；对概念形成的探求；对设计道德的觉悟……，乃至对空间造型、材质肌理、灯光照明、色调配置、陈设运用等专题的深层研究。这些都需要我们在新起点上对本专业范畴内的课题进行理性的回答。

读完作为中国本土优秀室内设计师集体亮相的08年鉴后，发现以70年代为分界线的两代中国室内设计师，当下正成为室内设计舞台上的主力军。相比第一代设计师作品所包含的文化使命感与沉淀感，以及理性、经典式的优雅气质，70后的新锐设计师更多轻装上阵，体现出对时尚发展的敏锐与新概念花样的追求，作品更趋前卫实验的意图。

最后，衷心祝愿《中国室内设计年鉴》作为当下强调严肃性、学术性的作品“年鉴”，能成为中国室内设计的一面崭新的旗帜。

Reflection after Elimination of Illiteracy

urses in these years, interior design in China has reached a new starting point, just like the rapid-rising Chinese economy. Standing on commanding height of this era, we have no difficulty in finding that this group of top interior designers in China can be fully confronted with any excellent designer abroad and can interpret and review their achievements from their own point of view. To Chinese interior designers, it is also a "starting point" that can have a dialogue with the past, present and future.

However, as for the development of the entire Chinese interior design, I still deem this starting point as an ending to a professional "elimination of illiteracy" movement. The whole generation, especially those contemporary Chinese interior designers who were born before 1970s, has spent almost two decades to basically complete an "elimination of illiteracy" movement in an unprecedented scale for the development of today's design. No matter from the varieties and complicities of the technologies, ideas and techniques or the functions, scales, and professional knowledge, the local interior designers all have experienced this "elimination of illiteracy" and enlightenment. In such case, it has made a professional foundation for the development of future industries. This achievement is reflected in "Yearbook of Chinese Interior Design 2008" everywhere. The yearbook, as a professional annual review, has presented us with the latest Chinese design levels. It is a collective demonstration of the entire strength.

Open "Yearbook of Chinese Interior Design 2008", we have no difficulty in finding that those more excellent design works have, in different degrees, reflected the understanding and maturity of the designers in the individual thematic fields and have consequently put this advantage into the various elements in space design entirely so that it can be presented more appropriately and the work tends more complete and orderly. All in all, this kind of ability is the ability of controlling the various space elements and enabling the application of formal languages and techniques more natural and smooth so that the space and atmosphere can be integrated together without any disturbance. Further speaking, the decorative technique only can demonstrate its aesthetic value through the restriction and coordination of the space arrangement. Otherwise, the excessive out-of-order decoration without space logic can only be deemed as a presentation of absent morals and laws as well as mediocre bearings. The aesthetics, that are character and morals, can be revealed just a little. In fact, anyone who has made the excellent achievements on various thematic design fields in the history always can make correct observation and explore the new orientations in the development of the design art and consequently developed a new world in the academic ideology so that they are highly appraised as the master of a generation. They all are the leaders of connecting the preceding with the following as well as carrying on the past and opening a way to the future and an entirety between the individual talents and collective knowledge. As far as this goes, the designer should more take the design for a learning, and this is also where the spirit of business exists.

The actual exploring road starts from the collective reflection after the elimination of illiteracy. In the yearbook 2008, it is easy to find out the latest works of the designers who come from different places with widely different styles. The various fashionable design techniques can be individually demonstrated here. What is behind this "abundance" is just a new starting point for the road we have to cross over after the "elimination of illiteracy". This shows that under the condition of lacking the talking right, more designers have had to make the helpless choices while entertaining the era. In our life, we often can see many design works, well speaking, which is deemed as "idea", but frankly speaking, is just "tricks". To degrade the design to a visually magic tricky performance is only an instant surprise. However, after the surprise, nothing can be kept any more, just like "being replaced by another product" rapidly after entering the ruthless market. Its main reason is the deliberate copy of the ideas and forms, even without full understanding of its meaning. The forms are deemed as a selling point, which has lost the inherent tastes. If it continues in this way, after the imported subjects are rapidly imitated and consumed one after another, its inherent lacking can fall in perplexities of creation immediately and become lost. The form of inherent tastes is based on a good design idea while the idea is base on the spiritual side of the entire cultural background of the era and is also based on the logic track of the developing rule of the design space itself from the point view of interior architecture. It can be demonstrated as a continuable development mode with inheritability and continuability. However, the "trivial" deliberate copy of visual forms is only a reflection of lacking inherence - its nature is "vanity", which is not only in spiritual level, but also included in the professional technologic level. We often clearly come to realize the problems that exist in the development of the two big levels in the future. One difficult task is now presented in front of all Chinese designers and intellectual stratum.

stes. The visual conception is only the physical form of this inherent cultural taste. In review of the Chinese and foreign histories, in fact, any strong times has its individual strong cultural spirits as its background. This kind of example is too numerous to be counted. Visually speaking, this taste may be romantic, elegant, grand, peaceful, violating, heroic, genteel, transcendent, enlightened, gloomy, mysterious, wild, rational, nimble, great…… and this visual feeling can constantly reflect the cultural features and spiritual connotation of that time, that place and that person while these are presented and spread through certain specific form and conception. It is a vigorous visual form and a form of connotation. Indeed, it demands a new cultural movement. It is an emergence of the entire socialistic and humanistic spirits. This kind of emergence will bring a more powerful integral cultural background. No matter what kind of culture and art, the visual form in this background will get perfectly reflected in the humanistic spirits. It is not only the responsibility of Chinese local designers, but also the task of the entire intellectuals. As a part of social knowledge and the creator of the environmental culture, the space designer is duty-bound.

If the problem at this level is a deep profound background problem, the problem at the second level is purely the professional accomplishments of interior designers and an academic subject. In the repetitive shouts for the demands of developing the academic theories in recent years, some local pioneer designers and educators are simultaneously devoted to this work. The research results aim to surpass the limits and continue infinitely in a limited confinement of individual talents through a more rational research method. This approximately scientific attitude first requires the interior designer to be confronted with the field of the rational and perceptual interior spaces as a scholar as well as study and explore in compliance with the rules of design self-discipline to enrich this space design study so that the research results of interior design can really get accumulated and innovated. For example, the acknowledgement of space constitution; the discovery of abstract relationships; the conclusion on thinking methods; understanding on the procedures and expressions; perception of design sequences; the exploration on the forming of ideas; awareness on the design ethics…… and even the further study on the themes such as space modeling, material textures, lights and lighting, color arrangement and ornament application etc. All these require us to make a rational answer to the subjects within this specialty at a new start point.

After reading through Yearbook 2008 - a collective presentation of the excellent Chinese local interior designers, it is discovered that the two generations of Chinese interior designers line with 1970s as a division now become the main force on the stage of interior designer. With comparison to the cultural feelings of tasks and precipitation as well as the rational and classic elegant tastes revealed from the works of the first generation of designers, the innovative designers after 1970s play their role in a simpler way and show their perceptivity to the development of fashion and their pursuit to the new ideas and styles. Their works tend more advanced and experimental.

In conclusion, I sincerely wish that "Yearbook of Chinese Interior Design" - a "yearbook" of the products emphasizing on seriousness and technicality, can become a brand-new stand for Chinese interior design.

设计-重新构筑我们的价值观

评述2008中国室内设计“喜”与“忧”

满登 Man deng
特约资深设计评论人
上海满登空间设计事务所创意总监，
上海市装饰行业协会环艺设计专委会副主任
中国建筑装饰设计专业委员会专家委员
IFDA国际室内装饰设计协会上海分会会长
中国室内设计研究院特聘专家

《中国室内设计年鉴》是一部记录中国室内设计发展成长的大型文献，她客观记载着业界设计师每年的作品，她更像一部设计“辞海”的百科全书为业界所关注。我们不仅从里面可以观摩中国室内设计界最优秀最杰出最齐全的作品“大汇集”，同时还能就其中有关优秀作品的概念、构思、空间构成、语言色彩、个性化特征和有关议题等进行深入的解读。用“解读”评论的方式去阅读《中国室内设计年鉴》在历年年鉴历史上还是第一次。她无疑是中国室内设计界最值得“收藏”的一本纪实性年鉴。所谓最值得“收藏”是因为里面藏有不少有研究价值的作品。同时也代表着中国室内设计界向世界展示中国设计师的设计风采和智慧。《中国室内设计年鉴》不仅是中国室内设计的编年史，同时还是中国室内设计界的一面旗帜……

总的来看，客观地说08版的《中国室内设计年鉴》和去年相比，整体水平有小幅度“上扬”，有些版块设计项目显然是找到了自己的DNA血缘设计密码，其设计语言的构思及原创语境已接近或达到国际先进水平。年鉴作品分类9大版块中，设计创意发展较为迅速的则是老建筑改造、会所、茶楼、酒吧、餐厅和售楼处。也许这些空间性质藏有潜在最易激发的跳跃挑战性元素缘故。从空间构成运动的自然轨迹上看，很多设计师已摆脱走出了传统模仿阶段，有些作品语言的特征创意似乎已进入空间构成运动的“高速轨道”，其设计语言已初显“空间个性”。

空间设计构成好像是一个设计“围城”，所谓“圈内人”想走出来，所谓“圈外人”也想走进去。“圈内”“圈外”在一个地方待久了都想“挪挪窝”，寻找最适合自己的空间栖息地。突破“围城”给我们带来思想禁锢的“解禁”是设计师必闯的一关。笔者注意到年鉴也涌现了一批优秀作品，他们有的在市场需求构思上找到了突破口，有的在空间构成上找到自己的表达方式，有的则在空间属性上练就了自己特有的语言，最大变化是很多设计师学会了如何改变自己的思维设计模式而适应新的市场需求，去寻找或调整自己的系统定位。

今年的年鉴不论从规模上，还是从质量上都比往年有所提高。尤其在空间构成意识上普遍也有所提高，这也是今年年鉴收纳作品的最大收获。我甚至认为有没有空间构成意识是衡量设计师正不正宗的首要条件。不论你是否有这种意识，也不论你承不承认，空间构成的最佳表述方式显然不以我们的意志为转移的，问题是你能不能发掘和发现。我们在浏览和阅读这本书时，会对每位设计师在不同层面上感受他们对空间的感悟和认识。我们在肯定《2008中国室内设计年鉴》的同时，在肯定中国室内设计取得不俗的成绩的同时，笔者也存有一些疑问，在此与大家共同探讨……

坦诚地说，世界是精彩的，和发达国家相比，我们的设计语言仍然不够丰富多彩，尽管“国门”打开了，各种设计资源、资讯和信息潮水般地向我们涌来，由于“封闭”太久，压抑已久的精神空间突然释放，我们等待已久的渴望使我们饥不择食。于是，我们在设计舶来品的菜单上挑三拣四，尽捡时尚和奢华的饱食一顿。当大家都拥挤在时尚和奢华的空间里，我们终于发现过于时尚奢华的设计浪潮波及全国室内设计界。我们认识设计思维世界的那扇大门还未真正打开，需要我们去探索、去了解、去学习的地方远远超出我们现有的知识结构。时尚和奢华尽管很流行，但并不是我们的唯一选择。我们的固有的设计思维是否已走到了终点？我们的设计语言表达和传播还相对比较单一。旧的“库存”不清空？新的“内存”岂能进来？对于空间构成表述的内在走向规律和空间设计的形态的理解我们还处于初级模仿摸索的阶段。系统的概念设计构思模式还仅仅停留在表面的解释阶段，有的甚至仅仅停留在空间的功能和材质模仿上，缺少一个构思框架的进入或进口方向，有时候即便进入了好的构思渠道，却苦于找不到设计系统的出口方向。设计思维的“进口”与“出口”是当下我们要解决面对的。此外传统与传承之间的关系仍然未理顺好，传统与现代、新与旧之间设计语言的转换，常常走向极端，也就是西方文明与东方文明，古代文明与现代文明的语言对接，延伸艺术和高度融合方面的“钥匙”还未找到。空间体量和尺度，形态与比例之间的辩证关系，空间与装饰的驾驭能力和控制能力尚欠火候。急于求成，急于表现的太多，稳扎稳打，深思熟虑、精雕细刻的太少。轻理论、重案例，轻设计、重装饰的多；既重理论，又重案例，既重设计、又重装饰研究的太少。对节能环保的考虑太少，对奢华耗材资源考虑太多，浪费了很多建材资源，同时也难免“污染”了空间。当然，这仍是笔者“一家之言”，其发现和观察是否真实客观，还有待于大家的指正。

借此年鉴编辑阶段，我们也邀请了美籍华人著名境外设计机构——如恩设计研究室的郭锡恩和胡如珊，日本SAKO建筑设计工社的迫庆一郎等境外设计师作为特邀观摩作品进入年鉴。目的是通过观摩学习，我们能从人家那里感受到什么？能接收到什么有价值的信息？

中国目前处于经济转型期的过渡阶段，经济发展过快势必会导致设计过热的“泡沫”现象，缺少设计“基因含量”的设计作品大量涌现，设计价值贬值现象仍未得到有效的扼制，室内设计评估市场机制还未建立起来。假如因为设计资源已达到饱和状态，势必会造成廉价设计“转嫁”高成本的投入，两个极端最终必然会造成因设计资源廉价而投入的资金成本“低走高开”。也就是说，我们因设计太滥、太廉价造成了太多缺少设计含金量的“视觉污染”。室内设计师居然成了中国资源消耗最大的“户头”之一。我们肩负的历史责任和使命任重道远，我们是否都应“掂量”自己？扪心自问：我的设计资源是否有价值意义？还是有可能被视为“视觉垃圾”？我们的设计应有的商业价值和文化艺术价值是否得到完整的体现？用现代科学观来解读我们的设计，是否将我们的设计意识，设计创意和设计原创放大到一个合理的机制空间？和日韩比，我们缺少一个尊重设计、普及设计、提高设计的“国家意识”和氛围。当然，面对中国室内整体设计发展的不平衡的实际情况，面对欧美等发达国家和境外设计师及设计机构蜂拥中国与我们争夺设计资源时，外来的和尚好念经以及收取高额的设计费优势必然要“统治”中国相当长的一个阶段。与“狼”共舞，作为只有20多年的室内设计历史的中国，与强者共筑一个平台，共同拥有一个市场起跑线的尴尬，而我们只有“沦落”配做设计费最少系统深化设计的“配角”，空间设计资源价值最大化效益将被境外设计势力划入囊中，这是我们作为一个中国设计师最不愿意看到的……

因此，作为一个有责任、有抱负和有志向的中国本土设计师，我们可以包容境外设计师在国内设计抢占设计资源制高点，也可以见证他们将世界最先进的设计理念带入中国，但我们决不能袖手旁观而仅仅做一个旁观者和模仿者。

Design–Reconstitute Our Value

Comments on "Pleasure" and "Worry" on Chinese Interior Design 2008

"Yearbook of Chinese Interior Design" is a large literature recording the development and growth of Chinese interior design. It has objectively recorded the works of the designers in the industry every year and is more like a design "encyclopedia" to receive the attentions from the design industry. We not only can study and emulate "a huge collection" of the best, most excellent and most complete works in the industry on Chinese interior design, but also can make further interpretation on the ideas, conceptions, space constitutions, language colors and individualized characteristics etc. It is doubtlessly an on–the–spot yearbook most worth "collection" in the industry of Chinese interior design. It is not only a chronicle for Chinese interior design, but also a banner in the industry of Chinese interior design……

In general, objectively speaking, compared with last year, the entire level of "Yearbook of Chinese Interior Design" in 08 Edition has "ascended" to a small extent. The design projects in some areas obviously have found their own DNA kindred cipher code. The conceptions and original contexts of their design language has been close to or reached the advanced international level. In nine blocks as the works in the yearbook is classified into, the design creativities that have developed quite quickly are the modification of old buildings, clubs, tea houses, bars, restaurants and real estate sales pavilions. As seen from the natural track of the activities constituted by space, many designers have broken away from the traditional imitation stage while the characteristic creativities of some work languages seem to have already entered "high–speed track" of the activities constituted by space and its design language has begun to show "space individualities".

The constitution of space design looks like an "enclosed city" of the design. The insiders want to go out while the outsiders want to come in. To break through the "enclosed city" and "liberate" from the idea confinement are the barrier the designers must overcome. The author has realized that many excellent works have appeared in the yearbook, which has found their breakthrough in the market demands or their expressive way in the space constitution. Some have formed their individual languages while the bigger change is that many designers have learned how to change their own design style to meet the new requirements in the market and find or adjust their systematic orientation.

The yearbook of this year is improved both from the scale and the quality, especially the ideas of the space constitution. I even think that the best expressive way of space constitution obviously will not change with our wills. The problem is whether you can explore and find out. When we are looking through and reading this book, you will feel every designer's comprehension and understanding on the space at different levels. While we are positively evaluating "Yearbook of Chinese Interior Design 2008" and acknowledging the outstanding achievements made in Chinese interior design, the author also has some doubts for discussion together with you here…….

Frankly speaking, the world is wonderful while our design languages are still not rich and colorful. Although "the door of the country" has been opened and the various design sources, information and data have purged towards us like the tidewater, the door that we come to know the world of design ideas is still not really opened and the things requiring us to explore, understand and study have largely outstepped our present knowledge structure. As for the interior trend rules of the expression on space constitution and the understanding on the forms of space design, we are still in the primary stage of imitation and seeking. The thinking mode of the systematic conceptual design still stops at the stage of superficial interpretation and some even at the imitation of the space functions and materials. Although sometimes even it has entered a good conceptive channel, it can not find the exit of the design system. The "entrance" and "exit" of the design conception is what we shall be confronted with and shall be settled at the moment. Too many people are over anxious to succeed and show off while there is too few people who can go ahead steadily and surely, be considered deeply and handled carefully. It neglects the theories and emphasizes the cases. There are too many that neglect the design and emphasize the decoration while there are too few that emphasize theory and case as well as the design and decoration. There is little consideration on environmental protection and too much consideration on the luxurious consumables and resources so that many construction materials are wasted and the space is also "polluted". Of course, it is only the author's "own opinion". Whether these findings and observations are authentic and objective is still pending for your correction.

When editing this yearbook, we also have invited the works of some foreign designers including Guo Xi'en and Hu Rushan from the famous foreign design institution held by American Chinese – Neri & Hu Design and Research Office as well as Keiichiro Sako from SKSK Architects as the special works for study and emulation in the yearbook. It is purposed to find out what we can feel from the works and which valuable information we can receive from the works through the study and emulation.

China is now in the transitional stage of the economic restructuring period. The over–speed of the economic development certainly will result in the "bubble" phenomenon – overheated design. If the design resources have reached the saturation status, it will definitely cause the cheap design and high–cost investment and two extremities certainly will result in the "rally" of the invested capital. The interior designers have become the biggest "maker" on the consumption of the resources in China. The historic obligations and tasks we undertake have a heavy burden and long road to go. Of course, it seems a little bit exaggerated, when we are faced with the real situation in china's unbalanced interior design and development. However, when we are confronted with the situation that the designers and design institutions from the European and American developed countries and abroad are swarming into China and compete with us for the design resources, the fact that the outside monks can chant scriptures better and the advantage that the high design fees will be charged certainly will "govern" China for quite a long time. However, we have to "come down to" "a minor role" to make the detail designs of the least design fees. The maximized benefits in the value of the space design resources will be included by the foreign design forces, which is what we don't like to see this as a Chinese designer……

Therefore, as a Chinese local designer with senses of responsibilities, ambitions and ideas, we can tolerate that the foreign designers have seized the commanding point of the design resources in the domestic design market and also can witness that they have brought the most advanced design ideas in the world to China, but we can never stand by to be only an onlooker and imitator.

目录

办公 Office

商业展示 Exhibition

SPA
美容

Spa & Hairdressing

售楼处
样板房

Sale Pavilion

CONTENTS

非办公设计
Non-office Design

沈雷 Shen lei
内建筑设计师事务所合伙人
高级室内建筑师
英国注册室内建筑师

谷腾
内建筑设计事务所
吴宗勋、竺春光、莫意波
张雷
王善祥
吴君华
姜峰、陈礼庆、袁晓云
宋国梁
曾建龙
于强
黄斌
李越
Keiichiro SAKO、Akiko MIYAZAKI
王闲
罗灵杰、龙慧祺
谈星火
史永杰
殷艳明
陈骏
张晔、盛燕
MARGAUX LHERMITTE、方钦正
贺钱威

吉禾
超锐体育用品发展有限公司
中石化宁波工程有限公司设计科研大楼
南画廊艺术家工作室
陈心懋画室
友邦家居总部办公楼
姜峰室内设计有限公司写字楼
[V]室内设计咨询有限公司办公室
概念办公空间
招商银行VIP贵宾厅
泉峰贸易有限公司办公楼
井冈山革命博物馆新馆
MATERIALITY
世新传播机构总部办公室
壹正企划办公室
北京体育大学国家队训练基地综合馆
香港华侨城集团写字楼
创域艺术设计有限公司办公楼
乔艺塑胶有限公司
大兴文图馆影剧院
NACO上海中心
HED Office

随着体系庞大、层级分明、工作固定的金字塔组织结构的解体，现代生活及工作借由E化逐渐融合。组织与工作流程的改变以及信息工具的不断更新带来作业形态的变革，形成了不限时间、不限地点、不论何时、不论何地皆可工作的新形态。这让使用者对办公地点的选择具有更大的弹性空间，办公空间也得以从纯粹的工业机能中解脱出来，与个性、自由、休闲、娱乐等发生联姻，成为一个混合体。

像机器一样运转的工作场所只能让员工每天面对工作存在感带来的压迫，因而成为办公设计的过去时。而当创意二字掀起的洪流迅速地推动起全球经济的发展时，如何寻找到有效的方式，让办公空间也可以担当起创意启发的功用，最大限度地挑战和开发这些潮流性行业员工的创造力，产生出非同凡响的商业价值，成为后工业时代业主与设计师们一次次实践推演的中心主题。

突破局限是建立起创意空间平台的重要方面，既有对刻板的办公空间格局的颠覆，也有对单调办公功能的拓展。非传统的办公设计的目的正是将最能点燃灵感之火的自由气息引入空间，让员工以最轻松的方式享受工作。在这样的理念引导下，办公空间趋向复合，以完成多重身份定义，最大限度地满足各种不同需求。在办公空间复合化的过程中，原本单一的空间肌理中掺入更多地非办公设计元素。当酒吧、咖啡厅、茶座、运动场、会所等多元化的场景和功能串接于办公空间内时，设计带来的空间错位感，如同拼贴的蒙太奇效果般，展现出冲破线性束缚的丰富空间体验。

随时可以变化的平面配置，是更为动态的办公机制的基本需求。空间的转向使明确的功能定位不再成为建构空间格局的唯一方式。笼统的区域划分、子空间的引入、切换模式的预设等非传统形式的表现，扩张了工作场所的概念，使之变为一个场域。模棱两可的功能性与非功能性空间呈现出自由、开放的态势，毫无距离感地欢迎任何人进入其中。

创造随意性交流空间的企图，从更大程度上体现了办公空间内对人性化追求的认可与鼓励。模糊化的空间边界与功能定义，以通达的环境、良好的视线、令人愉快的细节等消解识别带来的意识拘泥，通过对超越物质层面的关爱和兴趣的表达，在倡导共享与交流沟通的基础上提高工作效率。

在这一系列办公空间的设计中设计师都以突破常规办公模式的思维打破了概念语义的约束。在让人产生酒吧错觉的V室内设计咨询有限公司办公室中，办公与休闲的双重空间定义在光线切换与材料的质感组合中轻而易举地实现。同样是对空间的色相处理，吉禾与壹正企划则分别从暗调与无色两个截然不同的方向诠释了各自业主对自由的理解，让空间个性跃然体现。在G&D办公空间及世新传播机构总部办公中，设计从营造轻松舒适的工作氛围出发，清晰地传达了对人性化设计的考量。

通过这些非传统办公空间的展示，不难看出在后现代社会混杂、多变的外衣下，办公室形态已不再是单一的形式及功能。办公空间纯粹意义上为办公而设的有形规模在不断缩小和弱化。工作场所的设计行为也将在空间个性化和多元化的双重引导下，从功能、视觉效应、空间感受、实施手段等各个方面深入探求具有高包容度的非办公设计可能性。

With the disaggregation of the pyramid organization structure with the gigantic system, clear classes and stable jobs, the modern life and work are gradually integrated together through the computerization. The changes of the organization and working procedures as well as the constant updates of the information tools have brought with the reform of the working ways to form a new way in which the work can be done regardless of times and locations. In this case, the user has a more flexible space for the selection of the working locations. The office space also can be librated from the pure industrial function to connect with individualities, freedom, recreation and amusement etc. and become a mixture.

The working place running like a machine only can bring the staff to confront with the oppression due to the existence of the work every day and thus becomes the history of office design. However, when the current that is caused by the word "creativity" is rapidly promoting the development of the global economy, how to find an effective way so that the office space can also inspire the creativities and challenge and develop the creativities of the staff in the fashion industry to create the unusual business value has become the central theme practiced and performed by the owners and designers again and again in the post industrial era.

Breaking through the confinement is an important aspect of setting up a platform for creative space. There is not only an overthrow of the rigid office space arrangement, but also an expansion to the dull office functions. The non-traditional office design purposed to introduce the free sense that can best ignite the fire of inspiration to the space and make the staff enjoy the work in a most comfortable way. Under the direction of such idea, the office space tends to the complexion in order to accomplish the definition of multiple identities and maximally meet the various different requirements. During the complexion of the office space, the originally dull space texture is mixed with more non-office design elements. When the multiple scenes of bars, coffee houses, tea houses, playgrounds and clubs etc are linked to the interior office space, the sense of space displacement brought by the design, just like the pasted montage effects, has shown the experience of wonderful spaces that have broken through the linear confinement.

The plan arrangement that can be changed at any time is a basic requirement for a more dynamic office system. The change of the space makes the orientation of specific functions no more a sole way to constitute the space arrangement. The demonstrations of the non-traditional forms such as general division of the areas, introduction of subspaces and presetting of the changing modes etc have expanded the conception of a working place to a field. The spaces of ambiguous functionalities and non-functionalities have shown a free and open posture to welcome anybody in without any sense of distance.

The creation of the casual communication space is attempted to reflect the recognition and encouragement on the pursuit to humanization by the office space to greater extent. The indistinctive space boundaries and function definitions have dissolved the constraints of the ideas caused by the recognition through the clear environment, good vision and pleasant details etc. and improved the work efficiencies on the basis of advocating the share and communication through the expression of the concerns and interests surpassing the material level.

In this series of office space designs, the designers have applied the idea that breaks through the conventional office mode to break the confinement of the conceptual meaning. In the office of V Interior Design Consulting Co., Ltd. which brings an illusion of a bar, the dual space definition of work and recreation has been realized quite easily through the combination of light changes and material qualities. The same is applied to the arrangement of color and shape of the space. Jihe and One Plus Partnership Limited have interpreted the understanding of the freedom by the respective owners through the two completely different directions - dark and colorless to vividly demonstrate the space individualities. In G&B office space and the Office for Headquarters of Shih Hsin Media Group, the design has begun with the creation of a relaxing and comfortable working environment to clearly deliver the consideration on the human-orientated design.

Through the exhibition of these unconventional office spaces, it can be easily found out that the office type is not a dull form and function any more under the farraginous and variable overcoat of post-modern society. The physical scale designed for the work in a pure meaning of office space is constantly reduced and weakened. Under the dual direction of the individualization and diversification of the space, the design behaviors on the working places will also further explore the possibilities of non-office design with high compatibilities from the aspects such as functions, visual effects, space feelings and implementation methods etc.

设计:谷腾

吉禾
Ji he

面积:110m² / 坐落地点:上海 / 主要材料:水晶自流平、钢结构、玻璃 / 摄影:谷腾

生活中充满各种各样的颜色，每种颜色所带来的生活也不相同。在这充斥着缤纷色彩的世界里去寻找一个无色的社会，一种自由的、安静的生活，一个过去与未来交织的空间，无色，无限……来重新审视这世界上的每个生灵。

Our life was full of all kinds of color,Which bring us different kinds of life,We try to find a no color society,a free and quiet life style,or a space which contact between past , and future and so on in this colorful word,Colorless, unlimited……And review this world from every misery in the world.

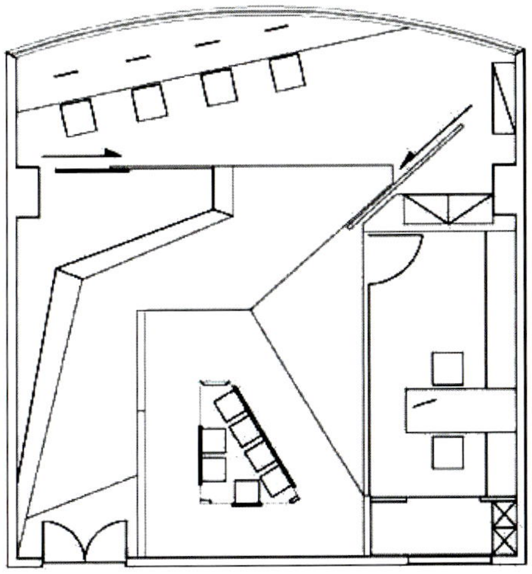

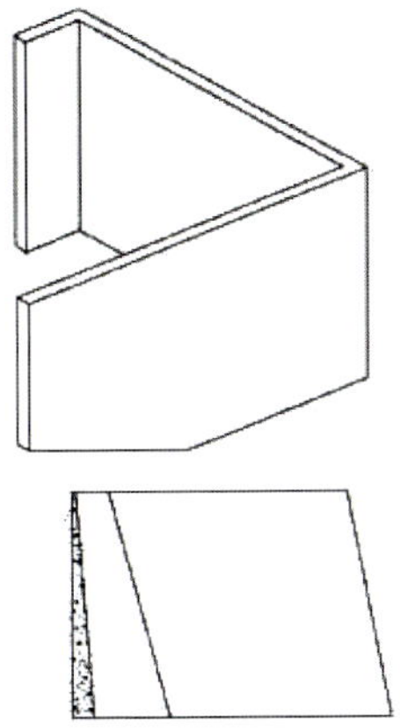

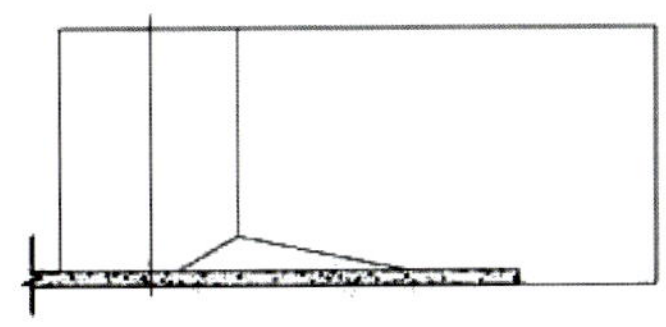

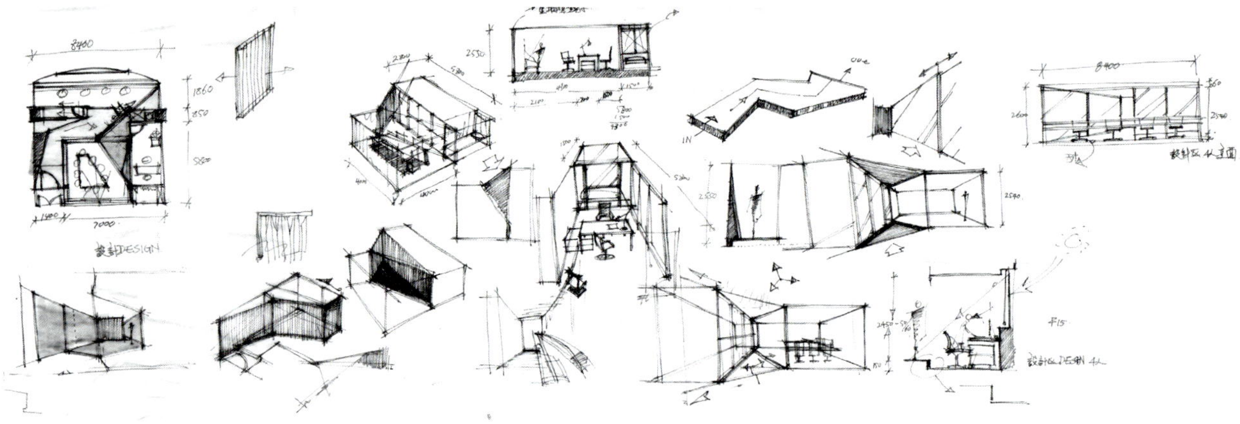

设计单位：内建筑设计事务所

超锐体育用品发展有限公司
Chaorui Sports Products Development Co., Ltd.

面积：650m² / 坐落地点：杭州 / 摄影：陈乙

超锐体育用品发展有限公司从事着国际知名运动品牌“kappa”和“RUKKA”的特许经营，在它的办公空间中，运动是一种发展的状态，也是一种创新的精神，经由设计师透过灵动开放的布局及明快的色彩搭配投射在空间中，便转化成活泼生动的场域气质，契合着企业运动品牌代理的行业特点。

设计极力营造具有动感的空间体验，突出活力亲切的空间氛围。入口接待区即以意大利的国旗颜色红、绿、白的色彩搭配呼应公司代理的“kappa”品牌。异型的螺旋造型形成半包围的等待区域，跃动的线条配合温暖的红色立刻让空间变得鲜活而富有冲击力，使客户在进入公司的第一时间就能直观而深切地感受超锐的热情与品牌认知程度。

主办公区采用开放格局，斜向布置的办公桌椅打破规矩的四方平面，扭转了办公空间的刻板印象。阳光的橙和灵性的蓝两色新型亚克力板如同旗帜般自上空垂下，顶部的滑轨让它们可以方便地按需要移动变化，不断适应瞬息万变的环境，拓展出空间变化的无限可能。变化带来乐趣，时而开启时而闭合的空间游戏，让群体空间不再枯燥乏味，也为个体的创意发挥提供了自由的典范示例。

大小会议室、总经理室等要求相对密闭的空间沿走道右侧错位参差排布，以白色的圆角盒体呈现，柔和的线条消解了立方空间的坚硬与凌厉。盒体当然绝不会是压抑的全封闭型，两面以弧线连接的玻璃墙面保持了空间通透性，再分别用橙色、蓝色、绿色为主基调的丝网印刷薄膜为玻璃覆面，除了可以清楚明晰地标识出的各自包含的不同功能外，也是走道一侧亮丽的风景。盒体内天花、地面和墙面都以竹胶板饰面，给人以整体感和一体性，亲切的质感也营造出轻松舒适的空间氛围。

Chaorui Sports Products Development Co., Ltd. is engaged in the franchised business operation of the well-known international sports brands “kappa” and “Rukka”. In its office space, the sports are a developing status and also an innovative spirit. It has been reflected in the space through the vivid and opening arrangement as well as the arrangement of vivid colors by the designer - Yao Lu, which is then transformed into a lively atmosphere to match the industrial characteristics as an agent for sports brands.

The design tries to create an dynamic space and stress on the lively and friendly atmosphere. The reception area at the entrance is arranged with red, green and white to go with - three colors in Italian national flag to echo with the brand “kappa” engaged by the company. The irregular spiral modeling forms into a half-enclosed waiting area. The active lines in combination with the warm red have made the space fresh and impressing so that the customers can directly and deeply feel the warmness and recognition on the brands of Chaorui once they enter the company.

The main office area is in an open style. The inclined arrangement of the office tables and chairs has broken the orderly square plane and changed the stiff image of the office space. The new acrylic boards in two colors - sunshiny orange and vivid blue are handing down from the overhead like flags. The sliding rails on top can move and change them conveniently as required to be constantly adapted to the ever-changing environment and explore the infinite potentialities of the space change. The changes have brought with the pleasures. The space games that sometimes open and sometimes close have made the integral space boring no longer and also provided a typical example of freedom for the development of the individual creativities.

The spaces that requires relatively closed including the big and small conference rooms and general manager’s office et are arranged along the right side of the corridor in an alternated and irregular way. They are demonstrated in the white cases with round angles. The soft lines have dissolved the hardness and sharpness of the cubic spaces. The cases are certainly not the depressive totally enclosed type. The glass wall to connect the two sides in a curve has retained the transparency of the space. The screen printed films separately with orange, blue and green as the main color are covered on the galls. In addition that the individually contained different functions can be clearly and obviously marked, they are also a bright scene in the corridor. The ceilings, floors and walls in the cases are all covered with bamboo-glued boards so that the people have a sense of integrity and entirety. Furthermore, the friendly qualities also can create an easy and comfortable space atmosphere.

S
K
kappa
TRA

主持设计:吴宗勋、竺春光、莫意波

中石化宁波工程有限公司设计科研大楼
Design and Research Building of Sinopec Ningbo Engineering Co., Ltd.

面积:31997m² / 坐落地点:宁波 / 主要材料:微晶白麻、深灰色微晶石、白玉石、壁布

一个企业的发展的文化信息，可以通过其建筑空间的设计呈现出来。其映射出的文化信息，正是一个企业发展的脉搏。现代、科技、智能化和国际化是中国石化集团宁波工程有限公司科研设计大楼的室内设计主题。设计源自生活、服务于生活，在整体空间中离不开人性化的细节关注。空间的形与色，就像人类的“血与肉”，只有丰满的体魄，“灵魂”才能得以体现，雕塑般的结构美感通过黑、白、灰系列，及纯净的色块表现，现代、科技、智能化空间的“灵魂”才呼之欲出。

在整个设计中，大量采用亮度适中的漫反射光槽和反射光槽，并充分利用自然光，减少灯光对人眼的直射，降低对人体的污染，绿色植物的采用增加了人与自然的亲和力，降低了环境对人体的污染。微孔铝板、铝挂板及吸音材料的使用，降低了声音对人体的污染。整个设计中无一不体现出“以人为本”这一永恒的法则。

The cultural information on the development of an enterprise can be displayed through the design of its architectural space. The reflected cultural information is the pulse of the development of an enterprise. The interior design subject of Sinopec Ningbo Engineering Co., Ltd. is modernity, science and technology, smart and internalization. The design originates from the life and serves life. The integrated space can not be without the concern on humanitarian details. The forms and colors of the space are just like human “blood and flesh”. Only with the well-developed physique, the “soul” can get better displayed. The sculpt-like structural aesthetics are expressed by the colors such as black, white and gray as well as the pure color blocks so that the “soul” of modern, scientific and technologic as well as automatic space can be vividly portrayed.

During the whole design, the appropriate diffuse reflective and reflective light channels are used in a great deal and the natural lights are fully utilized to reduce the direct lights on the eyes and the pollution to human bodies. The application of green plants has increased the affinity between human and nature and reduced the environmental pollution to human body. The microporous aluminum plates, suspended aluminum plates and the acoustical materials are used to reduce the noise pollution to human body. The whole design has shown the everlasting principle of “people orientated” everywhere.

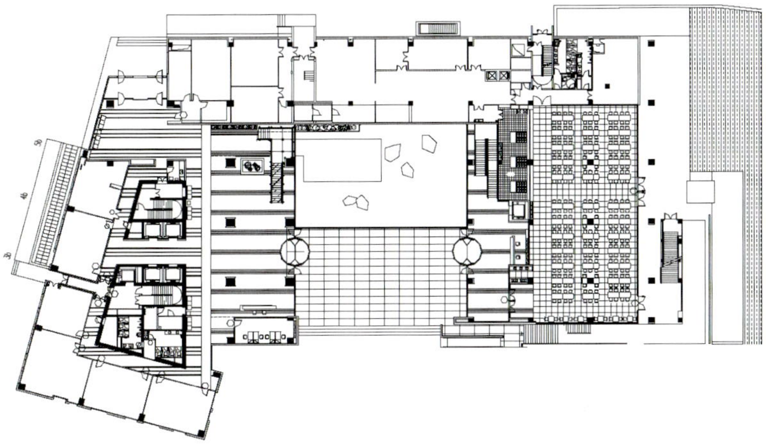

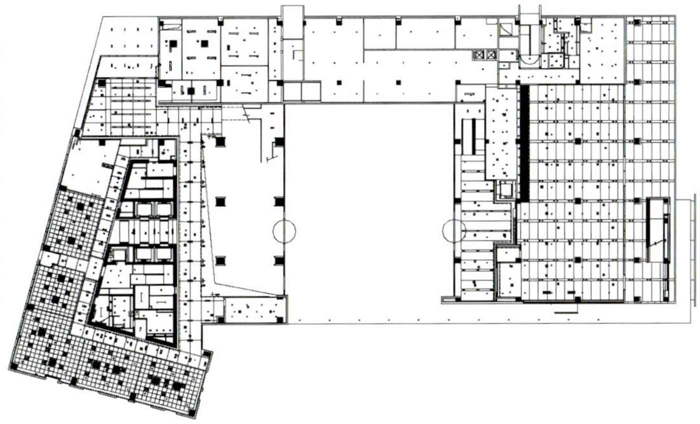

设计:张雷

南画廊艺术家工作室
RCM Gallery

面积:200m² / 坐落地点:南京 / 主要材料:瓦楞石棉板、红砖 / 摄影:Nacasa & Partners Inc

通过石棉瓦和锈铁门这两种原始材料，南画廊的表皮将场地的前符号现实的后工业气息挖掘出来。它不动声色地保持了建筑底色的连续性和场景的历史感。这一反视觉的做法并非只是为了让某种本来的场所气息得以延续和再生，它是一种无意识的偏移，某种难以启齿的东西（死亡驱力）吸引着它无意间闯入一个禁区。

和外层相比，内层处理显得低调得过分。它以原砖柱为基准，一斩齐做成连贯的大白墙面。原有的二层空间保留不变，在西侧增设了一个开敞的夹层，作为接待用途。一墙之隔，建筑师设置下了一个巨大的落差。首先是材质的落差，也就是符号落差。外皮的后工业废墟气息，在内部没有任何反映。虽然锈铁门在内部也还可见，但是在7米高的几面宽窄不一的白墙、白屋顶和暧昧的灯光的烘托下，它更像一个当代艺术作品。

Through two kinds of raw materials such as the asbestos tiles and rusted door, the surface of RCM gallery has dug the realistic post-industrial atmosphere out of the previous symbols in the field. It has unperturbedly retained the continuity of the architectural grounding and historical sense of the scene. This contra-visual way is not only purposed to continue and regenerate the certain original atmosphere in the field. It is a kind of unconscious deviation and certain unspeakable thing (drive of death) has attracted it to accidentally crash into a forbidden area.

Compared with the external layer, the internal layer is arranged over low-profiled. It takes the original brick columns as the benchmark to make a big continuous white wall in level. The existing two-story space remains unchanged while an open mezzanine is added in the west side for the reception. The architect has set up a huge difference through the partition by one wall.

The first is the difference of the materials, which is the difference of the symbols. The post-industrial relic atmosphere of the exterior surface does not have any reflection in the interior. Although the rusted door still can be seen from the interior, it looks more like a contemporary artwork in contrast to several 7m high white walls in different widths, white roofs and ambiguous lights.

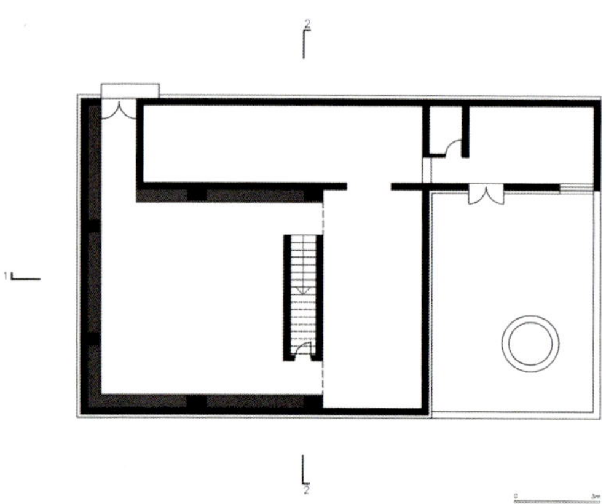

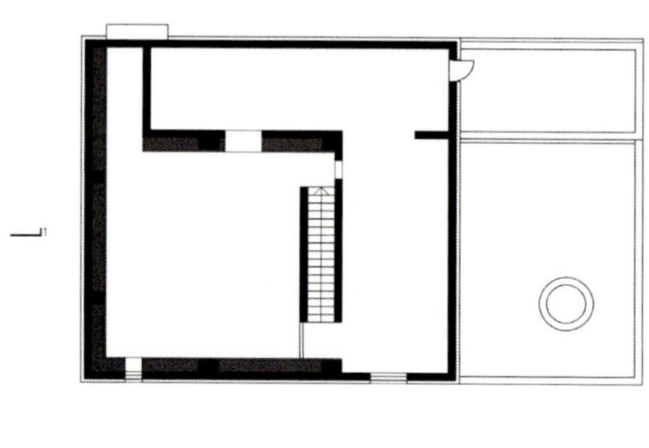

设计:王善祥 / 参与设计:王伟、管鹏、郑锦、岳勇飞、王善辉 / 结构设计:李春桥

陈心懋画室
Chen Xinmao's Studio

面积:392m² / 坐落地点:上海 / 主要材料:钢筋混凝土、乳胶漆、竹胶板、旧木地板、竹篱、型钢、旧青砖 / 摄影:王善祥

为了最大限度地制造使用空间，方形是最佳选择，建筑体量可以做到最大化。房形的外观体现了建筑典型的现代主义特征，为了呼应当地的旧时记忆，同时也是体现画家作品强调材质特色的主要特征。建筑外立面的主要墙面采用了白色拉毛粉刷，局部两个凸出的体块采用了建筑竹胶板做为饰面，建筑内部也采用了这两个主要材质与色调。建筑勒脚使用了旧青砖砌筑，南部院落也大部分使用了旧青砖铺装。院墙只在门垛及转角处立了少量实墙，大部分面积以钢框为骨架，上部安装了一种建筑工地上搭脚手架用的竹篱板（借鉴了王澍设计的中国美院象山校区的一些护栏的做法）。朝向南院的一层设计了高度4.5m的几扇铁制大门，打开后使室内外联为一体。

室内部分。一层，画家进行大幅绘画的创作，于是层高定在了5m；二层，以作品展示为主，同时安排了两间卧室和一间储藏间。作品展示的区域采用了低窗，为的是有更多的墙来展示作品；三层，主要为画家进行小型作品的绘制，屋面采用了天窗，并有一个阳台和一间休息室。一部小钢梯可以通至屋顶。为了最大限度争取空间，卫生间与厨房安排在了楼梯的休息平台部位。由于业主对内部一些详细功能一直不太确定，所以内部格局有些部分一直在变动。室内材质与色调也是同外立面以白色粉刷和竹胶板为主，大面积的楼板、楼梯板在浇铸时的粗糙痕迹被保留下来。画家收藏了大量旧家具和古董，以粗犷的材质、简洁的色调为背景，营造了室内非常轻松随意又极其浓郁的儒雅氛围。

To produce and use the space to the furthest, squareness is the best choice and the building size can be maximized. The visual appearance of the building has caused the typical modernism characteristics of the buildings to echo with the local old memories and also reflect the main characteristics of the materials emphasized by the drawer. The main wall facade of the exterior building is painted by white stucco and two protruded blocks in certain places are decorated with the architectural bamboo glued boards as the veneer. The interior building is also applied with these two main materials and colors. The building plinth is built up by the old gray bricks. The south courtyard is also mostly paved with the old gray bricks. As for the courtyard wall, few solid walls are built up at the buttress and corners while the most area is applied with the steel frame with bamboo fencing boards installed above, usually used for building the scaffolding at the construction site (which has been referred to the way of making some guardrails for Xiangshan Campus of China Academy of Art designed by Wang Shu). The level facing the south courtyard is designed with several iron gates with a height of 4.5m and the interior and exterior will become an integrated one after the doors are opened.

As for the interior part, the large-scale drawings are done by the drawer at the first floor, so the level height is set as 5m; the second floor is mainly used for exhibiting the works and two bedrooms and one storage room are also arranged here. The exhibition area is applied with the low windows. So that more walls can be reserved for exhibiting the works; the third floor is mainly used for making the small pictures by the drawer. The roof is applied with the skylights and one balcony and one rest room are arranged here with one set of small steel staircase up to the roof. To maximally strive for the space, the toilet and kitchen are arranged at the resting platform of the staircase. As the client has not been very sure about some detailed interior functions, some interior arrangements have been changing all the time. The interior materials and colors are also mainly the white stucco and bamboo blued boards. The rough traces on the large areas of floor slabs and the stair steps during pouring are kept. The drawer has collected a large amount of old furniture and antiques with the rough materials and simple colors as the background to create a very relaxing, free and strong cultured interior atmosphere.

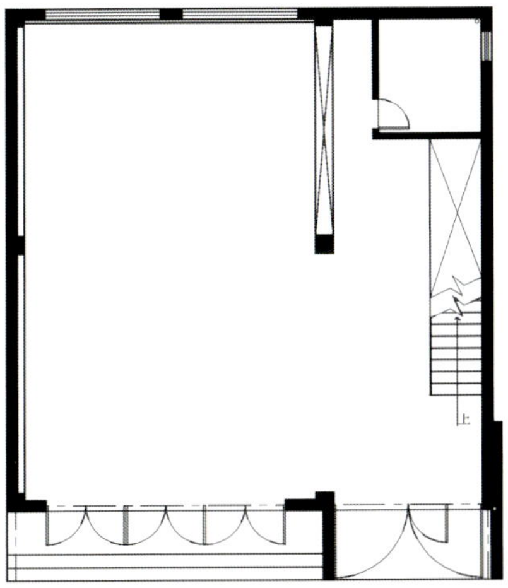

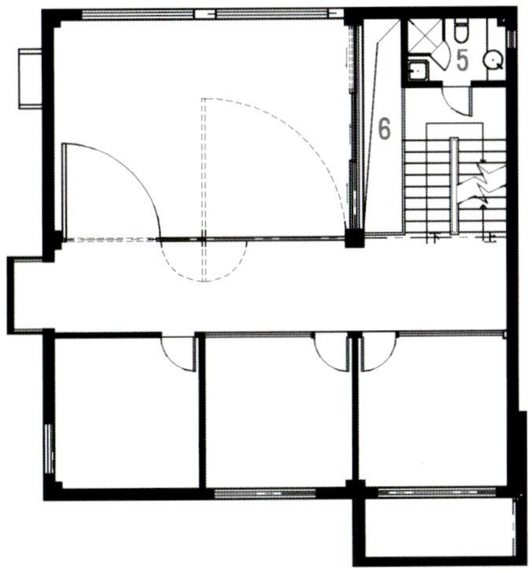
5
6

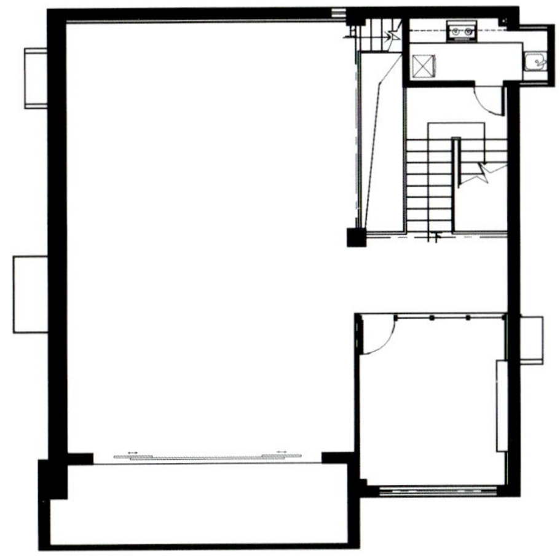

设计单位：翰禹·印加空间设计机构 / 设计：吴君华

友邦家居总部办公楼
Office Building of Youbang Home Supplies Headquarters

面积：600m² / 坐落地点：宁波 / 主要材料：微晶石、大理石、白橡木、发光膜、不锈钢、特殊玻璃 / 摄影：张学泉

运用最简单的素材，创造空间的质感和层次深度，是本案主要的设计主轴。

办公区以简洁实用为原则，在办公室的设计中力求突出现代感，同时融入艺术的内涵。

办公区公共部分采用玻璃隔断的形式，木制造型在生硬的走廊注入了生动感的色彩，线条则体现了理性和进取的方向，使得空间更有现代感。

总裁办公室在延伸整体简约风格的同时注重了空间使用效率和人性化的体贴设计，空间布置增加了私人空间的趣味性，多而不杂，少而不缺，背景墙的整体设计，最大限度地释放了空间的进深。

在空间规划上特别注重穿透性，希望坐在办公室里的人，不再感觉办公室只是一个包覆的盒子，而是无限延伸的空间。设计师用黑色的铝合金挂片给平淡无奇的吊顶做效果，深色的、冰冷的金属挂片就像深邃的眼神，怎么看也看不透眼神后面所隐藏的秘密。

触目所及皆大量采用能予人温暖感的白橡木，并以贯穿整体的几何造型元素及隐藏式收纳的概念，维持视觉的清爽度。设计师以无负担的设计语汇，极力营造出一个没有压力的舒适环境，将所有材质及色系单纯化，打造了一个简约的办公国度。

The main design focus of this case is to apply the simplest materials to create the sense of space reality and the profoundness of the levels.

The office area is designed on principles of simplicity and practicality, trying to lay stress on the sense of modernity and mix with the connotation of the arts.

The public part in the office area is separated by the glass and the wood style has filled the stiff corridor with vivid colors while the lines reflect the rationality and aggressive directions so that the space is displayed with more sense of modernity.

The president' s office is laid with stress on the considerate design of the efficiencies of the space utilization and individualities while it continues the entire simple style. The space is arranged with the interests of the private space, numerous but not messy, little but not insufficient. The design of the entire background wall has released the space depth to the farthest.

Special attention has been paid to the penetrability in space planning. It hopes that the people sitting in the office will feel the office is an infinitely outspread space instead of being a wrapped box any more. The designer has applied the black aluminum alloy hanging pieces to create certain effects on the humdrum ceiling. The dark cold metal hanging pieces are just like the profound eyesight. You can never see through the hidden mysteries behind the eyesight.

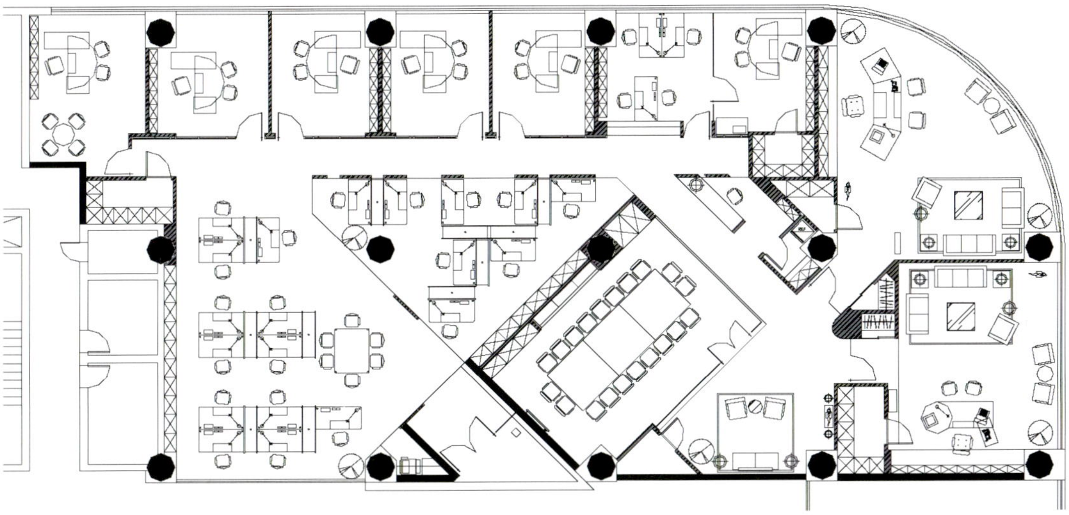

设计单位:深圳姜峰室内设计有限公司 / 设计:姜峰、陈礼庆、袁晓云

姜峰室内设计有限公司写字楼
Office Building of Jiang and Associates Interior Design Co., Ltd.

面积:1200㎡ / 坐落地点:深圳福田区 / 摄影:莫尚勤

红、白、黑是姜峰室内设计有限公司的标志色，公司写字楼室内设计以此为基调，在门厅的设计中加以深化，使之相呼应。红色醒目的背景墙面，深色木地板45度铺装，内藏LED的亚克力圆管成为竹子的化身，节节向上，极具雕塑感的墙面彰显个性。大办公室管道下局部吊顶，镂空为公司的座右铭。男女卫生间以黑白为基调，门上分别以1和0加以区别，女卫生间墙面为雪花白，镜面和手盆曲线柔美，家具统一为北欧风格。

Red, white and black are the VI symbolic colors of Jiang and Associates Interior Design Co., Ltd. The interior design of the company’ s office takes this as the basic colors and deepens in the design of the hall so that they can echo with each other. The distinguished red background wall is covered with dark wood boards in an angle of 45 degree. The PMMA round pipes with LED installed inside become the embodiment of the bamboos with each section uprising. The wall with strong sculptural sense has conspicuously demonstrated its individuality. The ceiling has been made below the partial pipes in the big office with the hollowed-out motto of the company. Black and white is the basic colors for the male and female toilets with 1 and 0 separately on the doors as the distinction. The wall in the female toilet is snow white. The mirror and wash basins have beautiful and gentle curves while the furniture is unified to be north European style.

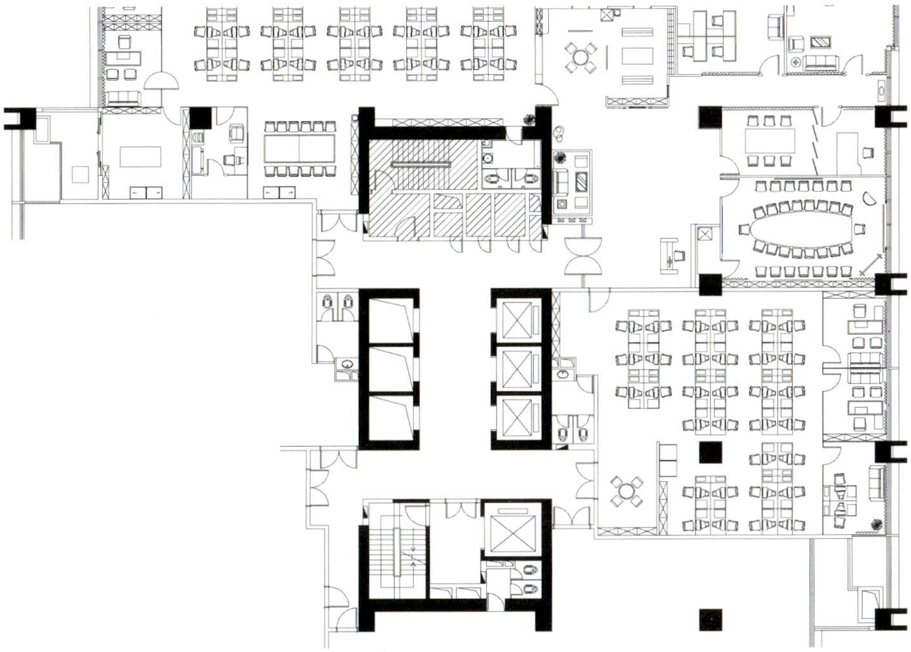

J&A
83416062

设计单位:宁波[V]室内设计咨询有限公司 / 设计:宋国梁

[V]室内设计咨询有限公司办公室
Office of [V] Interior Designer Consulting Co., Ltd.

面积:260㎡ / 坐落地点:宁波市 / 主要材料:橡木饰面板、大理石、镜面不锈钢、松木板

在人们的印象中，办公空间往往注定就是单调、理性、缺少情趣的模式。为什么不能改变这一成不变的工作方式和环境呢，生活需要更多元化的价值理念，本案的设计也许能给人带来一些启迪和思考。

当你推开高高的玻璃门步入办公室前厅的时候，会让人产生短暂的错觉，醒目的吧台、洋酒、飘渺的音乐让人仿佛置身于一处都市里的风尚酒吧，风格独特充满新意。公共空间的立面、顶部大胆采用了镜面不锈钢材料，把原本高挑的景深进一步放大了，设计师很好地把握好了设计的尺度，色彩纯净、手法干练，粗糙的砖墙、光滑的金属、铁锈板、大理石之间不同质感的组合搭配自然和谐，格调现代时尚又不失淳朴人文。楼梯运用力学支点采用仅为10mm厚的钢板，从侧面的角度看仿佛消失了。原有水泥结构柱经过毛面处理后，不施粉黛的裸露着，和2层的老树桩、旧地板一起散发着淡淡的自然风情，时光的沉淀就那样不经意地写在那里，看是随意却独具匠心。楼梯间墙壁上的涂鸦画和一些陈设品也出自设计师之手，在这个DIY的制作过程中，又有谁不认为它本身就是一种最vogue的行为艺术呢。关于2层悬挂的“蚕茧装置”，作为[V]设计机构主持人的宋国梁先生谈到该创意理念来源于设计工作本身，因为设计创作的过程就是一个思想蜕变的过程，在经过痛苦的蛰伏后，破茧化蝶的时刻就是对设计人最好的精神褒奖。

本案让人印象最深刻的是设计师在方案之初就考虑到它的变化多重性，办公室的共享空间设计了不同的灯光环境模式，在周末闲暇，工作疲劳的时候或邀三两好友开办小型派对，办公室就变成了休闲酒吧、浪漫秀场，2层的投影机亦能以墙壁为背景播放新潮影片，当你在迷离的灯光下与朋友、同事品享醇厚美酒、翩翩起舞的时候，又怎会想到这一切都是发生在办公室里呢！好的空间创意设计可以把有限的空间功能变得更宽广，生活中魔术师往往在有限的条件下变化出超越范围的事物，优秀的设计师也一样可以对空间施展魔法，二者具有异曲同工之妙，从本案具有魔幻色彩的办公室空间创意中，设计师做到了这一点。

In people’s mind, the office space is definitely doomed to be a monotonous, rational and unappealing mode. Why not change this invariable working method and environment? The life needs the ideas of more multicultural values. The design of this case may be able to bring you some inspiration and consideration.

When you open the high glass door and enter the antechamber of the office, you will have a short illusion. The striking bar counter, wines and illusory music make you as if in a metropolitan fashion bar with unique style and full of innovation. The facade and top of the public space has been boldly applied with the mirroring stainless steel materials. The originally high depth has been enlarged. The designer has properly grasped the design scale. The colors are pure and the techniques are concise. The combination of different textures including rough brick wall, smooth metal, ferruginous plates and marbles are arranged harmonically and naturally. The style is modern and fashionable, but loses no simplicity and humanitarian. With the application of the dynamic supporting point, the staircase is just a 10mm thick steel plate and seems disappeared as seen from the side. The original concrete columns are exposed without any decoration after being roughly treated. Together with the two-layer old stub, the old floor gives off a faint natural atmosphere. The settlement of the times just accidentally stops here, which seems unconscious but intentional. The doodles on the wall in the stair half and some displays are also made by the designer. During this DIY procedure, no one denies itself a most vogue behavioral art. As for the “silkworm cocoon device” suspended in the second floor, Mr. Song Guoliang, the owner of [V] design organization, has mentioned that this creative idea comes from the design work itself. He thinks the procedure of the design creation is a procedure of idea transmutation. After the painful hibernation, the time to break out of cocoon and become a butterfly is the best spiritual award to the designers.

The most impressive of this design is that the designer has considered its multiplicity of the changes at the very beginning of the plan. The common space in the office has been designed with different modes of light environments. If there is time during the weekends, when you feel tired after work, you can invite several good friends to have a small party. At this time, the office becomes a recreational bar and a romantic show place. The projector in the second floor also can play the latest fashionable films with the wall as the background. When you enjoy the good wines and dance with your friends and colleagues under the misted lights, how you can imagine that all these are taking place in the office! A good creative design on the space can extend the limited space functions wider. The magician in real life can turn out the objects exceeding the scope under the limited conditions. The excellent designer also can make magic on the space. Both are different in approach but equally satisfactory in result. The designer for this case has realized this from this magic creation on the office space.

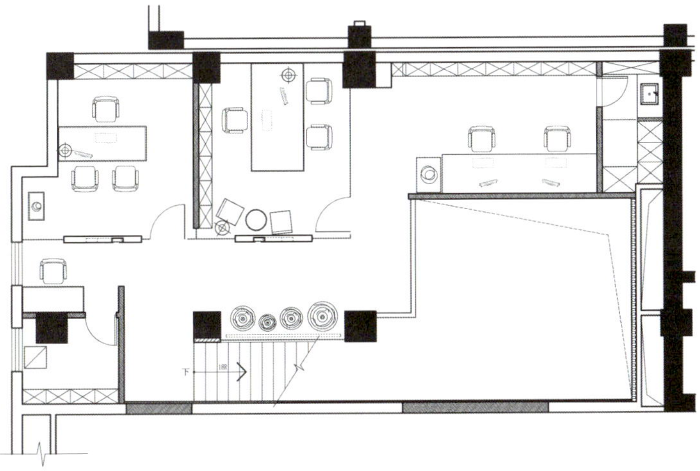

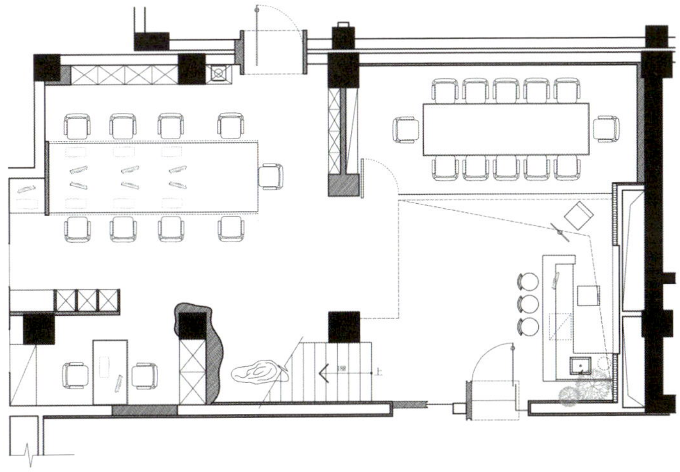

设计单位:新加坡PDI国际设计 / 设计:曾建龙

概念办公空间
Conceptual Office Space

面积:150m² / 坐落地点：温州市 / 主要材料:地面地毯、石材、木地板、橡木面板、涂料 / 摄影:香港ATA摄影机构

本案区别于传统的办公空间设计。在整体概念设计中采用会所的形式，虽然空间并不是很大，但是在空间结构层次上运用半开放的形式。玻璃和木饰面的运用结合，大理石的艺术处理，茶色镜面的运用以及地毯的动向运用，都围绕和体现空间的连贯性。阳台休闲区的设计带有东南亚风格的氛围，在这里可以看书、喝茶、聊天，可以去感受一份轻松和谐的心境，从而传达新办公空间的人性化。

This design is different from the traditional design of the office space. In the integral conceptual design, the form of the club is applied. Although the space is not very big, the half-open form is applied to the hierarchies of the space structure. The combination of the glasses and wooden veneers, the artistic arrangement of the marbles, the application of tawny mirrors as well as the dynamic application of the carpets are all surrounding and reflecting the continuity of the space. The design of the recreational area in the balcony has an atmosphere of Southeastern Asian style. Here, you can read the books, drink teas, chat and enjoy a relaxing and harmonious feeling so that the humanitarian of the new office space is delivered.

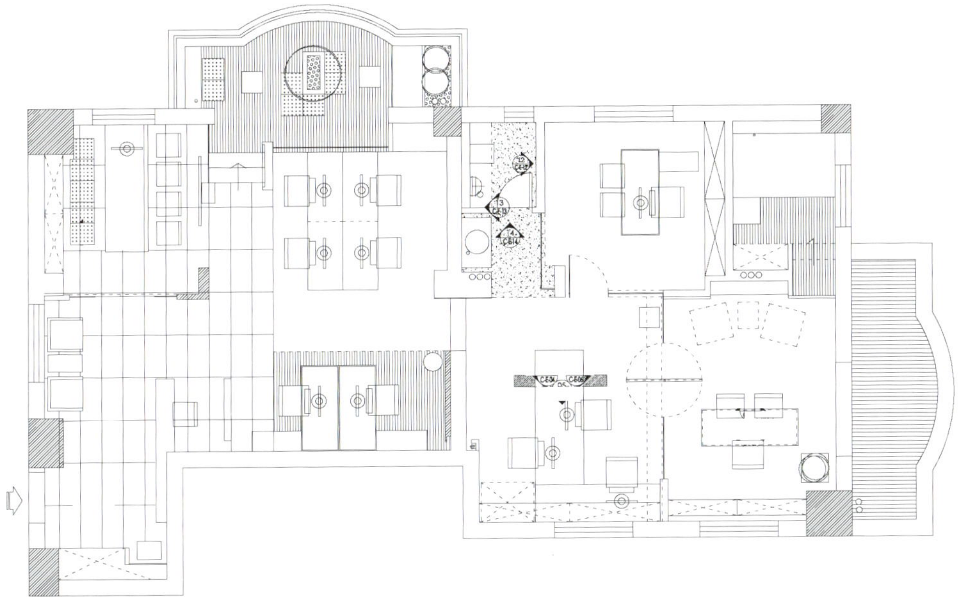

ARCHITECTURA
DESIGN FIRM
PDI
PHITA
DESIGN
INDUSTRIES
SDN
BHD

G&D
SINGAPORE
INTERIOR
ARCHITECTURAL
DESIGN FIRM
PDI
PHITA
DESIGN
INDUSTRIES
SDN
BHD

设计单位：于强室内设计师事务所 / 设计：于强

招商银行VIP贵宾厅
VIP Hall of Merchants Bank

面积：119m² / 坐落地点：深圳宝安机场

在“与时俱进，因客户而变”的新的招行企业精神指导下，时尚、活力成为本案的定位方向。设计以弧线为基调，把原棱角较多的单空间划分为若干个可自然过度且相互关联的小功能区域，以立面及天花层次形成各个小空间的独立感，视线上则是互通的，以重新巩固空间的整体感，亦散亦聚。弧线的流动感与随机性营造闲适而不惰怠的气氛，巧妙满足了各个功能需求。整体的装饰风格在表达招行“务求创新”企业精神的同时，追求品质、品位的高端形象。

Under the direction of new Merchants Bank' s corporate spirit of "keeping up with the times and changing with the customers", this design is oriented in fashion and vigor. Based on the arc lines, the original single space with quite many corners is divided into the numerous small functional areas what can be transferred naturally and interlinked. The independent sense of each individual space is formed by the layers of the facades and ceilings. It is interlinked in the visions sot that the integrity of the space is reinforced again, both scattered and collective. The flowing sense and randomness of the arc lines have created an easy but not lazy atmosphere and artfully meet the requirements of each function. The integral decorative style strives for a high-end image of pursuing qualities and tastes while expressing Merchants Bank' s corporate spirit of "the pursuit of innovation".

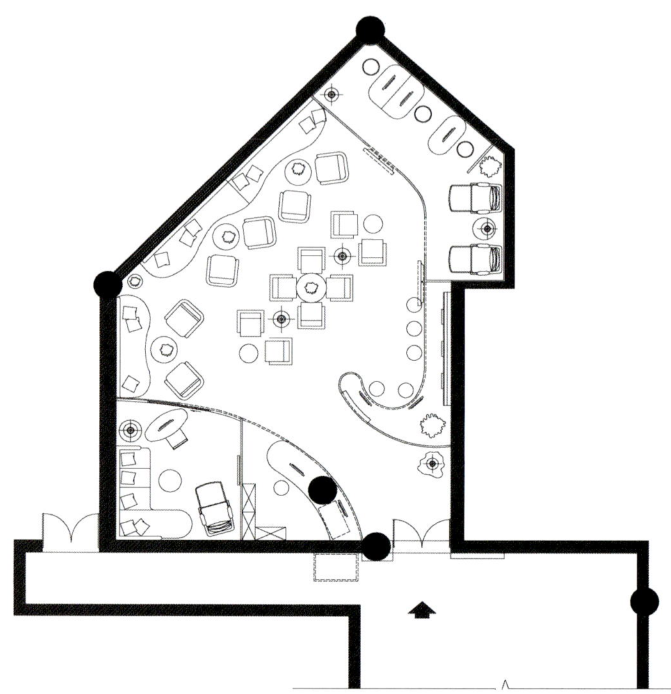

金葵花
SUNFLOWER

设计单位：上海现代建筑设计（集团）有限公司 / 设计：黄斌 / 参与设计：柯旸、施亮

泉峰贸易有限公司办公楼
Office Building for Quanfeng Trading Co., Ltd.

面积：26000m² / 坐落地点：南京 / 主要材料：枫木、石膏板、矿棉板、石灰石、石材、地毯 / 摄影：贾方

本项目建筑设计者为美国park&will公司，建筑师在充分考虑地域特点和时代特征的基础上，对该设计进行了一次大胆的尝试。整体建筑风格简洁而富有变化，将建筑、景观和室内综合考虑，加入中国园林空间特征的因素，使室内与室外取得了良好的对话与交流。在室内设计上秉持了建筑设计的思想与风格，采取大块面组织方式，强调室内与室外的交流，室内与建筑的协调，形成简洁、明快、结构清晰的室内空间环境。在铺底设计上，采用几种模数规格的石材混拼方式，简洁中富有趣味性。在主入口门厅设计上，局部设置绿化——竹子，既活跃了室内空间，又体现了中国古典韵味。在研发大厅中，采用高科技的展示方式，彰显了科技浪潮下的时代特征。

The architectural designer for this project is Park & Will Company from the United States. With the full consideration of the local features and the time characteristics, the architect has made a bold attempt to this design. The style of the integral building is simple but rich with changes. The building, landscaping and the interior have been considered comprehensively so that the interior and the exterior have had a good dialogue and communication together with the factors of the space characteristics of Chinese gardens. With the continuity of the ideas and styles of the architectural design, the interior design has applied the large-area arrangement to address the communication between the interior and exterior as well as the coordination between the interior and the building to form a simple and vivid interior space environment with clear structures. On the flooring design, the mixed arrangement of the stones materials of several modulus and specifications is applied, simple and interesting. As for the design of the hallway at the main entrance, the green bamboos arranged in some parts have animated the interior space and demonstrated the Chinese classic charms. In the research and development hall, the high-tech exhibiting way has been adopted to evidently display the times characteristics in the technologic tides.

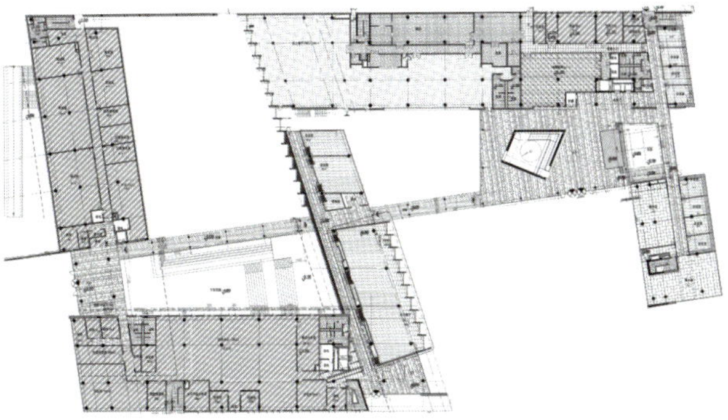

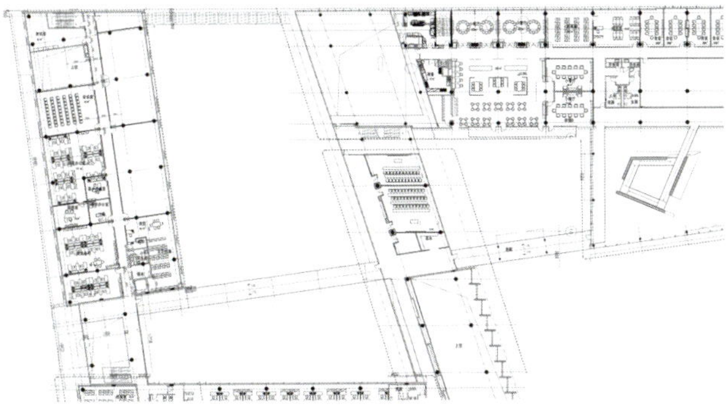

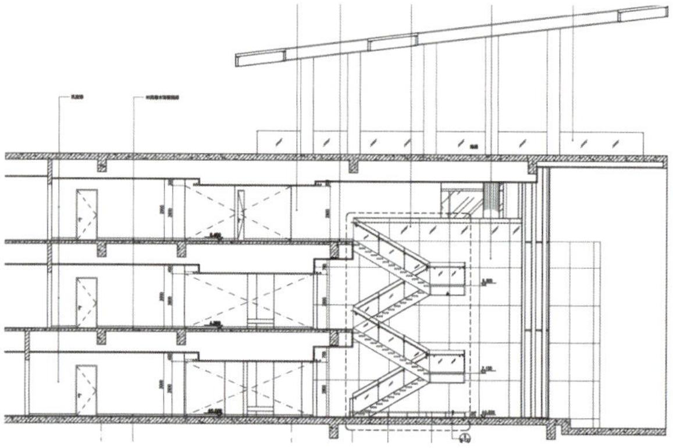

设计:李越

井冈山革命博物馆新馆
New Museum of Revolution in Jinggang Mountains

面积:7978m² / 坐落地点:江西井冈山市 / 主要材料:石材、氟碳喷涂铝板、玻璃、木质吸音板、涂料、地毯 / 摄影:张嗣华

井冈山革命博物馆新馆是一座全新的革命博物馆，位于井冈山红军南路，它是陈列展示井冈山革命历史、宣传井冈山精神的一个爱国主义教育基地。建筑设计上象征“革命之路”的红色砂岩大台阶由入口广场延伸至主题雕塑平台，一路分为四组梯段，象征井冈山革命历史中的四个阶段，坡屋顶采用砖红色金属板，与当地传统建筑取得呼应。室内设计延续建筑特有的“红色”主题，作为主要交通流线的U形展示走廊在整体白色基调的背景下，采用简洁的红色氟碳喷涂金属垂板吊顶，和室外的红色之路及红色屋顶进行对话。当参观人流通过大走廊进入各个展厅时，随时可以透过走廊一侧的大玻璃幕墙，感受到室内环境和整个建筑融为一体。位于博物馆地下室的报告厅进厅通过灯光的处理使这一没有自然采光的空间显得明亮开阔，白色素雅的格调是这个酝酿感情的空间最好的注释。报告厅的吊顶采用两组交错折行的金属波纹板和平板的特殊形式象征了中国革命道路的艰辛曲折，墙面用块面感很强而疏密有序的条形吸音板来表达朴实硬朗的感受，展现了革命博物馆报告厅的特殊气氛，红色座椅及舞台再次渲染了设计主题。贵宾厅的红色倒梯形吊顶造型象征了建筑层层红色屋顶的概念，方正的大型灯具也是为此空间度身定制。整个室内环境始终贯穿着建筑的主题，设计语言贴切而简洁有力，和建筑设计呼应延续，相得益彰。

New Museum of Revolution in Jinggang Mountains is a brand-new museum of revolution and is a patriotism education base where it displays the revolutionary history in Jiinggang Mountains and propagandizes the spirits of Jinggang Mountains. The big red sandstone steps that symbolize "the road of revolution" in the architectural design are extended from the square at the entrance to the thematic sculpture platform. It is divided into four sets of stages in total, which symbolize the four stages in the revolutionary history of JInggang Mountains. The sloping roof is applied with the brick red metal sheets to be corresponding to the local traditional buildings. The interior design has continued the unique "red" subject of the building. As the main passage, the U-shape exhibition corridor in the integral white background has been applied with the simple red fluorocarbon powered metal panel ceiling to have a dialogue with the exterior red road and red roof. When the visitors pass through the big corridor and enter each exhibition hall, they can feel that the interior environment has been integrated with the whole building at any time through the big glass curtain wall at one side of the corridor. The conference hall located in the underground of the museum, a space without natural lighting, looks very bright and wide through the arrangement of the lights. The white simple but elegant style is the best interpretation of this space filled with the feelings. The ceiling in the conference hall is made in a special form of two pairs of corrugated metal plates and flat panels that are interlaced and folded, which symbolizes the hardships and complications in the Chinese revolutionary road. The wall surface is covered with the stripe-type acoustic panels with strong sense of blocks in proper density to deliver a simple and strong feeling and display the special atmosphere of the conference hall in the museum of revolution. The red chairs and stage have again exaggerated the theme of this design. The red ceiling in the modeling of inversed staircase in the VIP hall has symbolized the conception of red roof of the building and the large square lighting fixtures are also custom-made for this space. The theme of the building has run through the whole interior environment all the time. The design language is appropriate and simple but expressive, which continues and corresponds with the architectural design, complementary with each other.

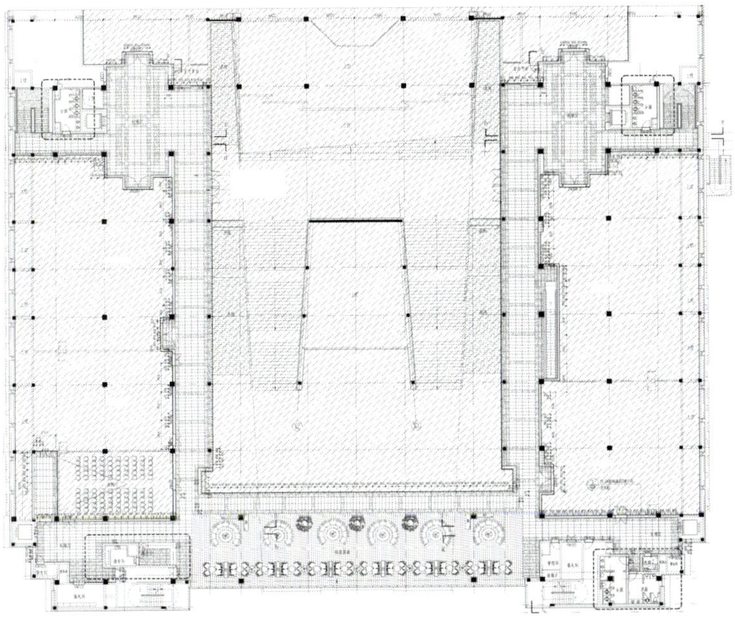

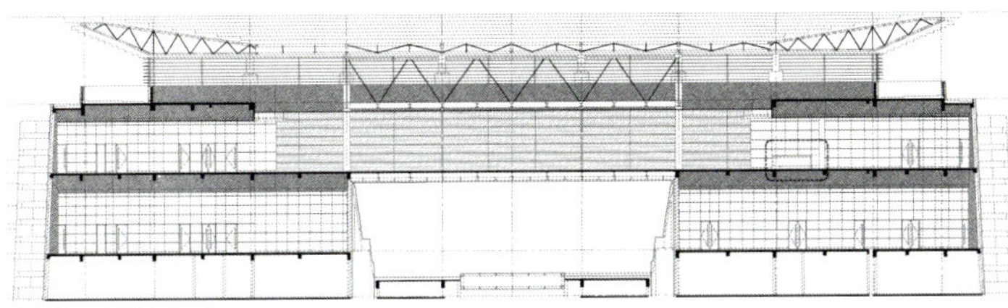

设计:Keiichiro SAKO, Akiko MIYAZAKI

MATERIALITY

面积:110000m² / 坐落地点:北京 / 摄影:Misae Hiromatsu

“三里屯SOHO”是大型城市开发项目，我们负责5座办公楼的室内设计。而办公室样板房的设计可视为整个办公楼室内设计的基础。

为了突出办公室的延展性，用不锈钢制成的细长钢架随意地布置在墙面上，显示速度感和活力。钢架的背板用四种彩色皮革覆盖，五彩斑斓，并使该空间成为环绕三里屯的氛围。透明的亚克力圆柱体作为隔墙，用圆柱体内反面的变形图像将办公楼逐步分割成不同的区域。

工作台沿墙以弯曲的方式布置。办公室被设计成“无领土”空间，并且新的移动线必须与未来的城市——“三里屯SOHO”的新工作方式匹配。

“SANLITUN SOHO” ,huge urban development project is just located at due south of “New Sanlitun” we took part in designing. We are responsible for interior design of five office towers. Firstly, we started to design of the office mock-up would be regarded as grounds for the whole interior design of real towers.

To emphasize the extendability of this office,slender shelves made by stainless steel are set in the wall at random, and then people gets speed feelings and activity of the space.

The back board of the shelves are covered with 4 kinds of colored leather, giving colorful feelings in chic and making the space become the atmosphere of surrounding Sanlitun area.Transparent acrylic cylinders are listed as partition. The office is divided into different areas gently by distorted images of contrary side in the cylinders.

Working table goes along the wall in curved way.We designed the office as non-territorial one, and the new moving line must match the new way of working in the futuristic city,” SANLITUN SOHO” .

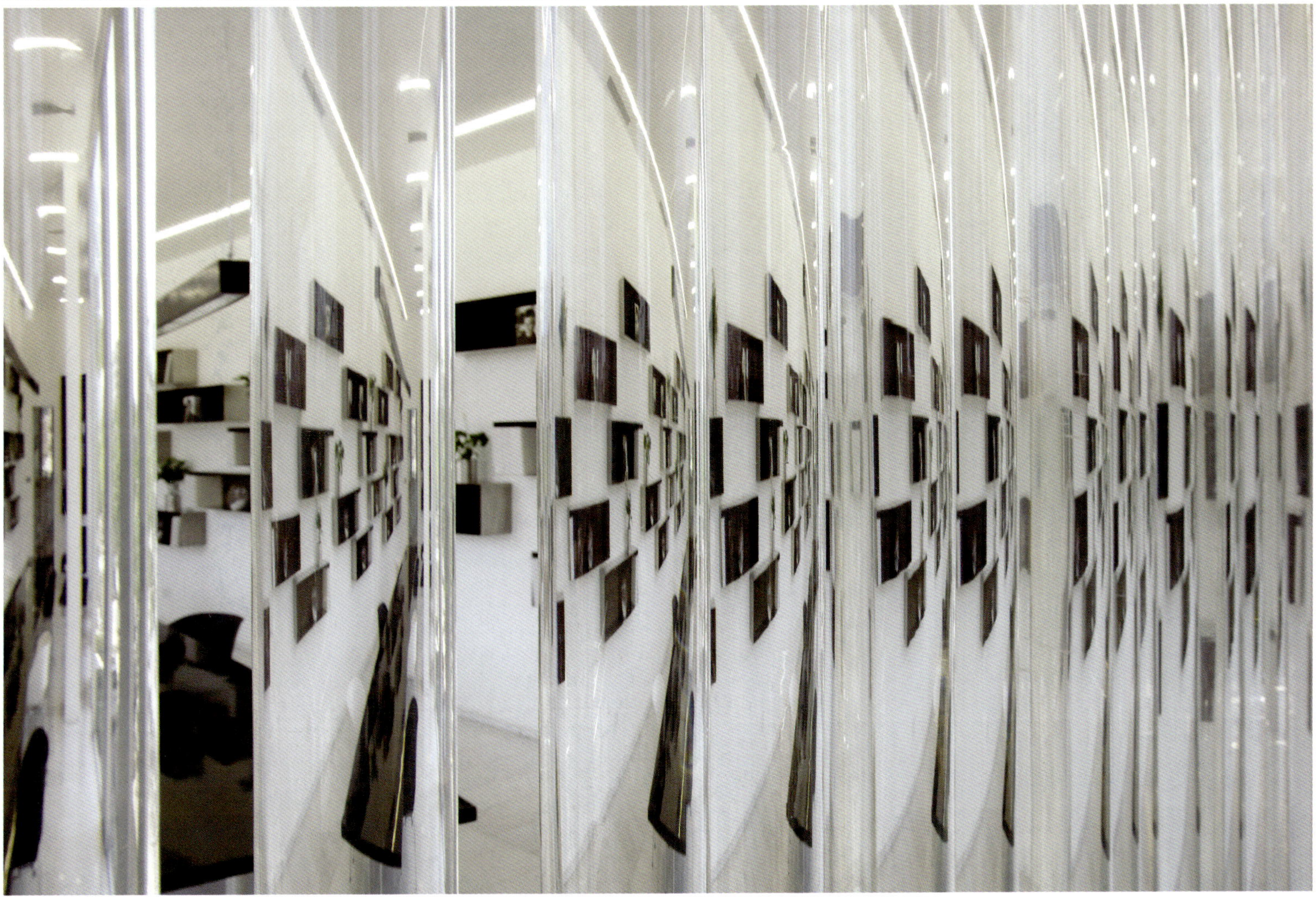

设计单位：北京包达铭室内装饰设计有限公司 / 设计：王闳

世新传播机构总部办公室
Office for Headquarters of Shih Hsin Media Group

面积：1000㎡ / 坐落地点：北京 / 主要材料：地毯、玻璃、金属制品 / 摄影：萧贵勇

在一个相对较大，较为宽松的环境里，用什么元素来营造办公的气氛，用什么指导思想影响办公空间的格调，同时鼓舞鞭策员工是现代办公空间所面临的核心问题。设计师在体现这方面有自己鲜明的观点，许多古典元素被输入进这间现代的办公空间。空间以白色花窗门为界面，一开一合间，空间的整合和通透淋漓尽致。两个场域的家具配置倾向也极具戏剧性，会议桌是意大利的，椅子却是官帽椅，可谓中西合璧。一辆哈雷摩托车的收藏居然成了员工工作之余的最爱。设计师在营造办公空间的舒适度和审美度方面也有自己独到之处，比如在员工办公空间的超大尺度和超大储物藏品上，让员工备感“受宠”和自豪。

In a comparatively large and quite relaxing environment, which kind If elements shall be applied to create the office atmosphere as well as inspire and urge the staffs is a key problem confronted with the modern office space. The designer has his own clear-cut viewpoint with respect to the expression. Many classic elements are input into this modern office space. The white patterned door is used as the interface in the space. During the opening and closing of the door, the integration and penetrability of the space are fully revealed. The furniture in two areas is arranged also very dramatically. The conference table is from Italy while the chairs are in the shape of the officer' hat, which can be called as a combination of Chinese and Western. The collection of a Halley motorcycle actually becomes the favorite of the staffs during their rest time. The staffs have felt extremely "being flattered" and proud of the super large size of the staff's office space and storages as well as the collections.

设计单位:壹正企划有限公司 / 设计:罗灵杰、龙慧祺

壹正企划办公室
One Plus Partnership Office

坐落地点:香港 / 主要材料:油漆、地毯

本案位于香港中央商业区的外围地带，对于这对近期炙手可热的设计师兼创办人来说，这所工房并不止于他们发展意念的地方，其本身也是一件见证他们新一个阶段的杰出制成品。

因应设计师们的喜好，那扇黑色的大门迎接客人进入设计师深层的设计思维空间，被玄关的昏暗氛围拥抱时就能有所感受。他们从不觉得黑色代表沉闷，而是代表了创作的自由空间。那里的黑色墙上的石头漆和先锋派字体的商标看板就说明了其作品里的笔触。而在房间内的闪绿色布帘和绿色古典花纹座椅，与附近的环境形成光暗的对比，这也代表了这家年轻公司的成长。

除了思考，设计更需要沟通和合作，亮丽的办公室正好为此提供了合适的空间。光亮的爵士白云石会议台和前卫造型的透明坐椅反映着窗外英皇道的繁华气氛，为室内带来生气，却又免除了不必要的烦扰，营造一个使会议更顺利的舒适环境。

处于后台区域的工作区以开放式间隔完善地划分，设计师最喜爱的黑色依然是当中的主调，但适当的颜色比例缓和了黑色的吸光特性，所以由全高窗户进来的天然光并没有受到影响。最为夺目的陈设是工作间内订造的十字形光管组合，此外形代表着公司的标志，使外观及实用性合二为一。这正体现了壹正企划的创作理念。

Located just on the edge Hong Kong' s CBD, the new atelier of One Plus Partnership is more than just an idea factory of Ajax Law and Virginia Lung, the up-and-coming designer duo who established the firm, it is also one of their impressive sculpture that appears as the start-up of their new era.

As a reference of their usual preference, the big black wooden door welcomes people into the designers' deep thoughts of design craft, just as what people can sense while they are being embraced by the even darker reception area. Yet black never means dullness to the owners, instead it represents the freedom space where all the creations begin. The feature wall with rough paint coating and the bold avant-garde signage has explained what kind of touch is included in their design works. The shiny green curtains and the green chairs with classical patterns, on the other hand, formed a bright contrast with the surroundings, as well as its meaning of this firm' s youth and growth.

Besides brainstorming, design also requires communication and cooperation, the luminous conference room is an ideal venue for this purpose. The glossy white marble conference table and the cool looking transparent seats are reflecting the energetic atmosphere of the hustle King' s Road below, but without the unwanted disturbance. Meetings can be as smooth as it can be in such a cozy circumstance.

Noteworthy features do not stop at the back stage, the areas are well defined by open partitions, the favourite black colour is still the domain while it is not absorbing much of the sufficient daylight of the full height windows. But the most eye-catching feature here is the tailor-made cross fluorescent light set, which is converting the company' s logo into practical lighting usage. The principle of One Plus' design is visualized through actual setting.

ONE PLUS PARTNERSHIP LTD.

设计：谈星火 / 参与设计：张磊、李晨晨、冀勉、张然

北京体育大学国家队训练基地综合馆
Comprehensive Gymnasium of Training Base for National Teams in Beijing Sports University

面积：29900m² / 坐落地点：北京 / 主要材料：钢、木质 / 摄影：江波

本项目是国家队为备战2008年奥运会建设的集训练、科研、教学为一体的重要训练基地之一。是为国家田径、跆拳道、艺术体操、蹦床队集中训练建设的现代化综合训练基地。而且考虑到我国未来体育训练的需求，因此建设规模和功能设置具有一定的超前性。建设地点位于北京体育大学校园内. 综合训练馆建筑面积29900m²，设计力求达到与比赛条件相同的“实战”环境，同时对体能训练、恢复治疗、安全防护设施等要求更高于比赛场馆。建筑总长291米，宽72米，主体结构为钢结构，其中田径馆室内200米跑道，是国内第一条木质基底塑胶跑道，馆内各项设施符合国际田联规定，除满足训练要求，还可举行国际室内田径大赛，是具有国际领先水平的室内田径训练场馆。艺术体操、蹦床训练馆是目前国内标准最高的场馆。室内设计风格简约，色彩明快，采光照明充足，材料新颖，科技含量高，以人为本，满足训练要求。新材料的使用及新颖的组合构成是此项目的设计特点。

This is one of the important training bases integrated with training, scientific study and teaching built for the national teams to prepare for the Olympic Game in 2008 and is a modernized comprehensive training base built for the intensive trainings of the national track team, kickboxing team, and rhythmic gymnastics team as well as trampoline team. Together with the consideration of the requirements by the future sport trainings in our country, thus, the construction size and the function setting shall be of certain foresight. The comprehensive training gymnasium, with a floor area of 29900m2, is designed to reach the “actual” environment same as the competing conditions to the maximum. At the same time, it has higher requirements for the physical trainings, recovery treatment and safeguarding facilities etc. than the competition gymnasiums and stadiums. The whole building is 291m long in total and 72m wide with the main steel structure. The 200m long indoor racetrack in the track gymnasium is the first wood-based plastic racetrack in China and each facility in the gymnasium meets the requirements of International Amateur Athletics Federation. In addition to meeting the training requirements, it can also hold the international indoor track competitions. It is an indoor track training gymnasium of the international leading level, among which the rhythmic gymnastics and trampoline training gymnasium currently is the gymnasium of the highest standards in China.

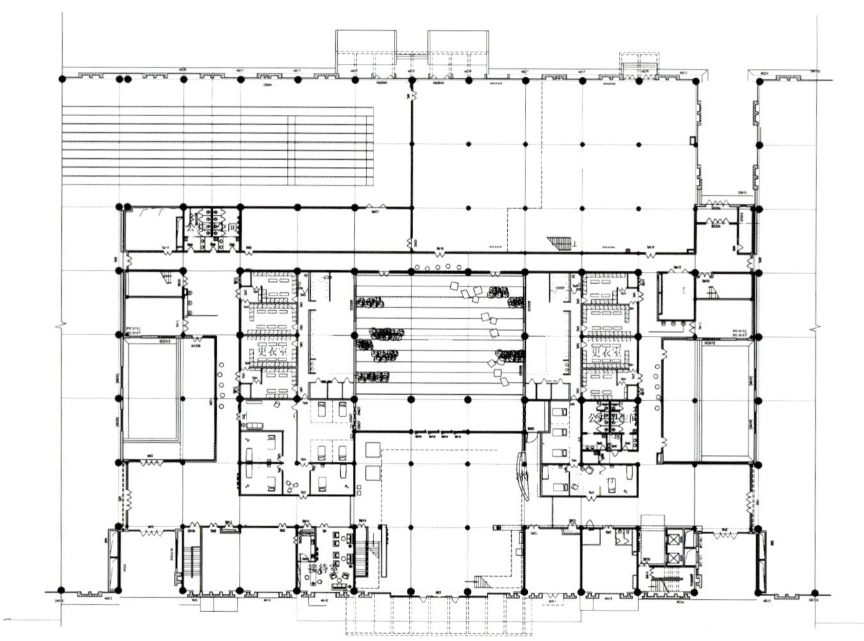

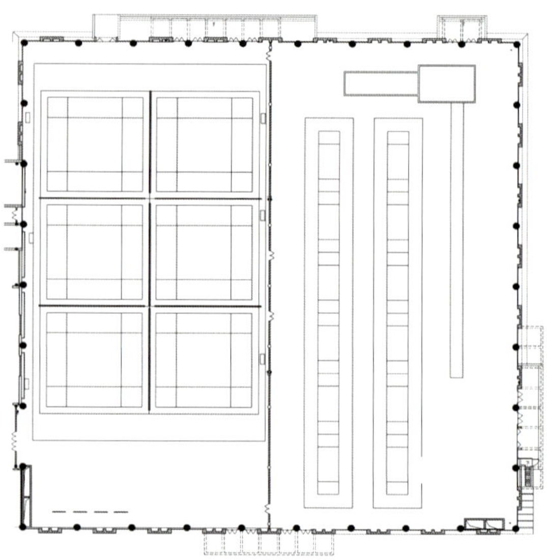

10
9
8

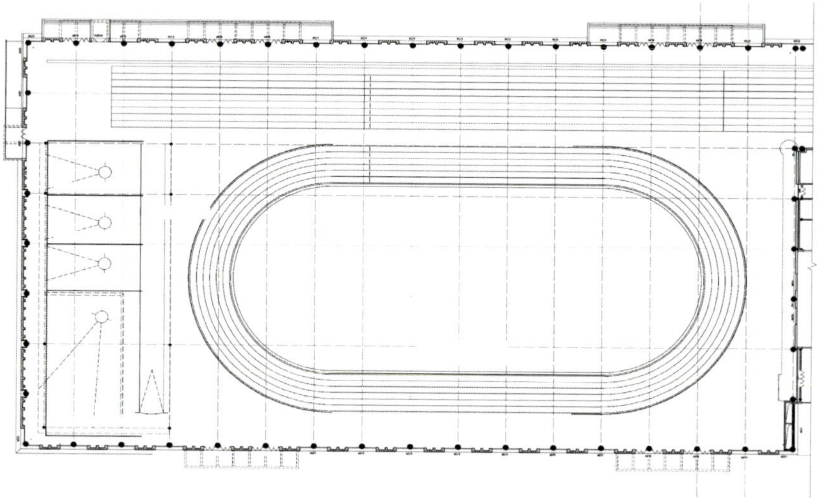

设计单位:深圳市鼎维设计有限公司 / 设计:史永杰

香港华侨城集团写字楼
Office Building of "Hongkong Overseas China Town"

面积:670m² / 坐落地点：深圳 / 主要材料:石材、玻璃、砂岩 / 摄影:郭一兵

入口门厅的主背景墙采用锯槽砂岩板与焗漆玻璃巧妙叠加，在材质与色调的对比上带给人轻松而不失稳重的气息，砂岩板钜槽处理使整面背景墙极具张力。浅色大理石地面由外向内延伸与暗藏光槽的玻璃接待台相接，三扇全部打开的通顶玻璃门使门厅的空间感得以扩展。中式家具与青铜雕像的装饰简洁，使现代的办公空间充满了浓厚的人文气息。现代双光照明系统大大减少了光污染，营造了一个明亮而又柔和的办公环境。

At the entrance of the hall, the great grooved sandstone slabs and the powdered glass are artfully overlapped together to form the main background wall. The contrast between the materials and colors has brought a relaxing but steady mood. The great grooves on the sandstone slab have endowed the whole background wall with great tension. The light marble ground is extended inward from the outside and connected to the glass reception desk with the hidden light channels. The three top glass doors that are completely open expand the space of the hall while the Chinese furniture and bronze statuaries have filled the concise and modern office space with the strong humanitarian atmosphere. The modern dual-light lighting system has greatly reduced the light pollution and created a bright and soft office environment.

设计:殷艳明

创域艺术设计有限公司办公楼
Creative-space Design&decoration Do.,Ltd.Office

面积:340m² / 坐落地点:深圳 / 主要材料:灰色涂料、灰镜、木材 / 摄影:臧兵

设计是空间的艺术，艺术的本质臻源于设计师内心的情感。作为设计公司的办公空间，不在于花哨的材料堆砌与复杂的造型展示，如何充分利用建筑空间的形态语言，从无序的空间中通过单纯的设计语言体现有序的比例尺度和形态美感，达到功能和形式的完美结合，这才是本案追求的重点。

空间形态方正，开放而友好。接待、交通、办公功能有机结合，完整高效，在延续设计公司一贯的灰色标准色系中，各种材质的运用恰到好处。灰色涂料传达着朴实、沉稳的语言，灰镜运用表现了虚实空间的共融，条形木材结构亲和了人的关系。各种设计元素在展现各自个性、张力与情感的同时，营造了丰富的视觉层次和结构美感。

现代感十足的屏风，具有独特品位的装饰画与镂空木雕传达出脱俗的审美气息，不经意间时光与情感已历炼出设计师的审美情结。灰色系空间里，暖色木与灰镜的对话，序列感、秩序感以及空间、人、物按照美学理念做出了丰富的诠释。

在虚实之间一切又回归到自然，那是一种建构在本质上的艺术享受。

Design is the art about space. As the office space of designer, it is important for us to think how make use of the language of building space to pass comparison dimensions and appearance pleasant impression in order to attain the perfect combination of function and form.

The office space which is square shows opening and friendly. The three function sections which include reception, transportation and work combine into a unified whole, and offers the environment of integrity and efficiently. We choose the consistent gray color in the office to inform simple and cool-headed feeling. Not only the usage of various materials, such as the grey coating, ash mirror and timber, but also the modern screen, the decorative paintings and the empty wood carving in the office construct abundant sense of vision layer and structure pleasant impression.

We have made use of all design chemical elements to do abundant annotation about the office space according to the esthetics principle.

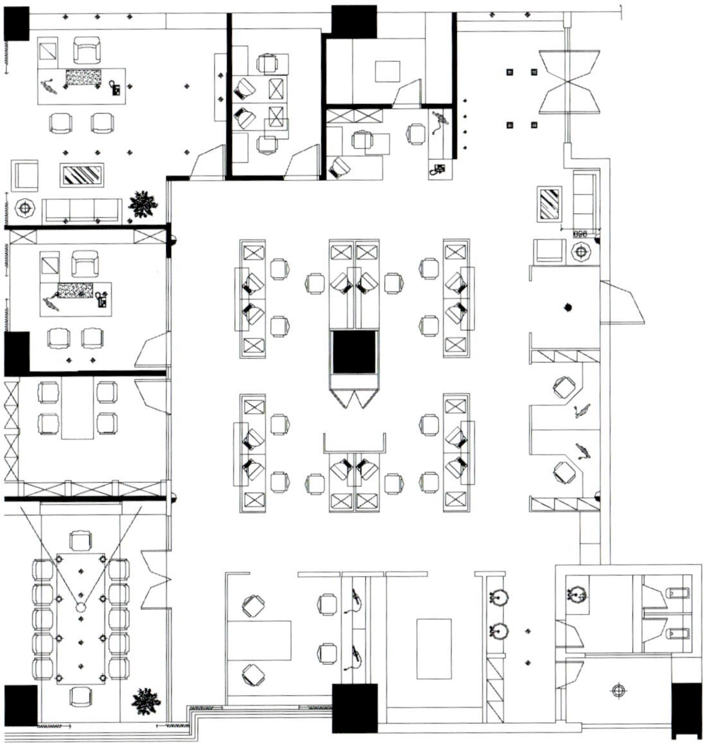

Creative space

设计单位:汕头市蓝鲸室内设计有限公司 / 设计:陈骏

乔艺塑胶有限公司
Qiaoyi Plastic Co., Ltd.

面积:1400m² / 坐落地点:汕头市 / 主要材料:抛光砖、吸音板、落地钢化玻璃、镜面玻璃 / 摄影:邱小雄

本次设计的焦点在于，原建筑挑空大堂进深不够，两条圆柱位于大堂中央，空间比例失调。所以着重在空间的拓展思维上去做文章，整个设计中没有多余的装饰，以期获得更好的空间使用效果。

首先，为了模糊室内外空间界限，设计师将室外两条圆柱纳入室内，采用清玻幕墙作为室内外分隔的材质，从而使得大堂向外延伸，空间进深扩展明显，在视觉上解决了圆柱位于大堂中央的尴尬。

为了营造纵深感及视觉焦点，采用了橙红色木纹吸音板和镜面，在形成强烈的空间导引作用的同时，也起到了延伸色彩及空间的感觉，弥补了一楼相对于二楼的空间缺失。

在色彩的衬托上，运用黑、白、灰原理处理空间中的各个面，大厅主通道、天花、圆柱的白色由于电梯间墙体周围地面、天花的深灰色的衬托，而形成大厅的主体空间。所有黑、白、灰的营造最终衬托出亮丽的橙红色主墙。同时为避免电梯间、墙面由于采用深灰色而形成的沉闷感，设计师加入不锈钢条去作横向分隔排列，弱化笨重感的同时营造了细节。

二楼办公区沿用公共挑空区域的思维，以拓展空间为主，采用清玻幕墙，利于采光的同时，加强了办公区域的联系，过道天花的悬浮感及墙面的暖色墙纸，解决了通道的沉闷及狭长。处于另一楼层的VIP客户接待区，采用六边形这一中国传统图案进行大胆的变形，天花板的黑色镜面将这一主题墙向上无限延伸，获得华丽、尊贵的感觉。

The focus of this design lies in that the depth of the overhanging lobby in the existing building is not enough and two round columns are located in the center of the lobby so that the proportion of the space is unbalanced. Therefore, we put our emphasis on taking action on the expanded thoughts of the space. The whole design does not have any extra decoration with expectation to obtain the better effects on space utilization.

First, to obscure the boundaries between the interior and external spaces, the designer has included the two exterior round columns into the interior and applied the clear glass curtain as the material to separate the interior from the exterior so that the lobby can be extended outwards and the depth of the space is expanded obviously. In this way, the embarrassment that the round columns are located in the center of the lobby has settled visually.

Furthermore, to create a deep sense and visual focus, the salmon pink grainy acoustic boards and mirrors are applied, which also has a sense of extending the colors and space while forming a strong function of space introduction and has offset the space lacking of the first floor relative to the secondary floor. In the setoff of the colors, the principle of black, white and gray is applied

to arrange each surface in the spaces. The white on the main passageway in the lobby, ceiling and round columns become the main space in the lobby as set off by dark gray on the ground surrounded by the walls in the elevator room and the ceiling. The arrangement of black, white and gray has finally set off the main bright salmon pink wall. At the same time, to avoid the tediousness due to dark gray applied in the elevator room and wall, the designer has added and arranged some stainless steel bars transversely and separately to create the details while weakening the heaviness.

The office area at the second floor has continued the idea of the public overhanging area. With emphasis on space expansion, the clear glass wall is applied to strengthen the contact in the office area while being good at day lighting. The overhanging senses of the corridor ceiling as well as the warm colored wall papers on the wall have solved the tediousness and narrowness of the passages. The reception area for VIP customers at another floor has been applied with hexagon the traditional Chinese pattern, which has been changed boldly. The black mirror on the ceiling has extended this wall of subject infinitely to achieve a magnificent and noble feeling.

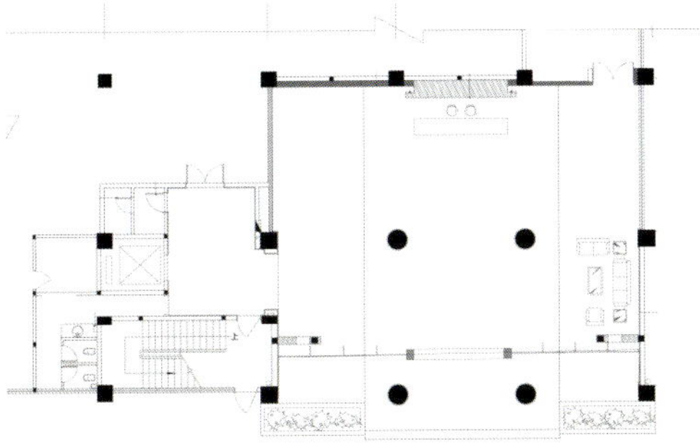

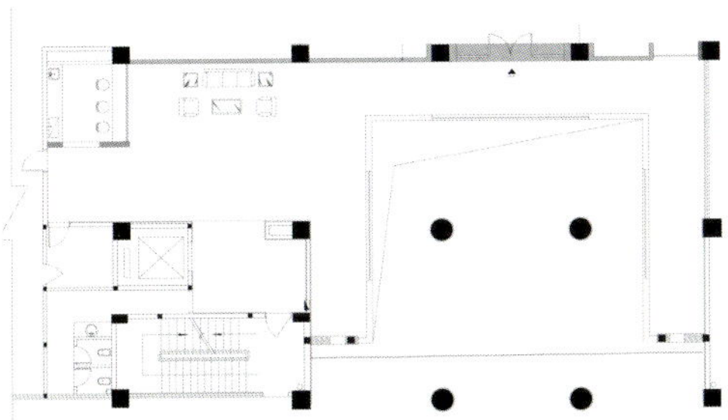

设计单位：中国建筑设计研究院环艺院室内所 / 设计：张晔、盛燕 / 参与设计：饶劢、邓雪映、刘烨、李晨晨、尤林、杨鑫、刘立洋

大兴文图馆影剧院
Daxing Culture Center and Cinema

面积：30000㎡ / 坐落地点：北京

大兴文图馆位于北京大兴区中心区。在建筑方案后期室内设计便予介入，目的是帮助建筑设计完成从外至内的一体化设计，完善建筑内部空间，达到内外的和谐统一，并使各室内空间满足使用功能要求。室内设计中贯彻建筑的空间形体和逻辑关系。通过建筑形体、建筑表皮向室内的延伸、建筑模数的延用，以及细部构造工法的统一，来实现建筑内部空间的整合。

Daxing Culture Center is located in the central area of Daxing District, Beijing . In the later stage of the construction plan, the interior design has got involved in the work with the aim to help the architectural design complete the integration design from the exterior to the interior and improve the interior space in the building so that it can reach the harmony and unification between the interior and exterior as well as meet the requirements for the functions of each interior space. The space form and structures a well as logic relationship of the building have been carried out throughout the interior design. The arrangement on the interior space of the building is realized through the building form, structure, the extension of the building surface to the interior, the continuity of building modulus as well as the unification of the structural workmanship for the details.

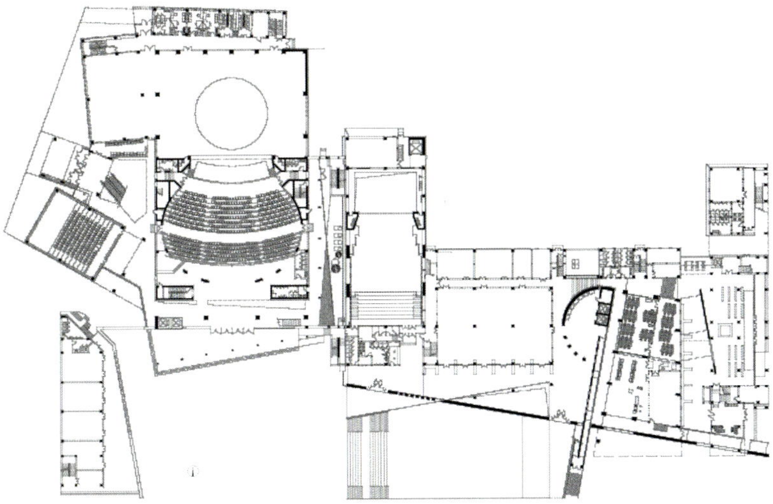

设计单位:Shanghai Infix Design / 设计:MARGAUX LHERMITTE、方钦正

NACO上海中心
NACO Shanghai Center

面积:100m² / 坐落地点:上海 / 主要材料:铝板、不锈钢、玻璃、混凝土 / 摄影:胡文杰

与NACO极富创意和革新思维的设计理念相同，他们的工作室也处处洋溢着激情和灵感。整个空间如同一个透明的玻璃盒子，明亮而充满活力。主入口采用了整块压成波浪型的红色铝板，从地面直竖向天棚，从而完成了门的功能。门把手，门框以及其他附件的“无”恰巧带给人们一种“未知”的体验。

空间布局整洁、从容，简单、拙旧的办公用具反而让人窥见一丝中西方文化的包容和交流。红白色调富有秩序感的整合更显空间的井然有序。在材料运用方面，地面和一堵墙壁运用了具有镜面反射的不锈钢材料，空间景象出现有趣而丰富的变化。室内一角保留着原来留下的电灯开关和红色的消防箱，它们融入了设计的整体成为一种有趣的元素，营造了工业化的室内风格。室内立面的部分墙壁保留了未加修饰的混凝土粗糙表面，丰富的肌理尽显设计的大胆不羁。

每张桌子的底部都预留了几个挂钩，既方便员工挂包也节约空间。每个电脑旁都有一个印有naco标志的红色工具盒子，每个员工都将自己的办公物品整齐地收纳在红盒子内，并养成习惯。为了保持桌面和整个环境的整洁，员工们尽量减少将私人物品放置在桌子上，个人的参考资料、书籍等其他物品被放置在窗前的架子上，或前台区域开放式的大书架上；公共书籍、杂志以及材料样板则采取统一放置管理的方式，方便每个员工取阅。这些细节也令整个工作室时时呈现一种轻松愉悦的氛围。

The same as Naco’s design ideas of rich creativities and innovative thoughts, their studio is filled with enthusiasms and inspirations everywhere. The whole space looks like a transparent glass box, bright and vigorous. The main entrance is applied with a red aluminum board which is pressed into the wave form by a whole board, top to the ceiling from the ground to realize its function as a door. The “nonexistence” of the door handle, door frame and other accessories happens to bring an “unknown” experience to people.

The space is arranged clean and orderly. The simple and old office furniture instead is revealed with a sense of the comprehension and communication between Chinese and western cultures. The orderly arrangement of red and white colors has obviously shown the space in good order. In aspect of material application, the floor and a wall have applied with the stainless steel material of mirror reflection so that the interesting and wonderful changes have appeared in the space scenery. One interior corner is reserved with the original electric button and red fire hydrant box. They have been integrated into the entire design and become an interesting element to create an industrialized interior style. The partial wall of the interior facade has kept the bare rough concrete surface and the wonderful texture has fully shown the boldness and unruliness of the design.

The bottom of every table is reserved with several hooks so that the staffmember can hang their bags conveniently and the space is also saved. Beside every computer, there is a red tool box imprinted with the logo of Naco. Every staffmember can put all their individual office articles in the red box and forms a habit. To keep the desktop and whole environment clean, the staffmembers have minimized their personal belonging on the desk. The other articles such as personal reference documents and books are put on the shelves in front the window or the open big bookshelf in the reception area. The public books, magazines and material samples are placed and managed in a uniform way for easy reading by every staffmember. These details also provide the entire studio with a relaxing and pleasant atmosphere all the time.

global architecture

设计:贺钱威

HED Office

面积:300m² / 坐落地点:宁波 / 主要材料:素水泥、锈板砖、大理石、钢材、镜面、ICI

HED是一家专业设计公司，室内是6m的挑高空间，不规则的平面格局，利用空间优势，局部挑高，其他区域分割为二层空间，一楼为门厅、接待区、创意1区、经理和财务办公及吧台、卫生间。二楼在空间上处理成错落有序的错层空间，以抬高的“L”走道为分割界面，区隔出创意2区、会议洽谈区、材料样板区、设计总监室。三面为5.5m的大落地玻璃墙，面临绿化带和小河，户外景观得天独厚，为使室内外相互呼应，设计师充分利用户外景观，在室内空间格局上充分利用天然采光，材质上利用自然质朴的环保绿色材料、清水泥墙和大面积留白，来映衬户外的自然景观，让绿色与阳光充分渗透到室内每一个角落，仿佛户外与室内的界面已不复存在，让人最大限度的感受绿色，感受大自然。在空间处理上制造禅意——静态的美、光影的美、建筑的美、质朴的美、和谐的美，真正做到以人为本。

HED is a professional design company. The interior is a 6m high-rise space in an irregular plan arrangement. The space advantage is used to raise partially while the other areas are divided into a two-level space. The first level is for the hallway, reception area, creativity zone No. 1, and manager and finance offices as well as bar counter and toilets. The second floor is designed as an orderly split-level space with the raised "L" corridor as the dividing interface to be divided into creativity zone No.2, conference and discussion area, material sampling area and design director office. The three sides are installed with 5.5m high bottom-to-ceiling glass walls, confronted with the green belt and stream. The outdoor scenery is richly endowed by nature. The designer has made full use of the outdoor scenery so that the interior and exterior can echo with each other: the natural light is fully used for the interior space management while the natural and simple environmental protection materials, plain cement wall and large area whitewash are applied to set off the natural outdoor scenery and permeate the greenness and sun lights through every corner in the room as if there is no interface any more between the outdoor and indoor. In this case, people can enjoy as much the greenness and nature as possible. As for the space arrangement, it has created a sense of Buddhist —— a beauty of quietness, beauty of lights and shadows, beauty of buildings, beauty of simplicity as well as a beauty of harmony and can really realized people-oriented.

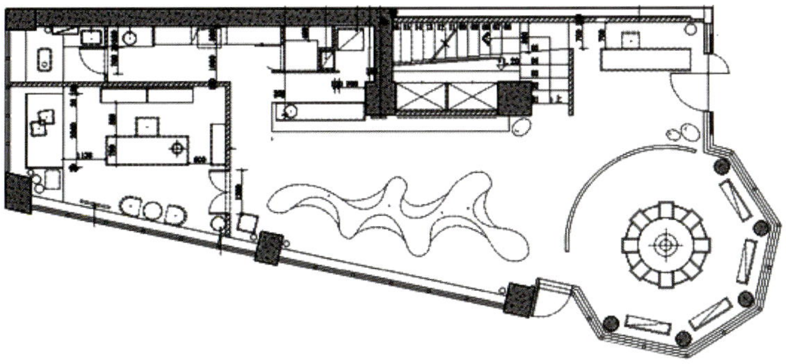

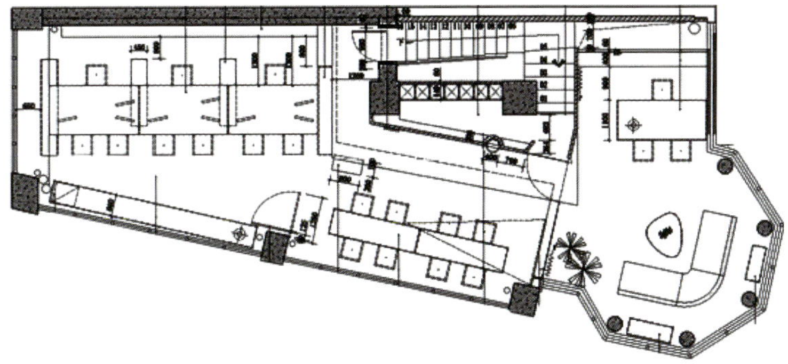

exhibition

商业・展示，不仅是一次合理的摆放
Business・Exhibition, not only a Reasonable Placement

陈耀光 Chen Yaoguang
杭州典尚建筑装饰设计有限公司创意总监、总经理
中国建筑学会室内设计分会副会长
中国建筑装饰协会设计委员会副会长
中国陈设艺术专业委员会副主任
浙江省建筑装饰行业协会设计委员会会长

郭锡恩、胡如珊

Keiichiro SAKO、Ken-ichi KURIMOTO、 Liang JU
陆希杰
曾建龙、林岳

隈研吾
Keiichiro Sako、Nobutoshi Hara
黄书恒
许志国
吕永中
孙天文
麦德斌
高立平
谢智明
陈耀光
杭州天澜建筑装饰设计院

鞋盒—PEDDER RED 香港旗舰店
圣诗得北京展厅
EIFINI
STEPHANE DOU
灯的摩登世界
圣诗得上海展厅
Z58
浪漫一身
驰骋劲驹的梦想
德国乐得福莱
纸围城 – 薇莳婚纱旗舰店
万德诺展厅
高陶洁具中国总部
亚寒带法国饰品店
伟祺玻璃建材有限公司办公室
浙江美术馆
亨利・戴家居

每时每刻，我们都无法选择地生活在一个被展示的世界里。

自然界的展示是宏伟无比的，它是以天地日月、春夏秋冬进行着千姿百态的时空现象，来展示地球生命空间的无限特征。这一点，人类无法与上帝的创意去相提并论。今天，我们在这里谈及的展示空间，首先是一种人为的环境创造，是一种传递信息的视觉场所，是一种建立在消费者和销售者之间的信息交换的有效过程。通过形态、色彩、材料、照明、声音和图文等诸多手段，将人们的视觉、听觉、触觉甚至嗅觉和味觉等一切感知、潜能调动在一个被围合的经过设计的空间内，构造一种人与物、人与人之间观赏和识别。

在我的理解中，将玻璃与钻石被同时放在一个环境里，玻璃毕竟像不了钻石，当小块玻璃被镶嵌在深色绒布之上，并在经过设计的光源下投射，此时的玻璃将会泛耀水晶般的闪烁。同样，棉纱天然的飘柔属性搁在厚重坚韧石块上，即使没有风，也仿佛会有被吹飘起来的感觉……这些通过设计被科技和艺术手法所展示的优秀案例，在我们这期商业·展示篇章里不乏其例：服装被悬吊于空中，向旗子一样高昂；鸡蛋的柔嫩润滑，与着装的肌肤需求混为一体；汽车的曲面弧线，在光反射的作用下，静态的空间里流溢着运动；被悬挂的灯饰都充分彰显着零件的工业化表情。其中一组服装和珠宝相存的白色展厅，能真正让人感受到白色的永恒魅力：它的清澈、剔透、纯净、高贵，使踏进展厅的所有观者在此不仅是一个有待消费的场所，更是一台艺术和创意的表演秀。此时此刻，艺术的视觉冲击已远远覆盖了产品的信息，传播与促销的直接意义开始淡弱。它的平面布置已远远超越功能和使用的定位，更像是T型台与观众席的布局……那一个纸折的墙体和晕眩纹理的立面，无不演示着设计师对自我独创境地的惊喜与自信，我们身不由己地被他们大胆、勇敢的感染力而折服……

展示空间艺术设计，是指特定物品、特定主题，在特定的场所做目标性摆设和演示的设计。它不仅是商业传达的形式，同时也包括博物馆、艺术馆、纪念馆等文化空间的一种形式。与商业展示所不同的文化空间展示，近几年我曾在关于美术馆设计的论坛上发表过自己这样的看法“看的场所，看不见的设计”，其室内设计是非常需要做到控制表现情绪的，尤其是艺术馆和博物馆，严格意义上讲，建筑和室内空间设计更应该像是一只画框，所有被展陈的艺术作品才是真正的画，如果画框被设计的过分精致耀眼，观赏画面的心情就会被打乱，人们之所以来到艺术馆主要还是看展品。室内设计应谐和建筑，构建一种低调的姿态，内敛的形态。所有具体和象征性的“积极”装饰符号在空间室内中出现，即便一点点，也是多余的泛滥，要有勇气让室内设计的“创意”消失在建筑空间和艺术展示之间。本人深深感到，艺术观赏是一种情绪交流的活动，但进行这样的活动必须拥有一个非情绪化的场所。近年来，我们做了些艺术馆、美术馆和纪念性的主题空间，这样的展示场所与许多商业室内设计有着许多不同的感受。艺术馆、美术馆不仅属于展示品，它还属于城市。许多城市许多民众到美术馆看作品也看环境。因此，这个场所需要充分体现它的大众性和交流性。咖啡店、音乐、艺术品商铺、图书、纪念品店等相关辅助空间，它们已构成艺术馆整体的一部分。它，是一个城市的停留点，人们在这里进行放松、悠闲，是一个养心养眼的过程。

无论何种展示概念，商业的、文化的、室内的、室外的，不论是产品还是艺术品，它的目的是给观者一次视觉的冲击。如果我们能做到不仅看得清楚、看得明白，而且还能让人看得有趣、看得心跳，那才是一件非常成功并值得庆贺的设计。商业·展示，不仅是一次合理的摆放。

We have to live in an exhibited world without any choice at any time.

The exhibition in the nature is incomparably magnificent. It has been changed with the space-time phenomena through sky, earth, sun and moon as well as spring, summer, autumn and winter to show the infinite characteristics of the lives and spaces on the earth. From this point of view, human beings can not be put on a par with the creativities of the god. Today, the exhibition space we have mentioned here first is a kind of man-made environmental creation, a kind of visual field to deliver information as well as a kind of effective procedure for information communication between the consumers and sellers. By the various means such as the forms, colors, materials, lighting, sounds and images and letters etc, all of people' s senses and potentials such as vision, hearing, touch and even smell and taste are mobilized into an designed enclosed space to constitute a kind of viewing and recognition between human and objects as well as among the humans.

In my understanding, glass and diamond are put in the same environment at the same time. After all, glass can not be like a diamond. When a small glass is embedded on the dark flannelette and projected by the designed light sources, it will be flooded with crystal-like flashes. At the same time, the natural waving characteristics of cotton yarns that are rested on the thick, heavy and hard stone blocks look as if it is waving as being blown even when there is no wind……for these excellent cases exhibited by technologic and artistic techniques through the design, there are a lot of examples in this chapter of Business · Exhibition; the dresses are suspended in the air, exalted like a flag; the tenderness and smoothness of an egg and the skin requirements for dresses are integrated together; the curve of the car' s hood face is flushed with the movements in the quiet space under the reflection of the lights; the suspended light ornaments are even fully revealed with the industrialize expressions of the parts. People can fully feel the everlasting charm of the white through a white exhibition hall with the coexistence of dresses and jewels: its limpidity, brightness, purity and nobility make this exhibition all not only a place for consumption to all the visitors entering it, but more a performance show of arts and creativities. At this moment, the visual impact of the arts has far covered the information of products and the direct meanings of distribution and promotion become weakened. Its plan arrangement has exceeded the orientation of functions and applications far, which looks more like an arrangement of T stage and auditoria....... that is a paper-folded wall and façade of dazzling textures and has demonstrated the surprises and self-confidence on the original creative conditions by the designer everywhere. We cannot help being convinced by their bold and brave appeals.

The artistic design of an exhibition space refers to a design that makes the targeted placement and demonstration in the specific location for a specific object and specific theme. It is not only a form of business delivery, but also includes a form of cultural spaces such as museums, galleries and memorials etc. The exhibitions of cultural spaces are different from the business exhibitions. In recent years, I have published my own opinion on it in a forum about the design of galleries as "visible place but invisible design". Its interior design shall be able to very necessarily control the expressive moods. Especially, in the strict viewpoint, for the galleries and museums, the design of architectural and interior spaces shall look more like a picture frame. All the displayed artistic works are the real pictures. If the picture frame is designed excessively delicate and brilliant, the mood of enjoying the pictures will be disturbed. The reason that people come to the galleries is mainly to see the exhibits. The interior design shall be harmonious with the building to constitute a low-profiled posture and introverted form. For all specific and symbolic "active" decorative symbols that appear in the interior space, even a little bit is also a redundant overflow. You shall have courage to remove the "creativity" of interior design among the architectural spaces and artistic exhibitions. I have deeply realized that the artistic view is a kind of mood-communicating activity, but such kind of activity shall be carried out in a non-emotional place. In recent years, we have made some art galleries and memorial thematic spaces. Such exhibition places have many experiences different from many interior business designs. Art galleries not only belong to the exhibits, but also belong to the city. Many citizens in many cities come to the art galleries for watching both the works and the environment. Therefore, this place needs to fully demonstrate its popularity and communicativeness. The related supplementary spaces such as coffee houses, music and artwork shops, book stores and souvenir shops etc have constituted part of the entire art gallery. It is a stop point of a city where people can get relaxed. It is a procedure pleasant both to the spirits and eyes.

No matter what kind of exhibition conception, commercial, cultural, interior or exterior, whether it is a product or artwork, its purpose is to give a visual impact on the viewers. If we can not only watch and understand clearly, but also feel interested and excited, then it is a very successful design worth congratulations. Business · Exhibition is not only a reasonable placement

设计单位:如恩设计研究室 / 设计:郭锡恩、胡如珊

鞋盒—PEDDER RED香港旗舰店
THE SHOE BOX – PEDDER RED FLAGSHIP STORE (HK)

面积:205m² / 坐落地点:香港 / 主要材料:橡木 / 摄影:莫尚勤

精心设计的零售店正逐渐成为一个品牌最好的形象宣传手段之一，有时甚至比产品本身更具影响力。在对Pedder Red这一新兴鞋类品牌零售商店的设计中，NHDRO希望创造出一个既符合零售店设计的规则，又能满足现代个性、理智的消费者需求的室内空间，从而塑造出充满魅力的品牌形象。

Pedder Red的设计理念是一个非常大胆的插入手法——商店两层运用了木块垂直交叠而成的橡木盒。这一木结构体将成为这一零售品牌令人印象深刻的视觉标识，它的灵感源于鞋带，以木材相互交错营造出一种来自家庭的温馨感觉。木盒表面各式各样的开口显露出木柜里浓厚迷蒙的红色，隐约间暗示着那是一个奇异而神秘的空间。木盒可以看作用以摆放鞋子的储藏室，一个被隐蔽起来的秘密空间。这一颠覆性的设计为零售空间的设计提供了一种新的思考方式，展现了零售业中普遍存在的、相当有趣而又重要的特征——储藏。这个红色“储藏盒”现在变成了整个商店脉动的核心，它增加了店内的交流与互动，成为店与店、消费者与销售者之间沟通的纽带。木盒的前面是一个精致而沉静的铜框玻璃窗。它在某种程度上减轻了威灵顿街道的狭窄感，使这城市长街显得更具活力。店铺玻璃窗的常规比例被打破，体现出一种人性化的尺度，与木头盒子的整体性形成强烈对比。精典现代主义的审美观感，对细节、材料及规模某种程度上的敏感将使购买者在店里寻找一双完美的鞋子时，感受到一种亲切感与温暖感。

Retail stores are becoming the most identifiable image maker for a brand, sometimes even more prominent and memorable than the brand’s product collections. For the design ofPedder Red, a new and refreshing shoe brand, NHDRO desires to create a captivating brandimage through its interior architecture by questioning some conventions of retail design and at the same time fulfilling the desires of today’s style savvy shoppers.

The design concept revolves around one bold insertion the stained-oak wooden box that stands vertically through two stories of the store. This structure provides a strong visual identity for the retail brand, while creating warmth and a sense of domesticity in its wooden weaving, which took cue from the stitchings on a shoe. Various punched openings in the box surface reveal a thick and luscious RED world inside, hinting at a strange inner sanctum. The wooden box functions as the main storage area for the shoe store, something that is often an after-thought, or at least tucked away behind hidden spaces. This subversive gesture offers a new way of thinking about retail space, and celebrates an extremely active and significant feature of the mundane retail business—storage. This red storage sanctum now becomes the pulsating heart of the store, pumping the ever consuming energy, negotiating between the shopper and the sales staff from shoe to shoe. In front of the wooden box is a carefully composed facade of brass-framed-window glass.

This is set back to allow some relief on the narrow Wellington street, attempting to reinvigorate this urban strip. The scale of the store window is broken down to be more informal, engaging the human scale as well as serving as a sharp contrast to the monolithic wooden box beyond. Although the total aesthetic is one of classic modernism, certain sensitive attentions to detail, material and scale contribute to inject a level of intimacy and warmth for the shopper to experience while searching for that perfect pair of shoes.

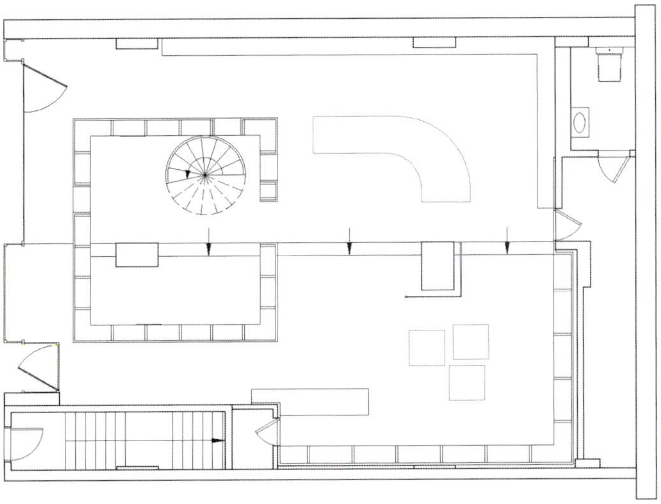

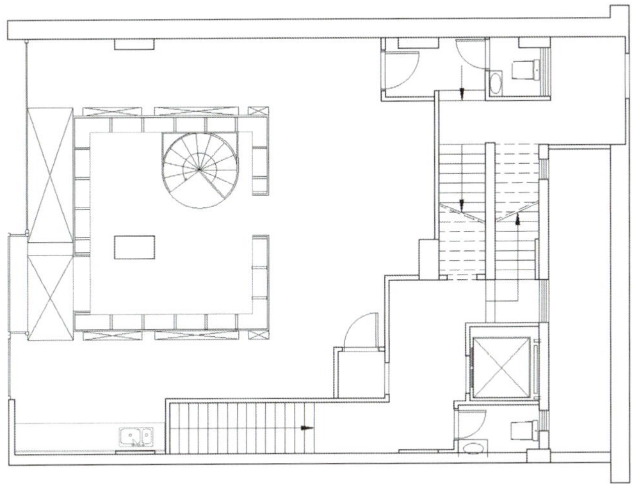

圣诗得北京展厅
Sunstyle Exhibition Hall in Beijing

坐落地点：北京市朝阳区南皋路东1号 / 图片提供：圣诗得

作为世界顶级户外家居品牌圣诗得的展厅，设计师因品制宜，选择户外展示这种做法。阳光、空气、天空，俨然是该展厅最佳的装饰元素，身临其境毫无建筑施予人的局限感，仿佛飘洋过海来到一个海滨之都。为了突出产品，不让建筑外观有喧宾夺主之感，设计师用色用材极为低调，纯白的墙壁、透明的玻璃主宰一切，只在少许不经意处配以蓝色马赛克做点缀。

As the exhibition hall for the world top outdoor household brand Suntyle, the designer just adjusts the measures to the products and has selected the way of outdoor exhibition. Sunlight, air and sky naturally become the best decorative elements in this exhibition hall. There is no sense of confinement imposed on the people by the building while being personally on the scene, just as you come to a seaside city after crossing the ocean. To emphasize the products and keep the building appearance from stealing the show, the designer is very moderate in the application of the colors and materials. Pure white wall and transparent glass is dominant and only blue mosaic is decorated in few casual places.

设计:Keiichiro SAKO, Ken-ichi KURIMOTO, Liang JU / 建筑设计:SAKO

EIFINI

面积:120m² / 坐落地点:北京 / 摄影:Misae HIROMATSU

EIFINI系列女装精品店内部的立面构成很简单，没有增设过多的隔断，一目即可尽览全景。设计的重点在于蜿蜒盘旋于室内的紫色荧光衣架，无论从哪个角度都能找到视觉的倚重点。因为处在纯白色这种特殊的背景下，紫色衣架还具有视觉引导性，人们的视线会随着线条的高低回环注意到不同的商品。设计满足功能需求这一理念贯穿始终。

The constitution of the facades inside EIFINI Series Suite-dress boutique store is quite simple without addition of excessive partitions. The overall sight can be viewed at a glance. The focus of the design lies on the purple fluorescent clothes racks circled inside the room in a wandering way. The relying point of the vision can also be found from every angle, because the purple clothes racks still have a visual introduction in such a special pure white background. The sights of the people will notice the different articles along with the height differences of the lines. The conception that the design meets the function requirements has run through the design.

设计:陆希杰 / 参与设计:陈宥铨

STEPHANE DOU

面积:148m² / 坐落地点:台北 / 主要材料:镀锌铁板、磨石子、布帘、铁板喷漆、钢浪板

全白色铁板喷漆的店面外墙，现代感十足的橱窗，创造简单强烈的视觉意象。空间中活动式轨道灯引领参观视线。在全开放式的空间中，感受流线与菱角的几何形体变化，金属浪板水平层架，陈列精品配件，在经过镜面包覆的柱子反射下，使得空间更具趣味。300多片大小不同的雷射切割冲孔铁网天花包覆，清爽简单及融合照明、空调、吊挂展示等机能于其中，流畅的线条展现数位科技的利落风采。天花铁网的吊挂功能、布帘后的收银台与储藏室，都同时满足功能性与许多空间想像。流线造型的布帘与展示柜，刻划出柔性的空间，原本坚硬的磨石子地板与展示台都因而形成软化但仍具几何感的形态。自然风味的磨石子与天花律动相呼应，工业素材经过精密设计呈现出高质感的展示空间。

A simple but strong visual image of a truly modern shopping window is created through the white colored, exposed metal external wall. Track lighting is designed as an exhibition guide for visitors to feel the changes of the geometric layout formed with flowing lines and angular curves. From the reflection of the mirrors that wrap the columns, horizontally aligned shelves of metal corrugated boards display luxury accessories, and enliven space. Together with over 300 pieces of laser-cut punched metal webs in the ceiling, the lighting, air-conditioning and hanging displays combine in clarity and simplicity to display the clean, smooth style of digital technology. In addition to the hanging metal webbed ceiling, the cashier counter and storeroom behind the curtain address functional needs and allow more space for imagination. The streamlined curtains and display cabinets connote a tenderness of space and soften the harshness of the marble flooring and display stands without sacrificing the geometric feeling of modern design. The natural tones of the marble flooring correspond to the rhythm of the ceiling, and a premier display space is thus presentedin both delicate and precise design using industrial materials.

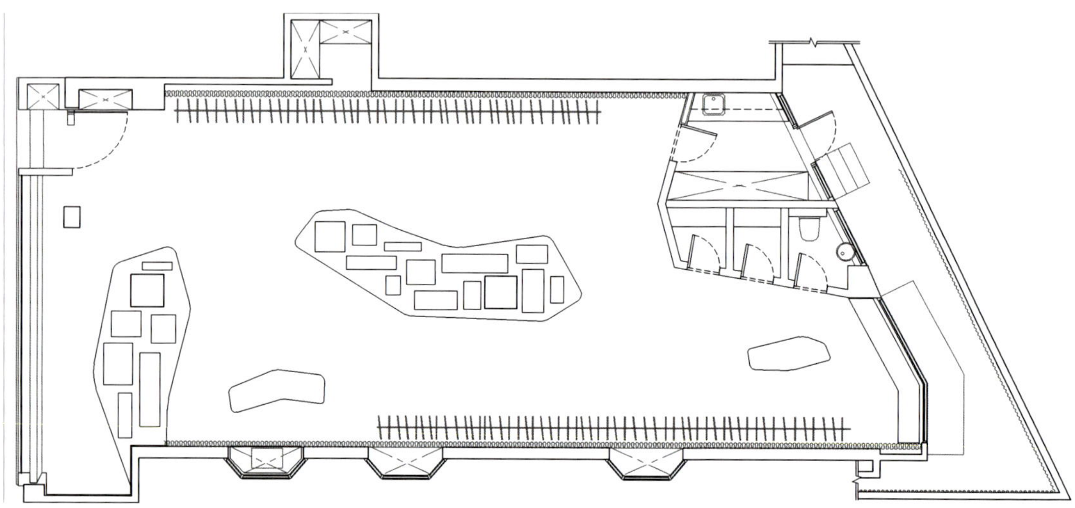

设计单位:新加坡PDI国际设计 / 设计:曾建龙、林岳

灯的摩登世界
MODERN WORID OF LIGHTS

面积:420m² / 坐落地点:温州市 / 主要材料:水泥、涂料 / 摄影:香港ATA摄影机构

本案是一个高档进口灯具专卖商业空间，在设计上难度比较大，既是产品的展示空间，同时又经营有很多不同风格的产品，而且本案空间夹杂在一、三层中间。一层是原有的展示厅，保持原来的布局，没有做太多的设计；三层主要展示欧式的灯具；而二层主要展示比较摩登和新古典的灯具，同时展示概念灯具。整体空间以金属感比较强的灰色度表现。在空间设计理念上以完全破常规的方式，从一层至二层的楼梯的设计，没有过多的修饰，以简约的线和面为主，在主体空间里与方圆结合，通过镜面和引导性的弧墙来增强空间的变化和动感。二层至三层的楼梯间保留了原来的旧装饰的部分造型面貌，通过改造加以修饰。同时传达一种多元化的视觉美感，以全新的设计理念来展示灯的摩登世界。

This design is for a business space engaged in the exclusive sales of the imported superior lighting fixtures. The owner has a lot of his own ideas. It has quite great difficulties in the design of this case as it is not only an exhibition space for the products, but also is engaged in the products of various different styles. Furthermore, this design space is between the first and the third floor: the first floor is the existing exhibition hall and its original layout is maintained without much variation; the third floor is mainly engaged in the exhibition of the continental-style lighting fixtures while the second floor is mainly engaged in the exhibition of the quite modern and neoclassic lighting fixtures as well as some conceptual lighting fixtures. The whole space is demonstrated by a grayness of strong mental senses. As for the idea of space design, a method that has completely broken through the conventions is applied. The staircase from the first floor to the second floor is designed without any excessive decoration. It is mainly applied with the simple lines and surfaces to combine with the squares and circles in the main space. The changes and the dynamics of the space are strengthened through the mirrors and introductive arc walls. The staircase from the second floor to the third floor is kept with the partial modeling appearance of the original decoration and finished through modifications to it to deliver a multiple look-out angle aesthetics. The lamps' modern world is presented with a completely new design conception.

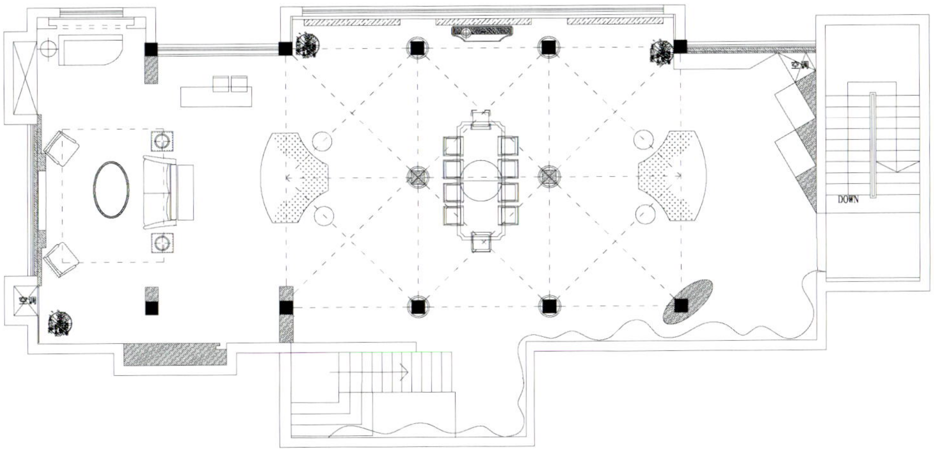
空调
DOWN
空调

圣诗得上海展厅
Sunstyle Exhibition Hall in Shanghai

坐落地点：上海市城市雕塑艺术中心 / 作品提供：圣诗得

作为圣诗得户外家居品牌的上海展厅，设计师本可以随心所欲，尽展设计才华，但现实的面积给他出了一个难题，怎样能让走进小展厅的人没有狭窄之感，并且在欣赏产品的时候会对空间产生不断的惊喜是这次设计的重点。镜面在合适角度的摆放不仅令空间层次感更丰富，还从不同的侧面用设计的语言介绍了产品各个角度的美。建筑结构做的有点类似挑高小户型，线条纵横，于开阔处开阔，狭窄处狭窄，空间便在一张一合中予人惊喜。

As the exhibition hall in Shanghai for Sunstyle outdoor household bran, the designer of course can show his design talents at his pleasure. However, the actual area has put the problem in front of him. The key point of this design is how to make the people who enter the small exhibition hall without the sense of narrowness and get constant pleasures on the space while enjoying the products. The arrangement of the mirrors in the proper angles not only can enrich the hierarchic sense of the space, but also introduce the beautifulness of the products at different angle in the language of design from the different sides. The architectural str ucture is made similar to the small overhanging apartments with the crossed lines, wide when wide and narrow when narrow. Thus, the space surprises the people in an appropriate way.

设计:隈研吾

Z58

面积:4000m² / 坐落地点:上海市番禺路 / 主要材料:玻璃、抛光不锈钢、格栅

当中泰照明，中国最重要的照明专家，想要将位于上海市中心一座空置的手表厂厂房建造一个中国最好的商业空间时，他们自然而然地找到了最擅长建筑用光的隈研吾。

为了创造一个充满特殊的“光的现象”的特殊空间，设计师设计了由盒子构成的新结构。这是一个探寻建筑与光的实验室，同时，也是一个“光的现象的大教堂”。

穿过梦幻般的中厅，你会发现，Z58本身不仅仅是一个卓越的商业空间，更是一个品质出众的交流平台。几乎完全中空的一楼大厅内，照明系统能够根据不同的活动对每个灯具做到迅速调校、精确控制，营造完美的光的氛围。二楼是中泰照明的展厅，陈设了工业设计有史以来最为经典的照明作品，展陈布置均出自大师手笔，让来访的每一位宾客置身艺术之光下。三楼的办公空间通透、大气并充满生机，给使用者创造了一种融洽、自然、心情愉悦的工作环境。中泰照明作为本身是提供照明解决方案的公司，除了充分引采自然光，北面的办公区域还补充了柔和、均匀的人工光，加上能满足功能性需求、又符合建筑结构的重点照明处理，完美诠释了自然光与人工光。四楼客房是最让参观者流连忘返的地方。可开合的天顶再次体现了Z58对自然的敬意——或在晴日望白云朵朵，或在雨天赏水波涟漪，或在夜晚享星光点点，谁又能不说，这不是现代都市丛林里最奢华的享受呢。

新的四层玻璃结构的大楼，是与不同事物对话的结果。首先，是与原有地块上现存结构的对话，在保留原结构的基础上，加入了玻璃，铝和绿意。是在结构中与现存物体的对话，是与为这幢大楼增添个性和内涵的时间的对话，还有与这个建筑所在的番禺路的对话。在街道与大楼之间创造水意和绿意，将水动起来，通过水与绿的对话，让绿色的叶子摇曳。

When Zhongtai Lighting, the most important lighting specialist in China intends to build into a best business space an empty watch factory building located in the downtown area in Shanghai, they naturally find Wei Yanwu who is best good at architectural light.

To create a special space filled with special “light phenomena”, the designer has created a new structure constituted by boxes. It is a laboratory to explore the architecture and light and also a “cathedral of light phenomena”.

Passing through the dreamlike middle hall, you will find that Z58 is not only an excellent business space and is more a first-class communication platform. In the almost completely hollowed-out hall at the first floor, the lighting system can make a quick calibration and precise control on every lighting fixture according to different activities to create a perfect light atmosphere. The second floor is the exhibition hall of Zhongtai Lighting, displaying the most classic lighting works since the industrial design. All the arrangements of the displays are done by the master to make every visitor staying in the lights of arts. The office space at the third floor is spacious, grand and vigorous in order to create a harmonious, natural and pleasant working environment for the users. Zhongtai Lighting is a company to provide the lighting solutions itself. Besides the sufficient introduction of natural lights, the office area in the north is also supplied with soft and uniform artificial lights. It not only can meet the requirements of the functionalities, but also comply with the key lighting arrangement in the architectural structure. All these have demonstrated the natural and artificial lights perfectly. The guestrooms at the fourth floor are a place where the visitors are indulged in pleasure without stop. The ceiling that can be opened and closed has shown the respect from Z58 to the nature again - you can see the clouds in sunny days, enjoy the rippling waves in rainy days or watch the dotted stars at night. Who can deny that it is the most luxurious enjoyment in the modern city forest?

The new four-story mansion in glass structure is the result of the dialogues among the different objects.

First, it is a dialogue with the existing structure in the original block. We have added the glass, aluminum and greens while retaining the original structure. The dialogue with the existing objects in the structure has added the time dialogue between the individualities and meanings to this building. Furthermore, there is a dialogue with Fanyu Road where this building is located. Watery and green feelings are created between the street and the building to activate the water while the green leaves are swayed through the dialogue between water and greens.

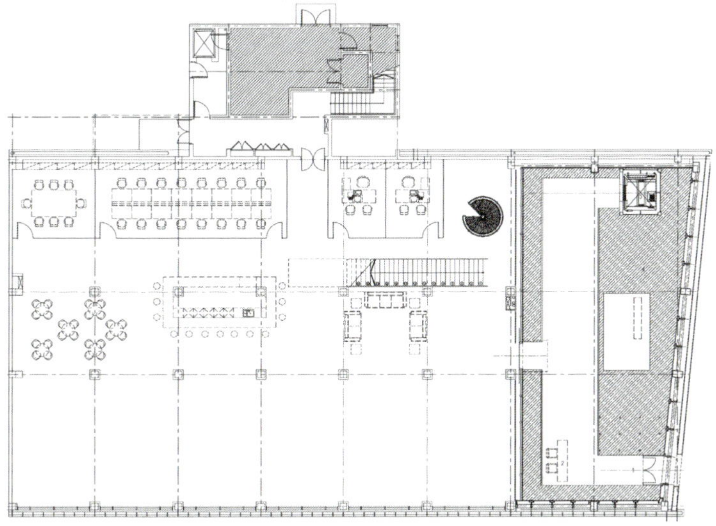

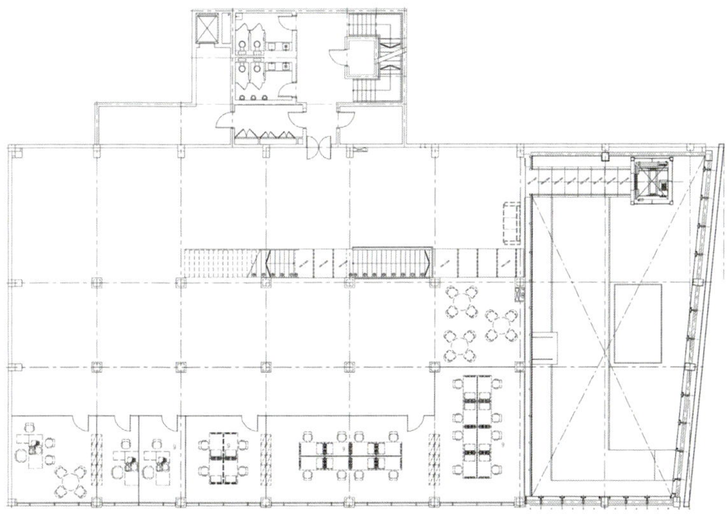

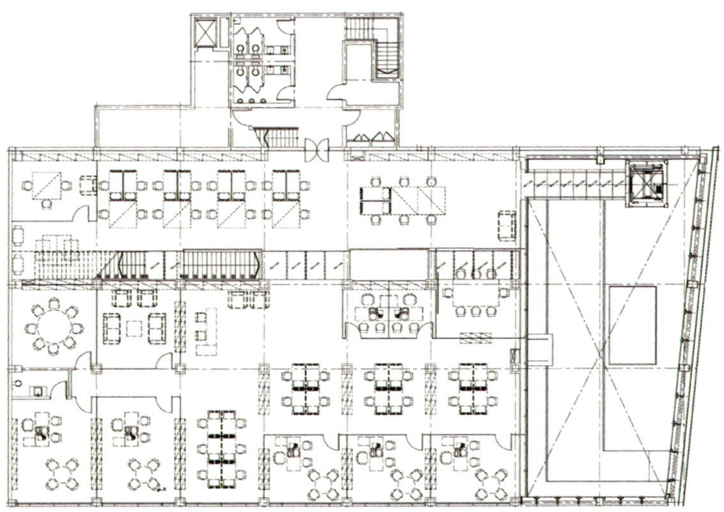

设计:Keiichiro Sako, Nobutoshi Hara / 建筑设计:SAKO

浪漫一身 Romanticism

面积:80㎡ / 坐落地点:杭州 / 主要材料：水泥砂浆、白色瓷砖、木材、纸面石膏板、胶合板

“浪漫一身”是在中国拥有500家分店的服装品牌。业主在委托做内装改造设计的同时，对于自己的服装品牌形象也制定了向更精炼方向转型的目标。设计对于此品牌的“新生”的构思，和“鸡蛋”本身所具寓意——诞生、未来、发展是互相吻合的。

“鸡蛋”通过两种手法来表现。一是把像南瓜马车一样的主题扩大化，二是将星空和亮晶晶的形象无限化。

店铺中央的巨大蛋壳被分为上下两部分。利用倾斜的剖面作出直对入口的服饰陈列，同时吸引店外的行人注意到巨大蛋壳，自然而然进入店铺浏览商品。这样，“鸡蛋”成为店铺的中心，客人围绕它转一圈，就可以看到所有商品。门面标牌对于店铺至关重要，无数个用内藏式灯箱表现的“鸡蛋”吸引了大家的视线。LOGO浮在灯箱的表面上。

各个位置不同形式的“鸡蛋”都有它自身的震撼力，也正是通过它们衬托出构成小型店铺的重要元素——商品LOGO。今后业主准备将其他类似规模的店铺进行同样的改装。“设计主题”和“商品・LOGO”的恰当关系，这就是做“浪漫一身”设计时考虑的关键问题。

“Romanticism” is a fashion brand that holding almost 500 of boutique. Client is aiming at image changing into elegance through our decoration design for their shop. We regard new-born of the brand as the initial meaning of Egg’s born, future and develop.

Egg was used as 2 means. One is making motif getting the hugest like pumpkin carriage, and another one is countless like stars or spangles in the sky.

The huge egg that standing in the middle of shop is separated into 2 parts. The slope section that is facing the entrance is not only an image of egg itself, but also able to display garments on it. Meanwhile stepping around the huge egg, customer could see all of the commodities once a time.

Logo is very important element to a shop. Countless inter-light type eggs draw attention of people, and we could also put logos on those eggs.

Every egg is not only impact existing, but also help to make important elements of the shop like logos as well as garments stand out. And we have reached an agreement, in the future, similar area chin will be re-decorating based on the design of this shop. Another thing I think it over on Romanticism Hangzhou shop is that “design motif” and “commodity logo”.

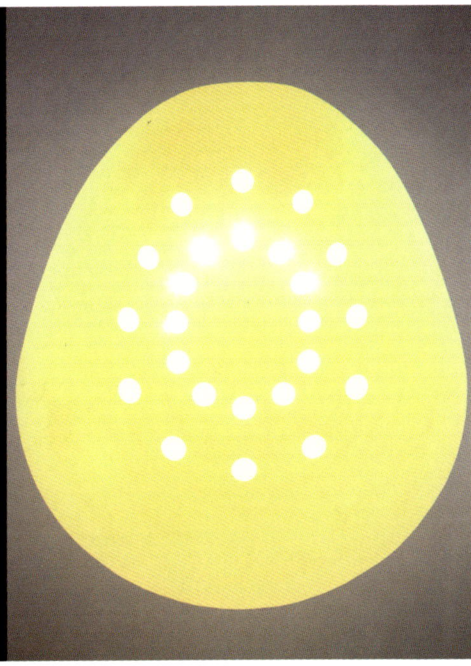

设计单位:玄武室内装修有限公司 / 设计:黄书恒 / 参与设计:欧阳毅、陈新强、林彩雯、陈佑如

驰骋劲驹的梦想
Mercedes-Benz Showroom

面积:455㎡ / 坐落地点:高雄市 / 主要材料:马来漆、抛光石、英砖、胡桃木皮、水泥板、喷砂镜、CDM复金属灯 / 摄影:王基守

Mercedes-Benz的百年造车工艺，不但兼具舒适驾乘感受与顶尖安全科技，更具有个人品位与身份的指标意义，如何延续一贯的优雅时尚及诠释品牌核心价值，成为设计者着手展场规划的首要考量。

位于高雄市的繁华中心，拥有三角窗的地理优势，以略高于人行道的高度塑造品牌顶尖价值，并配合品牌以飞箭银及深夜蓝两个颜色作为主题色。以线条作为架构空间的主轴，垂直与水平的线条放大外观幅宽，撷取量体的聚合与延展性，同时修饰建筑西晒与开窗面不一的问题。设计重点在于凸显五部展示车，并规划一处主题车款的展台，清楚地交代主从关系，空间因此区分为展示、洽谈、吧台、办公及交车等五大机能。一进门，接待区采用蓝白渐进色调与光源，表现空间中流转的速度感与力道，呼应整个展示空间的主题。

配合车主在赏车时可能环绕车款的动线，于车款周边安排洽谈区，特意保留洽谈区的独立性，同时展示汽车零件与造车工艺，采用金属板与胡桃木的组合衬托顶级车款与众不同的精致零件，让车主在赏车之余更实地参考零件、配备、内装等选择。天花板设计三个胶囊造型，搭配灯具结构形塑如同品牌标志的象征，考量展场的主角就是新颖车款，为避免喧宾夺主，在照明方面尤其重视均匀照度，借由向上投射的光源撞击天花之后蔓延的光晕，衬托车款的优雅线条与细致工艺。除了展示区域外，为服务车主而规划的吧台休闲区，以架高60cm的姿态提供足以俯瞰全场的视野。

交车区是品牌与车主进行第二次接触的地方，因此空间着重让车主能够仔细检视车辆的明亮光线与舒适座位，撷取黑白镂空造型勾勒如弹指琴键般的优雅情境，创造舞台聚焦效果，让每辆交到车主手中的劲驹戴上品牌高贵的光环，实现驰骋的梦想。

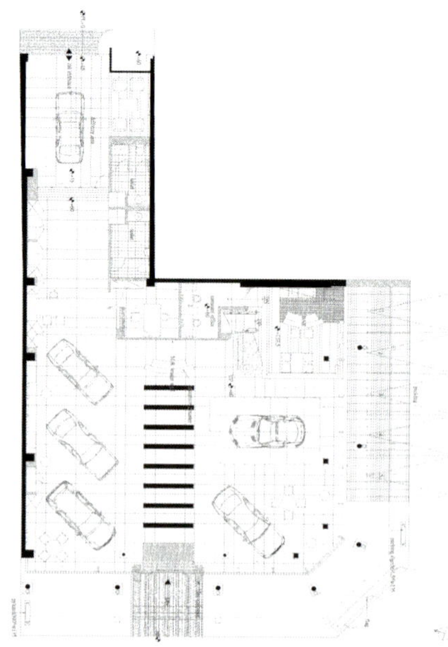

Mercedes-Benz has over hundred years of automobile manufacture workmanship. It not only provides a comfortable driving feeling and the top safety technologies and more has an index meaning of individual taste and status. How to continue the persistent elegance and fashion and how to interpret the core value of the brand has become the first consideration of the designer for the planning of the show room.

It is located in the prosperous downtown of Kaohsiung and has a geographic advantage of a triangular window. It is at a height slightly higher than the passageway to build up the top brand value and the two colors including arrow silver and dark blue are applied as the thematic colors to coordinate with the brand. The lines are applied as the main axis to constitute the space. The vertical and horizontal lines have amplified the visual breadth and pick the convergence and ductility of the objects. At the same time, a western exposure of the building and the different window openings are also complimented in this way. The design has put emphasis on showing five exhibited cars prominently and planning an exhibition place for a thematic car models as well as specifying the principal and subordinate relationship. Therefore, the space is divided into five functions including exhibition, discussion, bar counter, office and car handover. Upon entry, the reception area has applied a tone and light source of the gradual combination between blue and white to deliver a sense of speed and force moving in the space and echoes with the theme of the whole show room.

To match the active line that the car owners may walk around the cars while appreciating the cars, the discussion areas are separately arranged around the different cars with the deliberate maintaining of the independence of the discussion areas to display the car parts and manufacturing workmanships. The combination of metal sheets with the walnuts is applied to set off the top car models and the extraordinary exquisite parts so that the car owners can more practically study and select the parts, fittings and interior decorations etc while appreciating the cars. The ceiling is designed to three-capsule models and is like the symbol of the brand logo in arrangement with the lighting fixtures. The leading role that evaluates the show room is the new car models. To avoid being distracted by the other objects, the lighting has paid special attention to the uniform lux. The light halos spread after the light sources projecting upwards strike on the ceiling are applied to set off the elegant lines and delicate workmanships of the car models. Besides the exhibition areas, the casual bar-counter area planned for serving the car owners has provided a visual field to overlooking the whole area at a raised height of 60cm.

Car handover area is a place where the brand and the car owners have a second contact. Therefore, the space puts emphasis on the bright lights and comfortable seats by which the car owners can check the cars carefully. A black-and-white through-carved modeling is applied to outline an elegant situation like playing the pianos and create an effect of stage focusing. Every powerful car that is handed over to the car owner puts on a glorious brand aura so that the driving dream can be realized.

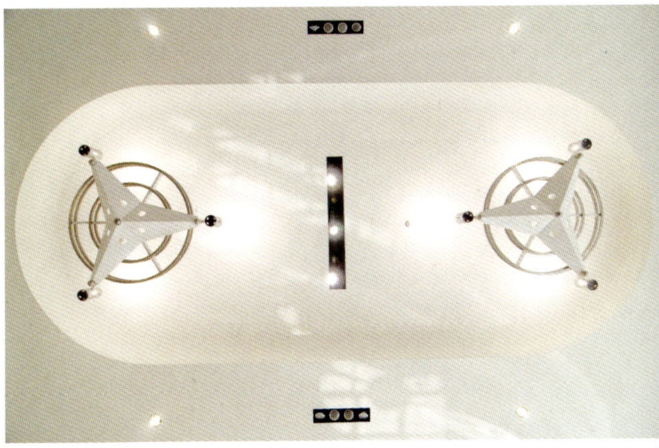

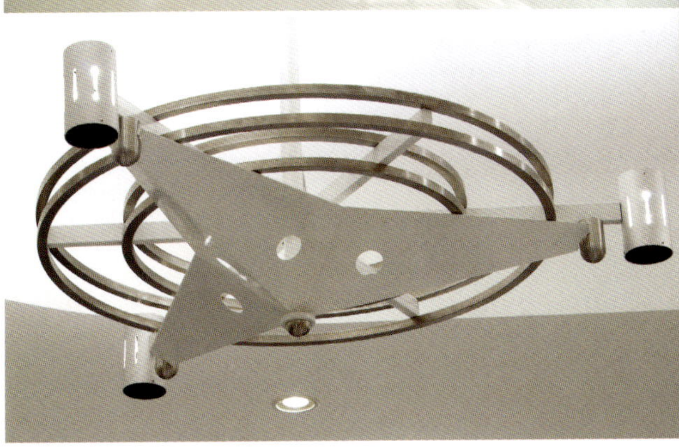

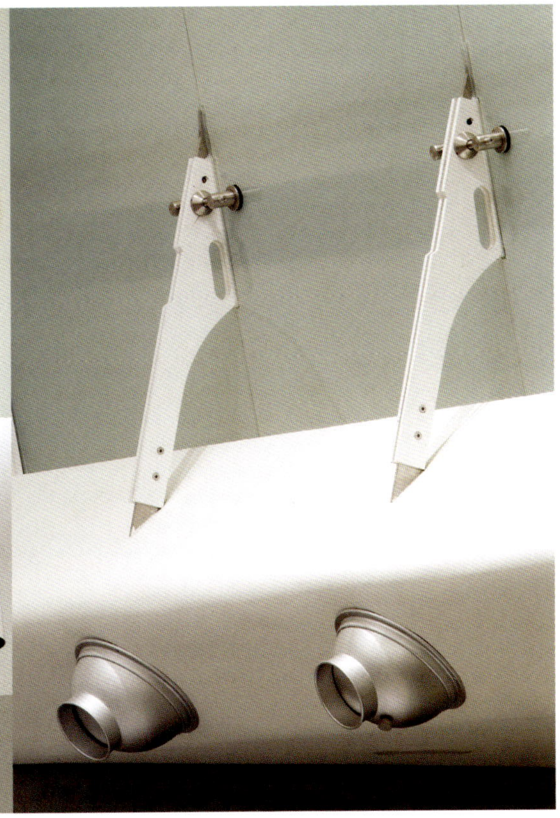

E 280

设计:许志国

德国乐得福莱 lattoflex

面积:400m² / 坐落地点:南京 / 主要材料:地毯、地板、纱帘、有色乳胶漆、橡木饰面板、烤漆玻璃 / 摄影:文宗博

本空间为一软体床品专卖店。照明设计上摒弃了直接照明而采用更适合客户停留与体验的漫反射光源。立面运用了大量的白色纱帘来软化空间，结合穿插其中的小型场景的点缀，避免了通常床品展示的单调罗列，营造了静谧、柔和而舒适的心理氛围。材料主要以橡木，地毯及有色涂料为主，亲切而温暖。

This space is designed for a soft bed franchised store. In the lighting design, the diffuse reflective lighting sources which are more suitable for the customers to stay and experience are chosen instead of the direct lighting. The large quantities of white gauze curtains are applied in the facade. In combination with the decoration of the small scenic spot, the monotonic display of the bed products is avoided and a quiet, soft and comfortable mental atmosphere is created. The materials selected are mainly oaks, carpets and colored paints, friendly and warm.

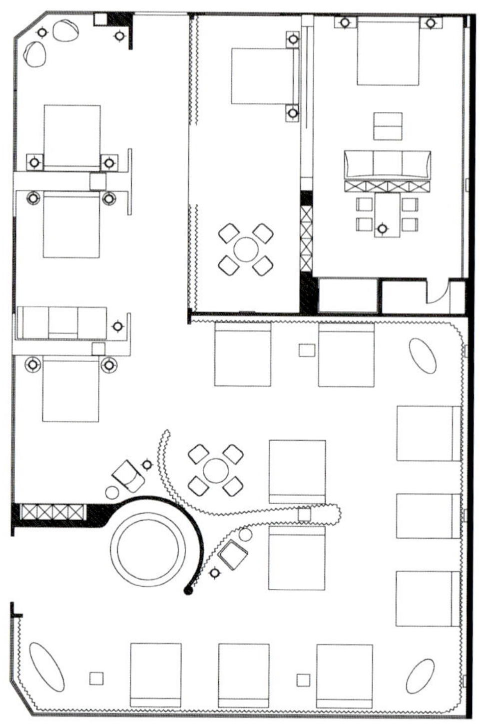

lattoflex 德国乐得福莱

lattoflex
COOMO

设计单位:唯品设计 / 设计:吕永中

纸围城—薇莳婚纱旗舰店
Paper-Besieged City - Flagship Shop of Watabe Wedding Dress

面积:330m² / 坐落地点:上海浦东 / 摄影:吴永长

日本华德培（上海）婚庆工作室位于上海浦东新区上海滩街区。室内设计欲创造一个崭新的个性化创意婚庆体验空间。室内空间规划面临的首要问题，是如何在一层狭小区域内同时安排三个功能区。经过多次比较，放弃了众多规整划分的功能分区，最终选择了较自由的空间布局。用围合、游离、分离、相望等空间组织手段，把不同功能区域加以整合。由于场地两面临街，因此设立了四个出入口，分别可独立出入各功能区域，互不干扰，同时提供更多潜在的商业机会。

一层中心部位是接待大厅，占据了近一半的面积，空间围合是采用定制的杜邦特种纸做成的“纸围城”，平面呈椭圆形状布局，控制着整个一层空间效果，成为整个空间的视觉印象。圆弧形城墙断开处成为开口，没有边界，并延展到另外两个区域。半透的纸围城墙外面左右两边分别是花店和婚纱区，上下两边是沿街的橱窗，人流动线由此展开，迂回曲折，步移景移。纸墙的最初想法是婚纱蕾丝的纺织结构，其半透明状的结果是由纺织的疏密关系形成，质感丰富而轻柔。由于3.5米的墙体高度，采用虚透的立体网状结构容易支撑墙体并保持平整度。经过工厂多次试验，确定立体网格的钢丝直径和间距，选择的纸带为杜邦原用于印刷包装的新型材料，具有强度高、不怕水、防尘等多种功能，这次通过防火处理首次用于室内建筑空间。编织的过程充满了乐趣和传统手工艺的特色，项目组设计师研究了手工编织的程序和方法，集体配合花费了近五天的时间完成。

第三层空间根据功能布局去安排造型及材质，为婚纱室内摄影提供了多种创意背景的选择。蛋形接待台暗示婚姻的诞生与圆满，同时为三层较方正的空间带来一丝柔性与浪漫，光洁圆滑的表面与纸墙丰富的肌理形成对比。深木色的咖啡书吧提供给新郎安静舒适的环境用以消磨时间，对面的新娘化妆间只一纸城墙相隔，二者位置的关系有如观众与舞台。墙内独立化妆间带给新娘更多明星梦幻般色彩。婚礼服装区是一个步入式衣帽间，好似传统意味上的木头衣帽箱，按总裁办公室布局设计的贵宾厕所同时具有化妆功能。两个摄影工作室可分可合，为以后摄影沙龙活动提供延展的可能。设计师更像在导演一出婚礼的故事片，场景在围绕故事的线索展开，通过室内及界面去纺织梦幻，追求梦幻。

Japanese Watabe Wedding (Shanghai) is located in Shanghai Bund Block, Pudong New Area, Shanghai. The interior design is supposed to create a brand-new creativity wedding space of the individualities. The chief problem confronted with the interior space planning is how to arrange three functional areas in this narrow space at one floor simultaneously. After multiple comparisons, the functional areas in the various orderly divisions are given up and a quite free space arrangement is finally selected. The techniques of space arrangement by enclosure, dissociation, separation and face-to-face etc. are applied to rearrange the different functional areas. As both sides of the shop are facing the streets, four entrances and exits are arranged to enter and exit independently from each functional area without the interference with each other, also providing more potential business opportunities.

The central part at the first floor is the reception hall with an occupation of about half area. The space is enclosed by the custom-made special Dupont paper as “Paper-besieged City”. The plane is arranged in a elliptical form to control the space effects of the overall first floor and become the visual impression of the whole space. The broken location on the circular arc city wall becomes an opening without boundary and also extends to the other two areas. You can see the different sceneries outside the semi-transparent paper besieged city wall whiling walking. The preliminary idea of the paper wall begins with the textile structure of the laces for the wedding dresses. Its semi-transparency results in the density of the textiles. It feels rich and soft. The applied fake transparent three-dimensional net structure is easy to support the wall and maintain the flatness due to the 3.5m wall height. After many experiments in the factory, the diameter of steel wires and the distance between them for the three-dimensional grid are decided. The selected paper tapes are the Dupont’s new materials originally used for print packages with various high-strength, waterproof and dust-proof functions. it is the first time to be used for the interior building space after fireproof treatment. The knitting process is filled with the pleasures and the characteristics of the traditional workmanships. The designers in the project teams have studied the procedures and methods of the manual knitting and group cooperation has taken them almost five days to complete the work.

The modeling and material for the space at the third space are arranged according to the layout of the functions so that the various kinds of the creativity backgrounds are provided for the indoor photographing of the wedding dresses for selection. The egg-shaped reception table implies the birth and satisfaction of a marriage and also brings certain gentleness and romance to the quite square space at the third floor. The bright and smooth surface has formed a contrast with the rich textures of the paper wall while the dark wood colored coffee and book bar provides a quiet and comfortable environment for the bridegrooms to kill time. The opposite dressing room for the brides is separated only by a paper wall. The relationship between the two locations is just like the audience and the stage. The independent dressing room inside the wall brings the brides with more star-like dreamy colors. The area for the wedding dresses is a step-in type locker room, just like the wooden chest in the traditional sense. The VIP toilets designed according to the layout of the president' s office also has the make-up function. Two photographing studios can be either separated or combined to provide a possibility of extension for the future photographing salon activities. The designer is more like to direct a feature film about the wedding and the scene is developed around the clues of the story. The dreams are woven and pursued through the interior and interface.

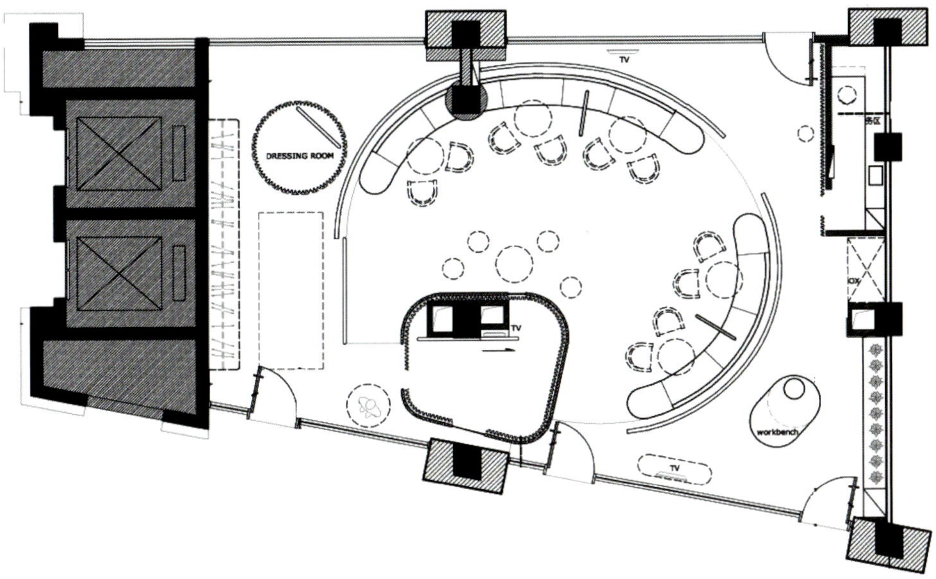

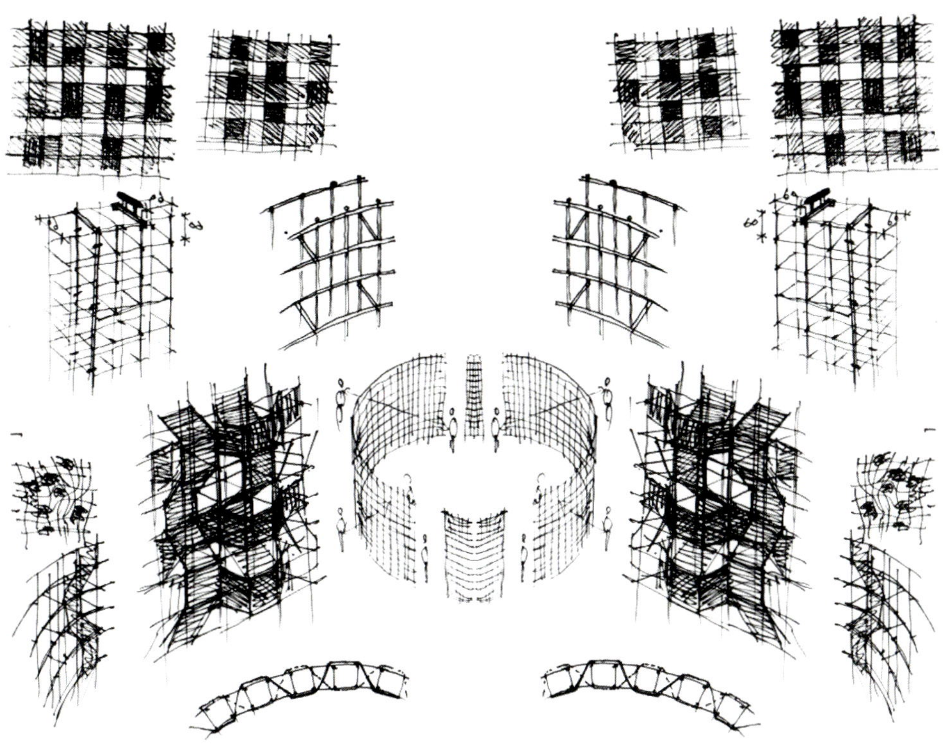

设计单位:上海黑泡泡建筑装饰股份有限公司 / 设计:孙天文

万德诺展厅
Exhibition Hall for Water Knoll

坐落地点:上海 / 摄影:陈乙

极简、纯粹，整体空间充满了大气磅礴的写意气度——这就是世界家具排行前五位的万德诺（上海）新展厅给人的第一印象。高低错落的大尺度空间拉近了建筑与室内的高度，使其协调统一。配合新材料与新元素的有机结合，体现了这一家具品牌锐意创新的产品设计研发理念。在体现展示家具功能的同时，又通过灯光的设计营造出了晚上组织PARTY的氛围，使人仿佛置身于一座超时空的艺术殿堂。

Very simple, pure and integral space is filled with a decent and majestic enjoyable bearing – this is the first impression that the new exhibition hall for Water Knoll (Shanghai) listed in the top five in the world's furniture industry. The large space of different elevations has closed the distance between the building and the interior to make it harmonious and uniform. Together with the organic combination of new materials and new elements, it has demonstrated the dauntless and innovative design, research and development conceptions on the products for this furniture brand. While respecting and showing the function of the furniture, the lights are designed to create an atmosphere in which party is held in the evening as if you are personally in an artistic palace going beyond the time-space.

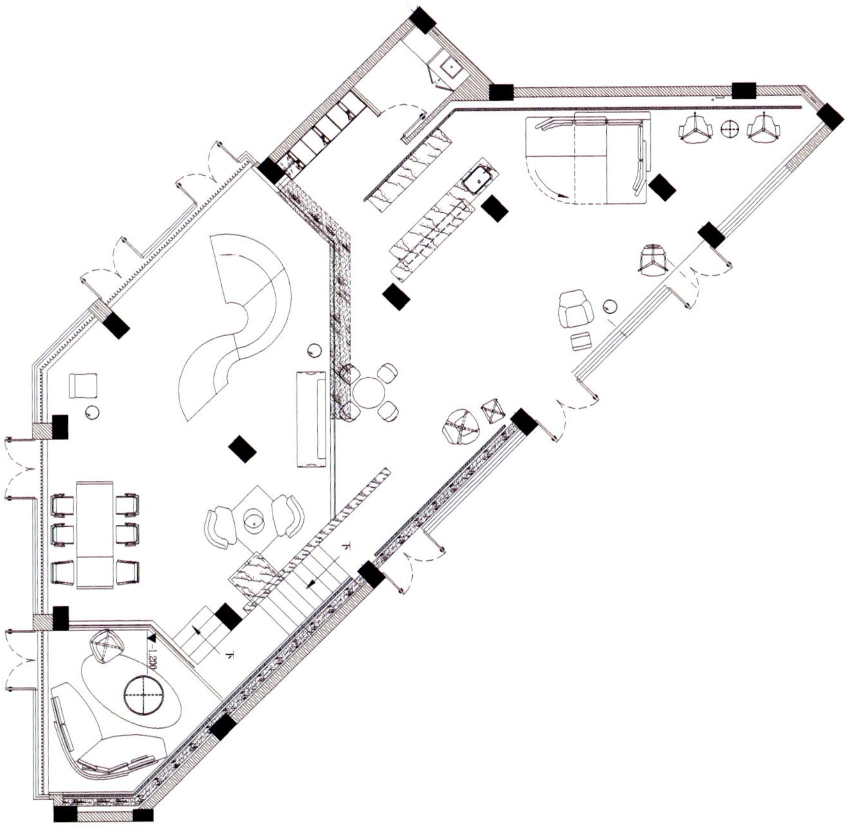

设计:麦德斌

高陶洁具中国总部
Cotto Sanitary Wares China Headquarters

面积:700㎡ / 坐落地点:广东省中山市 / 主要材料:水泥、玻璃、不锈钢、粗石 / 摄影:周敏强

作为中高档消费品的COTTO洁具，在强调品牌魅力的同时，如何对其品牌展示空间作一个完美的诠释，使人们对这个品牌产生一个深刻的视觉印象呢？设计师从这个基点出发，全面改写了人们印象中洁具用品潮湿、凌乱的画面。

展厅的设计糅合了展示效果和功能上的应用。展厅入口的左侧是产品橱窗展示区，这是一个吸引人们视线的平台，进而展区背景COTTO的标志也在不经意间进入眼帘，促使人们前去一窥其特色。通过宽阔的展厅入口，首先看到的是洗手盆展示区，纯白色的光滑理石底座上立起黑色的粗糙石柱，简约写实的洗手盆被展示在人们视野的高度，可以轻易看清展品的光洁质地。洗手盆两边不规则地摆放着马桶样本，看似随意的摆放，却是经过精心的设计。落地灯投射出的光束角度，最大限度地突出样本而不会令展示空间产生眩光，同时避免展品产生过重阴影，无论从哪个角度看，都能够达到良好的观看效果。顺势右行看到的是，模拟水流开启状态的洗手盆安放在大型装饰石块上，水流经洗手盆漫溢到周围的凹槽，在这里可以看到使用状态的真实演绎。在展厅北面墙面，设置了6个样板间，每一个样板间摆放着经过精心搭配的成套洁具，或简约，或豪华，或复古，在向人们展示着不同的视觉感受，靠墙展示不仅能够节省空间，还增加现实模拟空间的真实感。从各个样板间一一观看过来，你的目光一定会被潺潺流水所吸引，如瀑布般的流水从墙面流下，经过地面下层再沿阶梯流下到有公司网址石材的回流池，圆形的泡泡浴缸摆放在上面，仿佛假期在碧海沙滩上享受日光浴的闲情。一墙之隔的房间里面是完全高科技的展示区，或许，你刚进入，洁具会自动开启，而使用后这种无触摸式产品能够自动判断适量洗净水自动冲洗；又或许，拧开浴缸龙头，暗藏于洗手台下的电视会自动开启，让人们体验不同的享受。

For Cotto sanitary wares engaged in the medium and high grade consumables, how to make a perfect interpretation on its exhibition space and leave a deep visual impression on the people while emphasizing the charm of its brand? Based on this point, the designer has entirely changed the damp and disorderly picture of sanitary wares in people' s mind.

The design of the exhibition hall has incorporated the exhibition effects with the application of the functions: the left side of the entrance of the exhibition hall is the window exhibition area, which is a plat to attract the people. Then the logo of COTTO at the background of the exhibition area has accidently entered people' s eyes and urged them to step out to see its distinguishing features. Passing through the wide entrance of the exhibition hall, you will first see the exhibition area of the washing basins. The black rough stone columns are erected on the pure white smooth marble base. The simple true washing basins are exhibited at a height to the people' s visual field so that the people can easily see the smoothness and qualities of the exhibited products. On both sides of the washing basins are placed with the lavatories irregularly, which looks like a casual arrangement but in fact is delicately designed: the angles of the light beams projected by the floor lamps can maximally stress the samples instead of producing dizzy lights in the exhibition space and also avoid the impression of over excessive exhibits. It can achieve the good visual effects whichever angle it is viewed. What you can see while walking along the right side is that the washing basin simulating the on status of water flow is placed on the large decorative stone and the water overflows the washing basin to the surrounding grooves. You can see the real demonstration of the application status here. On the wall in the north of the exhibition hall are set up with six model rooms. Each model room is placed with the delicately arranged complete set of sanitary wares, simple, luxurious or antique, which has displayed the different visual feelings to the peoples. The exhibition against the wall not only can save the space of the wall, but also increase the sense of reality about the actual simulated space. After watching each model room one by one, your eyes surely will be attracted by the slowly flowing water. The waterfall-like flowing water flows down along the wall, pass through the underground floor and come to the circumfluence pond with the stone materials of company' s website along the steps. The round babble bathtub is put over it like the leisurely mood when you enjoy the sun bath on the beach beside the blue sea during the vacation. The room separated by a wall is a completely hi-tech exhibition area. Maybe the sanitary wares will automatically open once you enter and this kind of touch-free product can automatically judge the proper quantity of water for flushing after being used. Or maybe, once you turn on the tap of the bathtub, a television hidden below the washing table will automatically turn on so that the people can experience the different enjoyments.

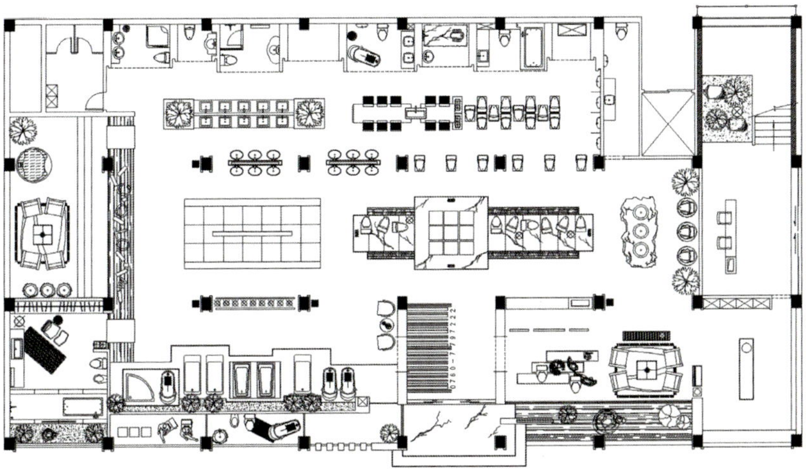

高陶洁具中国总部落成暨
五星酒店设计论坛
COTTO Sanitary Ware
China Headquarters Opening Ceremony &
5-Star Hotel Design Forum

COTTO

设计单位：北京优恩空间环境设计有限公司 / 设计：高立平

亚寒带法国饰品店
Sub-frigid Zone French Ornaments Store

面积：299m² / 坐落地点：苏州市工业园区旺墩路158号 / 主要材料：白水泥、木材、型钢、钢板网、石头、玻璃 / 摄影：文宗博、潘宇峰

环境赋予场所的诸多元素是在寻求环境新的意义过程中必须面对的，所有的连接都会因对诸元素的敏感性的不同，读取到让我们兴奋的使之意义扩大的状态。拓展或提升必然遵循通常的审美逻辑范围和智库储存的能量浓度。当然，过程是意义重大的，设计在一定程度上被看作是信息的编码过程，使用者则被看作对其译码。浓度的稀释与凝结，取决于"现代性"的批判方向选定的明晰性、心智的尖锐及对生存现状的关注程度。环境的意义已不只停留在一般的表象禁锢，它会潜入到事物的内在深度与之相连，传输意义以确定情境，转而影响行为和交流方式，以改变习惯和常规行为——寄予视觉且穿越表象，透视出真实。

In the process of seeking new significances of environment, we have to face many of the factors brought about by environment, and all the interlinkages will be so magnified as to make us excited due to the fact that our sensitivities vary at diverse factors. The expansion or uplifting must observe the common aesthetic logic scope and in proportion to the energy density in our intellectual banks. Evidently, processes are of great significance. To some extent, design is regarded as a process of encoding information, and it is the users' task to decode it. Density dilution or condensation is subject to the clarity of selecting critic direction of "modernity", the incisiveness of human mind, and the concern for the state of existence. The significance no longer stays confined by general representations. It insinuates itself into the inner depth of matters to form links and transmit meanings to define situations, and then exert impacts upon patterns of behavior and communication in order to change habitual and common behaviors ---by expressing visions, transcends the surface and reveals the truth.

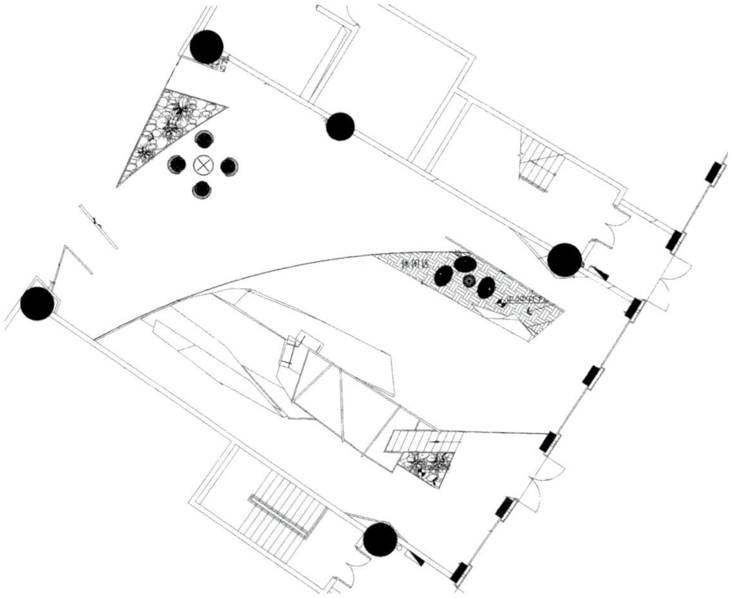

设计单位:佛山城匠设计院有限公司明威事务所 / 设计:谢智明 / 参与设计:霍律鸣、萧伟奇

伟祺玻璃建材有限公司办公室
Office Building for Victory Glass Building Materials Co., Ltd.

面积:3000m² / 坐落地点:广东四会工业园 / 主要材料:玻璃马赛克、乳胶漆、玻璃、钢架 / 摄影:钱翔

本案设计风格总体简约明快，整体规划从建筑到室内，随着空间功能划分（首层展示，二层、三层办公，四层休息）的同时，利用建筑的框架突出空间的视觉穿透力，同时满足业主展示作品的需要。面与面之间的材质、色彩、纹理、图案能相互协调，相互烘托是本案设计的特点之一，而大量使用的马赛克起到平衡的作用。流动的马赛克与椭圆柱子贯穿每层的空间。一体化的流线酒吧，增加了展厅的娱乐性。幽默的鞋子形状沙发为办公增添一份乐趣。雪丝状的玻璃马赛克灯柱使本案有了一份亲切感。设计既可满足办公楼的使用要求，又可体现业主产品的展示力。同时，建筑内外的结合，让办公楼有了生命力。

The overall design style of this case is simple and vivid. The frame as well as the interior and exterior compatibility of the building is applied to lay stress to the visual penetrability of the space while it is divided from the building to the interior according to the space functions with aspect of the integral planning (the first floor is for exhibition, the second and third floors for offices, the forth floor for rest). At the same time, it has shown the requirements of the owner to display the products. The materials, colors, textures and patterns between the surfaces can coordinate and set off with each other, which is one of the design characteristic of this case. The application of the large quantity of mosaics can balance their relationship between them. The workmanship, lights and decorative faces shall be artfully integrated. The flowing mosaics and elliptical columns have run through the space at every floor. The integrated streamlined bar has increased the fun of the exhibition hall. The humorous shoe-shaped sofa has added some interest to the office. The snow-shaped glass mosaic light poles have endowed it with some friendliness, which not only can meet the application requirements for the office building, but also can reflect the exhibitive abilities of the Owner' s products. Furthermore, the combination between the interior and exterior of the building has endowed the building with the vitalities before you.

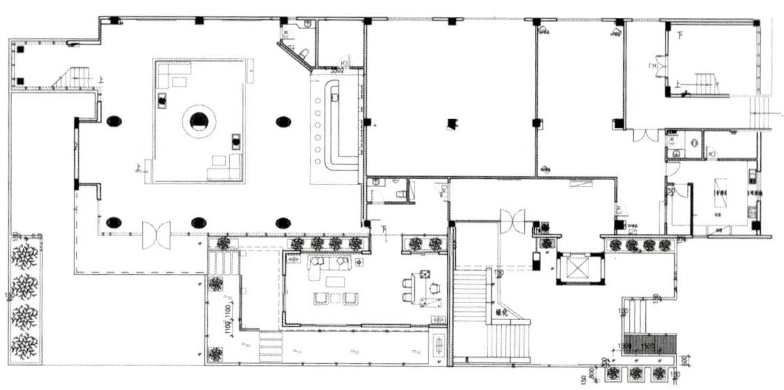

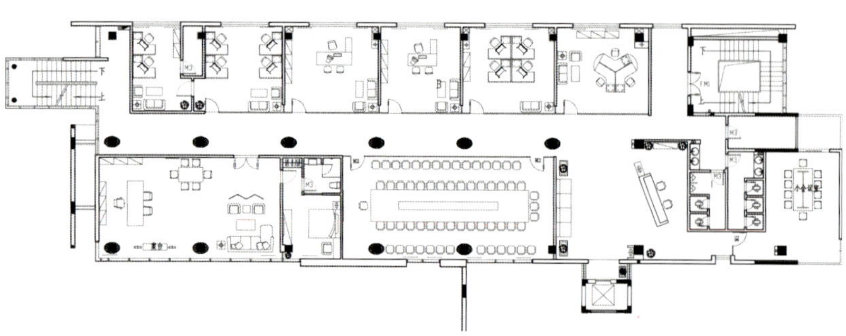

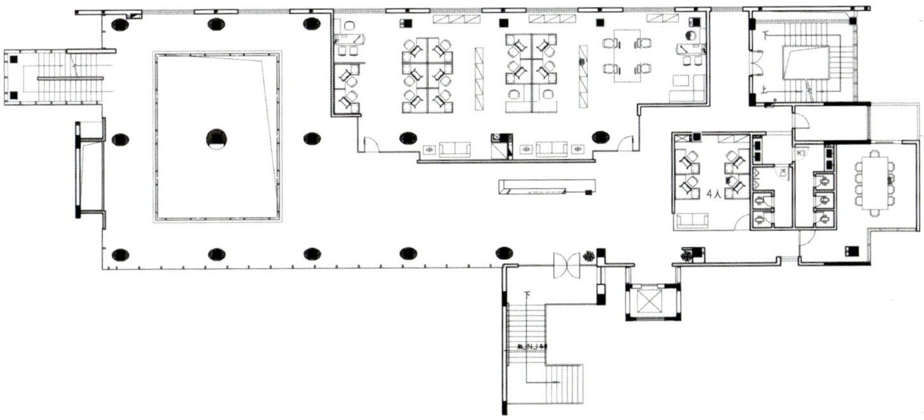

设计单位:杭州典尚建筑装饰设计有限公司 / 设计:陈耀光 / 参与设计:胡昕、施献峰、夏世英

浙江美术馆
Zhejiang Art Gallery

面积:32000㎡ / 坐落地点:杭州 / 摄影:陈乙

美术馆、艺术馆、博物馆尽管都是“看”的场所，但这次室内设计却做了很多看不见的设计。人们之所以来到艺术馆，主要还是来看展品，不是来看装修的，美术馆的室内设计应该建立在低调的姿态、平静的形态、努力积累这样一种较为克制的空间设计理念之上，“内敛、含蓄、克制”是这次的设计状态。始终建立以展品、艺术品为主角的主导地位，装饰设计退而求其次，让室内设计消失在建筑空间和艺术展示之间。

室内设计的理念则以浙江的江南传统、山水意境和黑、白、灰作为空间基调，达到与建筑空间、外部山水环境和谐共存的状态。形态上，空间浑厚而不笨拙、灵秀而不飘浮。大面积的室内墙面、地面铺以乡村绿的大理石，隐约捕捉雨后的青苔微绿，体现湿润的江南感觉。抽象的东方空间环境，虚实相间、室内与室外的界定交相融合。

“平静、从容、清淡”是空间设计的特征，几乎以一种材料、一种肌理、一种色彩的表现，让一种从容平静的空间感受贯彻始终、一气呵成。尤其是空间的立面和地面上几乎是用一“平”如洗的纯净手法，将空间的写意、大气、整体的建筑气势做了极致地刻画。之所以没有过多的不确定的描述，这完全来自于建筑师与室内设计师对预期设计结果的果断和自信。

该馆的建筑空间变化丰富、虚实重叠，特别是公共区域，充分利用了建筑空间的优势，在室内设计中着重梳理规整空间。如山峦起伏的大型玻璃采光顶，实体的墙面，敦厚的力度，几乎近似钝拙的手法，使空间犹如天然斧劈而成。以空间体量和尺度与人的心理感受产生共鸣，最大程度地反映了建筑空间的结构感染力及地方美术馆的权威性和规模感。抽象空间的审美给人更大的想象。

Although art galleries and museums are all “watch” areas, this interior design has done many invisible designs. The aim that people come to the art galleries is to watch the exhibits instead of the decorations. The interior design of the art galleries shall be based on a low profile and calm form and attempts to accumulate such kind of quite restrained space design conception. “Introversion, implication and restraint” is the status of this design. It shall always take the exhibits and artworks as the primary role and the decoration and design as secondary role in order to make the interior design disappear among the architectural space and art exhibition.

The idea of interior design takes the traditions in the south of Yangtze River, the artistic conception of rivers and mountains in Zhejiang as well as black, white and gray as the space keynote to realize the harmonious coexistence of architectural space and exterior river-and-mountain environment. In the form, the space is profound but not clumsy, delicate but not floating. The large interior wall and floor are covered with the green marbles, which indistinctly catches the slight lichen green after rain and deliver a feeling of the humid south of Yangtze River. The abstract eastern space environments are alternated between reality and delusion while the interior and exterior boundaries are integrated together.

“Peace, ease and moral loftiness” are the characteristics of the space design. It is almost expressed by one material, one texture and one color and the feeling of an easy and peaceful space has run throughout and is achieved in one instant. Especially the facade and ground of the space are almost applied a completely “flat” pure technique to depict the enjoyment and grandness of the space as well as the integral architectural atmosphere in an extremely detailed way. The reason that there are not excessive uncertain descriptions comes from the determination and self-confidence on the desired design results of the architects and interior designers.

This gallery has wonderful changes of architectural space with realities and delusions overlapped together. Especially the public area has made full use of the advantages of the architectural space to rearrange the whole space in the interior design. With the large glass lighting roof like the undulated mountains, solid wall surfaces, simple but strong force as well as the almost awkward techniques, the space looks like being naturally chopped into being. The space measurement and dimension has generated the resonance with the people’ s mental feelings to fully reflect the structural inspiration of the architectural space and the authority as well as scale of the regional art gallery. The aesthetics of the abstract space has provided people with more imaginations.

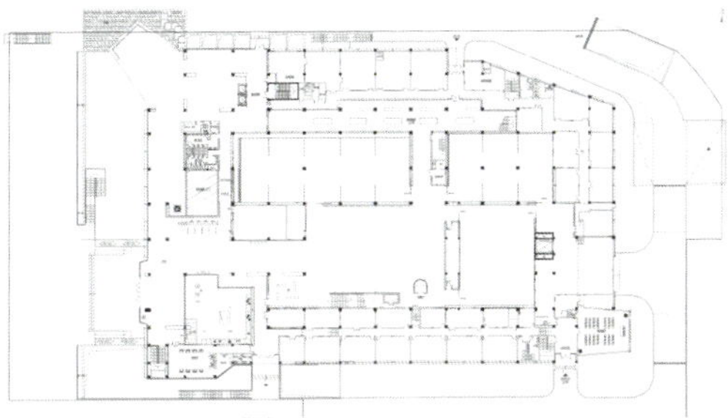

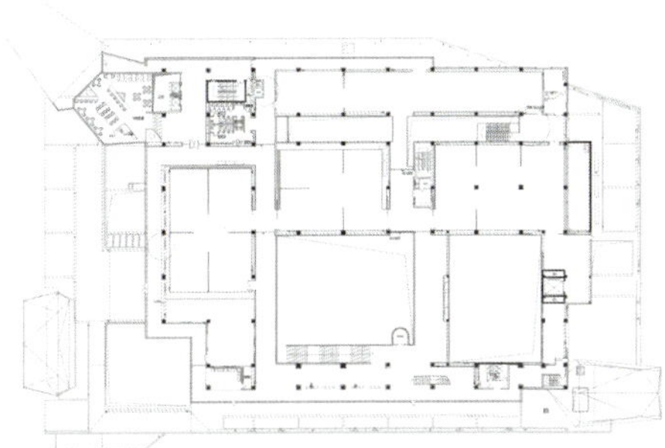

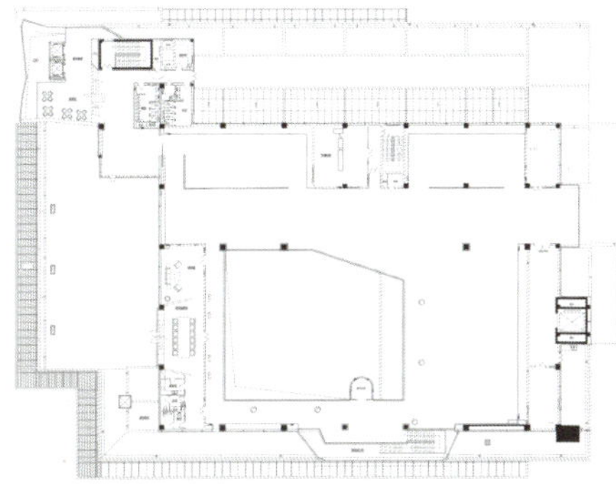

设计单位:杭州天澜建筑装饰设计院

亨利·戴家居
Henry Dai Furniture

面积:4600㎡ / 坐落地点:杭州 / 主要材料:不锈钢雕刻板、爵士白大理石、墙纸、地毯 / 摄影:陈乙

亨利·戴家居位于杭州西湖大道211号，总营业面积约4600㎡，分为地面及地下一层，其中地下层约3500㎡，占总面积的3/4，主营高品质家具及饰品。

本案设计的着重点将取决于其主营特性及商场营业面积的特殊分布性。因此将通往地下层的入口作为此次设计的重点，以斜向的隔墙、精致的连廊和高低错落的展示台来丰富空间关系，减少地下层所带来购物心理的负面效应，消除购物过程中的视觉疲劳。

本案主调为浅色，立面多处模拟户外窗景，多层次的灯光布局从不同角度更好地体现了家具的质感及其精致的工艺。由于建筑层高较低，且顶部交错的各类管道，局部的镜面不锈钢雕刻板是个不错的选择，它的镜面效果既能弥补层高的不足又可以带来很好的视觉感受。总之，任何的材料运用及色调的控制，都只是为了最终的一个目的：以人为本，营造一个让人愉悦的购物环境。

This case is located at No. 211, Xihu Avenue with a total business area of about 4,600m2. It consists of one floor on the ground and one floor under the ground. The underground floor has an area of about 3500m2, accounting for a 3/4 of the total area. It is mainly engaged in high-quality furniture and ornaments. The focus of this design case also depends on its business characteristics and the special distribution of the business area in the shopping center. Therefore, the entrance of the underground is stressed in this design. The leaning partition walls, delicate linkage corridor and the exhibition stage in different heights are applied to enrich the space relationship and reduce the negative effects on the shopping psychology as well as eliminate the visual tiredness during the shopping.

The case has mainly applied the light colors. The outdoor window scenes are simulated in many areas of the facade. The multi-hierarchic light arrangement can better reflect the realistic senses of the furniture and its delicate workmanship from different angles. Due to the low level height of the building as well as the various pipes crossed on the top, the mirror stainless carving panels in partial areas are a quite good choice. Its mirror effects not only can supplement the shortcoming of the level height, but also can bring good visual feelings. In general, the application of any material and control of colors are arranged for the final purpose – "human-orientated to create a pleasant shopping environment".

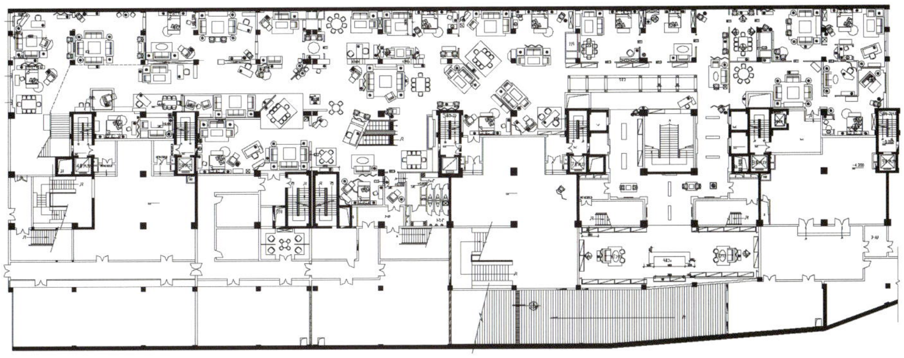

SPA，水的点滴

SPA, Drops of Water

王兆明 Wang zhaoming
哈尔滨唯美源装饰设计公司总设计师
高级室内建筑师
中国建筑学会室内设计分会副会长

SPA HAIRDESSING

朱永春	固足坊
孙志刚、大帅	英格玛SPA会馆
张奇永	美丽新娘
姜湘岳、吴海燕	龙景温泉会所
BEN LING	GREEN MASSAGE
孙志刚、徐慧	希尔斯池典
陈卫新	今生有约
上瑞元筑	班雅SPA会馆
陆文星	王春美容院——龙江风尚馆
高轶、张苗苗	卡伊造型
吕邵苍	名仕会SPA
韩冠恒	Lea spa

在每个城市都把自己打扮成水泥“怪物”的同时，我们的设计师悄然的设计着许多“自然”的水环境。为行色匆匆的人们营造着拥有不同风格、不同文化、不同装饰手法的沐浴休闲空间——SPA。

SPA，通常被认为是一些拉丁短语的字母组合，Solus par aqua、 Solus是健康，Par意为经由,aqua是水的意思。泛指用水来增强体质治疗疾病。我国是东方沐浴文化的起源，在中医理论中也将水视为健康的基础。强调水是人体内生命能量循环的媒介,水既是物质的也是神圣的。

当代的SPA更多地指可以使人们放松解压、恢复体力、振作精神的场所，并拥有相对完善的管理体系和专业化的服务流程。它包括各种保健因素，倡导健康生活方式和自我意识的发掘。在服务形式上更是多样化和地域化，充分利用地域的优势和地域历史的健康文化渲染着SPA的“色彩”。从阿尔卑斯山到巴厘岛，从国际品牌连锁店到专营店，他们有的在风景秀丽的山谷或海边，也有的隐于大都市某酒店的某一层；有门庭若市的万人浴场，也有亭亭玉立的街边美容美体店。它们在设计师的巧妙的装点下成为休闲空间的新亮点 ，并迅速的遍布到我们身边。人们大都在这种环境中涉足过、享受过、受益过。由于SPA行业发展较快，给设计师和投资者带来了更新更多的挑战。这要求设计师们在设计定位和市场投资定位等问题上都要有准确的选择。我们不仅要追寻着前沿的设计理念，还要考虑到人们现阶段的审美观念、地域文化、人文环境及城市历史的不同价值观趋向。更要考虑投资者的经济回报、服务项目与流程、空间的使用比率等因素。SPA空间的设计不仅仅注重单一的表象，更重要的是为人们提供一个适应行业要求的商业平台。

我们今天所看到的这些作品就体现在这几方面，探索、实践，设计师们付出了很多的努力，在做中国自己的SPA。从作品中可以看到用心的利用中国文化的神韵与自然的山石结合成一个禅境的幽深意境。SPA作品有的看到的是简洁、时尚，注重陈设的美容院。更有气势恢宏的大浴场，处理方法成熟、细节统一，才能达到这一水准。每一个空间处理、每一盏灯的选择、每件饰物的放置，都展示了设计师的责任心、文化艺术修养及对生活的感悟。随着这一行业的发展和生活水平的提高，从服务上国际品牌更多的引领这一行业发展，服务的规范化和专业化划分越加明确。我希望我们这些设计师要更加准确的划分专业SPA的个性设计重点，尊重自然、信守绿色健康的理念。挖掘出服务的特点，深刻的学习和掌握相关的行业知识，用我们的设计来推动中国SPA馆的行业发展。坚信将来会看到更多更好的中国设计师的SPA作品。

While every city is decorated itself into a cement “monster”, our designers are quietly designing many “natural” watery environments to create a shower and recreation space of different styles, cultures and decorative techniques for the busy people - SPA.

SPA is usually deemed as a monogram of some Latin phases: Solus par aqua. Solus means heath, Par means “by” or “through” while aqua means water. It generally means that water is applied to build up health and cure the diseases. Our county is the origin of the eastern shower culture and water is also deemed as the base of the health in the theory of traditional Chinese medical science, which has emphasized that the water is a medium for the circulation of life energies in human body. Water is both physical and sacred.

The contemporary SPA more refers to the place where people can get relaxed, resume the physical strength and cheer up. It also has the comparatively perfect management system and professional service procedures. It includes the various healthcare elements and advocates the exploration of the healthy life style and self-awareness. It is also diversified and localized in the service forms. The healthy cultures of the regional advantages and histories are fully used to embellish the “colors” of SPA. From Alps to Bali Island, the international brand chain store to franchised store, some of them are located in the valley or seaside with beautiful scenery while some are hidden in certain floor of certain hotel in a metropolitan city; there are bathhouses with a lot of visitors which can hold over ten thousand people and are also the graceful beauty and body shops along the streets. They have become a shining point among the recreational spaces under the tactful decoration by the designers and rapidly spread around us. Most of us have ever stopped, enjoyed and benefitted here in this kind of environment. As SPA industry has developed very quickly, it has brought more and newer challenges to the designers and investors, which requires the designers to have a correct selection on the design orientation and market investment orientation etc. we not only seek for the best design ideas, but also have to consider the current aesthetic ideas, regional cultures, humanistic environment as well as the tendency of different values in the histories of the city. More consideration shall be taken on the factors such as the economic returns to the investors, service items and procedures as well as the use rate of the space etc. The design of SPA space not only stresses on the single superficial phenomenon and more importantly, it is purposed to provide a business platform adaptable to the requirements of the industry.

The works we see today just have reflected these aspects. With the exploration and practice, the designers have made great efforts to make Chinese Own SPA. It can be seen from the works that the charms of Chinese cultures are combined carefully with the natural mountain stones to create a profound Buddhism artistic conception. What you can see from some SPA works are the simple and fashionable beauty salons with stress on the displays. There are even the gorgeous big bathhouses with the skilled arrangements and unified details. The arrangement of every space, the selection of every lamp as well as the placement of every ornament have all shown the responsibilities, cultural and artistic educations as well as the understanding on life of a designer. With the development of this industry and the improvement of living standards, the international brands are more leading the development of this industry in the service sense and the standardization and professional division of the services have become specific. I hope that our designers can classify the individual design focuses of the professional SPA more correctly, respect the nature and comply with the conception of green health. We shall explore the characteristics of the service, further study and master the knowledge of the related industry and promote the industrial development of Chinese SPA shops through our designs. I strongly believe that we can see more and better SPA works by Chinese designers.

设计单位:朱永春设计有限公司 / 设计:朱永春

固足坊
Goodjoy Health Club

面积:650㎡ / 坐落地点:盐城市双元东路 / 摄影:吴永长

接这个项目前，我对足疗业知之甚少，反而带着偏见。一接触，才知道，此业有源远流长的历史，博大精深的文化，以及与之不相匹配的亟待改善、提升的环境及品位。总之，其中大有我们设计师的用武之地。

遥想当时，业主在设计任务中强调最多的是文化。我理解，那多指装饰性的符号，并不入骨的。按照那种文化的界定，恐怕现在固足坊里唯独缺少的就是文化，但未见妨碍开门营业，也不折损品位档次。——文化，实在有更内在的东西，更含蓄的表达！这么说，但愿不致影响固足坊软装到位的进度，有些符号还是必要的，有些文化本身就是符号的，何况，我们的设计，软中含硬的。

I have known little about the foot massage industry, even with having a bias against this before I undertake this project. Instead, I have some prejudices on it. Once in contact with it, I come realize that this industry has a long history with the broad and deep cultures and its mismatching environment and taste are desiderating the improvement and upgrade; in a word, there is great margin for our designer to make.

Recalling that moment, what the proprietor has mostly emphasized on the design task is culture. I understand that it means the decorative symbols and but is not to the marrow. According to the definition on that kind of culture, I am afraid that what Gu Zu Fang is lacking now is only culture. But it will neither necessarily affect the operation nor decrease the taste and grade. ——Culture actually has more inherent things and more implicative expressions! Saying so, I hope it will not affect the progress of the soft decoration in Gu Zu Fang. Some symbols are still necessary and some cultures are the symbols by themselves. What is more, our design has softness in hardness.

设计单位:大展·孙志刚空间设计工作室 / 设计:孙志刚、大帅

英格玛SPA会馆 ENIGMA

面积:4500m² / 坐落地点:沈阳市和平区三好街 / 主要材料:白微晶石、沙曼沙兰、LED

ENIGMA（英格玛）是一家高级SPA会馆。共四层。分男女士馆。首层接待，二层女士馆，三、四层男士馆。主要经营项目是男女美容与SPA。

ENIGMA（英格玛）译为：迷，不可思议的东西。这样梦幻般的词汇正好拿来做为设计的主线。男女士馆分别以两种风格体现。男士馆的风格为“灰酷”；女士馆为“白白”。

材料上，男士馆大量运用灰色石材、灰色乳胶漆、白钢等材料，配上点光源与LED，显得神秘，硬朗，突显男性的阳刚；女士馆运用大量白色纱帘、白色微晶石、镜面马赛克、电脑影像图案及大面的镜子等反射材料，显得梦幻、轻柔，突显女性的柔美。

灯光是本次设计的一个重点。用现今流行的LED做为主要的光源。而以往用的筒灯、射灯等为之配合。如男士馆的接待前厅与浴区就采用了蓝色的LED做为主要光源，射灯作点缀。显得冷静，神秘，令人产生如海一般的影像。

本次亦有不少有趣的设计。如一层主入口高4m的双开大门，用了128个水晶烟灰缸堆砌而成，效果晶莹、梦幻；还有玻璃泡制成的置物架……总之，这是一次不同以往的设计，是一次转型的设计，也是一次快乐的经历！

ENIGMA is a superior SPA club with four stories. It consists of men’s and women’s clubs with the first floor for reception, the second floor for women and the third and fourth floors for men. It is mainly engaged in the beauty and SAP for men and women.

ENIGMA can be interpreted as: mysterious, an unimaginable thing. Such a dreamy word is taken as the focus of this design. The men’s and women’s clubs are demonstrated separately in two styles/ the style of men’s club is “gray and cool” while that of women’s club is “white and pure”.

In material, men’s club has applied large quantities of gray stone materials, gray latex paints and white steel etc. together with the spotting light sources and LED, it looks mysterious and hard with a prominence on the masculineness. Women’s club has applied large quantities of white yarn curtains and the reflective materials such as the white crystallized glass panels, mirror mosaic, computer imagine patterns as well as large-area mirrors so that it looks dreamy and gentle with a prominence on the feminineness.

The lights are a key point of this design. The currently popular LEDs are applied as the main lighting sources while the formerly applied down lamps and radiation lamps are used as decoration. For example, the front reception hall in the men’s club is applied with the blue LEDs as the main lighting sources with radiation lamps as the decoration, which looks calm and mysterious and has left a sea-like impression.

There are also some interesting designs this time, such as the 4m height double-swing door at the main entrance of the first floor built by 128 pieces of crystal ash trays with a crystal and dreamy effect as well as the shelf made of glass bubbles……. In a word, it is a quite different design from the past, a transforming design as well as pleasant experience.

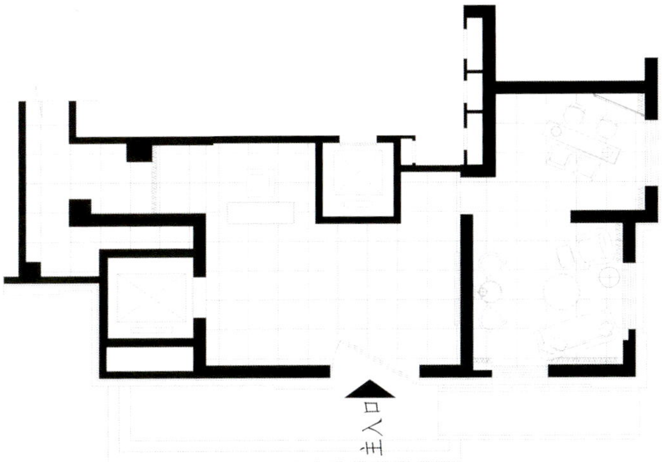
主入口

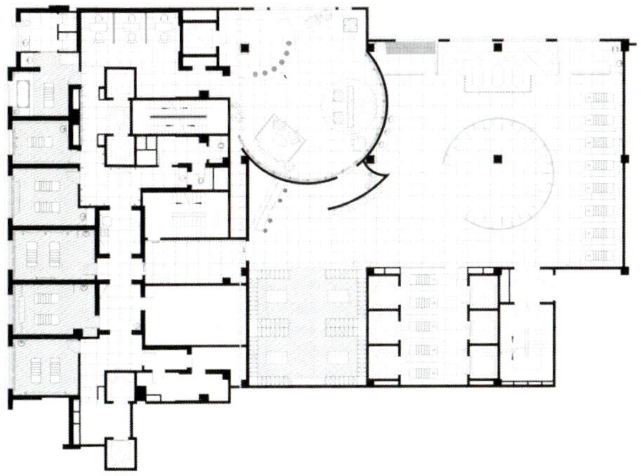

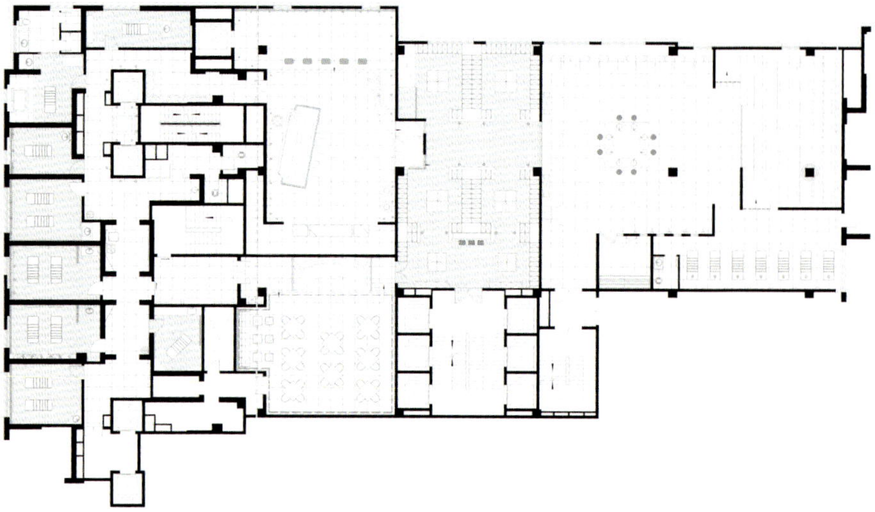

ENIGMA

设计单位：哈尔滨唯美源装饰设计有限公司 / 设计：张奇永

美丽新娘
Beautiful Bride

面积：1400㎡ / 坐落地点：哈尔滨市中央大街 / 主要材料：涂料、马赛克、镜子 / 摄影：张奇永

每个寻得幸福的女孩都会披上称为圣袍的白色婚纱，与心爱的另一半携手踏进礼堂。仅仅几码的织物，却编织着每个女孩心底里浪漫的梦。婚纱影楼是留住这个梦的地方，它是虔诚、圣洁、美丽、浪漫的浓缩。空间里经典的后现代古典家具，体现出的奢华。后现代茶金色镜框线配合茶色镜片，镶嵌于天花板上，时刻捕捉到每对新人甜蜜幸福的微笑瞬间，留下了无限的浪漫想象。银色科林斯柱头与幻彩马赛克的柱身，形成强烈的视觉冲击。古典与现代，庄重与浪漫，美丽与圣洁完美的结合，演绎出影楼整体空间的设计理念。

Every girl who has found the happiness will wear a white wedding dress called sacred robe and step in the church with her bridegroom. The textile only in several yards yet has woven a romantic dream in the bottom of every girl' s heart. The wedding studio is a place to keep this dream and it is a concentration of sincerity, sanctity, beauty and romance. The classic post-modern furniture in the space has revealed the extravagance. The post-modern golden mirror frame lines in arrangement with the brown mirrors are embedded in the ceiling to grasp the sweet and happy smiling moments of every new couple all the time and leave the infinite romantic imaginations. The silver Corinthian necking and colorful mosaic column shafts have constituted a strong visual impact. The perfect combination of the classics and modernity, sobriety and romance well as well the beauty and sanctity has demonstrated the design concept of the integral space in the wedding studio.

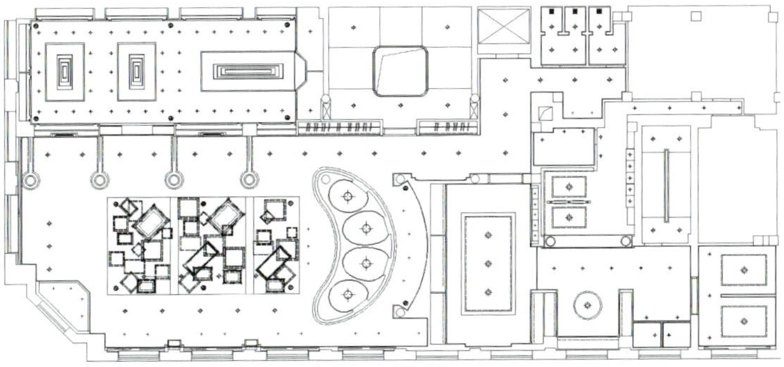

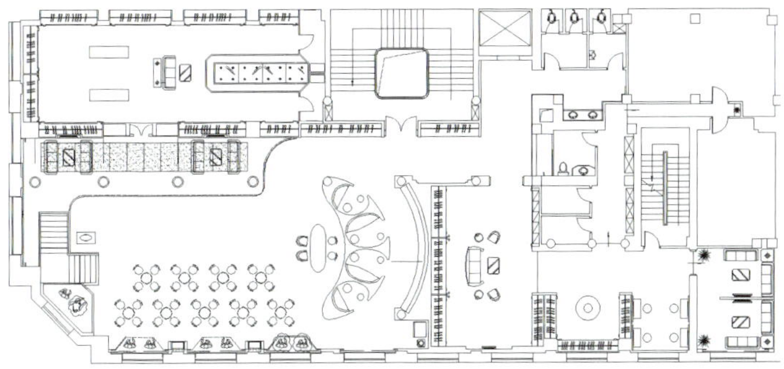

设计单位:江苏省海岳建筑装饰工程设计有限公司 / 设计:姜湘岳、吴海燕 / 参与设计:纪红、徐云春、钟铮、潘婷

龙景温泉会所
Longjing Spring Club

面积:1200㎡ / 坐落地点:南京中山陵 / 主要材料:日本特色墙纸、地毯 / 摄影:陈乙

想到SPA，我就会联想到天空失重状态下飘渺并且放松的感觉；想到香薰油在空气中轻松的弥漫、令人彻底地放松感觉。所以在设计立意时，我把“纯净”列为第一主题。希望在这里能够褪去外表的饰物，安静地享受宁静、自我的空间。

纯净的“白色”是空间的主色调，因为不要其他装饰，所以在大堂里我仅仅采用了投影的方式。在白墙上，投射出游动的红金龙，让空间有流动的感觉。一进入接待空间就能让来客强烈地感受到水，似乎我们也在水中。

进入水区，在纯净的白色空间里，中心便是一注湛蓝如海水般的水池，“蓝”与“白”是这个空间的色彩主题。进入休息区，灯光暗了下来，光线基本以间接光为主，让来客的视觉能安静下来。墙面设计的壁灯犹如幽暗的烛台。微微散发的黄光传达一种安静的暖意。在墙面的设计上，我们通过镂空木格背后的镜，路过的人感觉像是在两道围墙内，窗棂之外还有一个透明的世界，所以原来狭小的过道，让客人觉得宽了。进入休息间，还是安静的白。台灯的布罩都是黑的，控制着光线的散发，静静的躺一会疲劳感全消散了。客人路过楼梯，可以看到由上至下悬挂着的水晶球，错落无序地吊挂着，仿佛像是从水底冒出的气泡。楼梯的周围墙面，玻璃的水鸟，安静地停在悬挑漂浮的玻璃上，让人感受到安静的氛围。

Once thinking of SPA, I will associate it with the dimly discernible and relaxing feeling in the status of weightlessness and think of the essential oils that pervade comfortably in the air and relax you thoroughly. Therefore, in the conceptual design, I have listed "purity" as the major subject. We hope that here we can remove the superficial ornaments and enjoy the peaceful and individual space quietly.

The pure "white" is the main color in the space. As there is no need for other ornament, I only apply the way of projection in the lobby. The swimming red arowanas are projected on the white wall to endow the space with a feeling of flowing. Once entering the reception space, the visitors can strongly feel the water and it seems that we are also in the water.

When entering the water zone, in the pure white space, the center is a water azure blue pool as the sea. "Blue" and "white" is the subject of color in this space. Entering the rest area, it becomes dark and the lights are mainly the indirect so that the sights of the visitors can be calmed down. The wall lamps designed on the wall are just like the dim candle holders and the dim emitted yellow lights deliver a quite warmness. On the design of the wall, through the mirror at the back of the hollowed-out wooden grids, the passengers feel a transparent world between the two enclosed walls outside the muntin, thus, the visitors feel that the original narrower passage becomes wider. It is still the quite white after entering the rest room. The lamp cloth cover is all black to control the emission of the lights. The tiredness will completely disappear after lying down quietly for a white. The visitors can see the crystal balls hung up to bottom in disorder like the air babbles out from the bottom of the water when passing through the staircase. The wall around the staircase and the glass birds quietly stationing on the suspended floating glass endows the people with a quiet atmosphere.

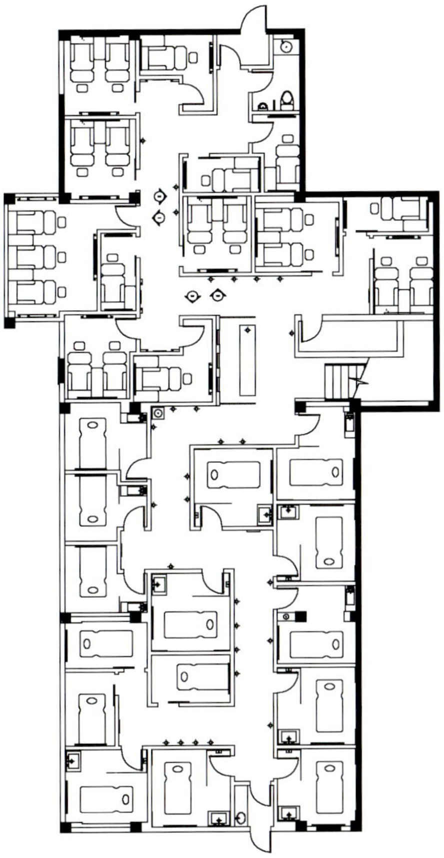

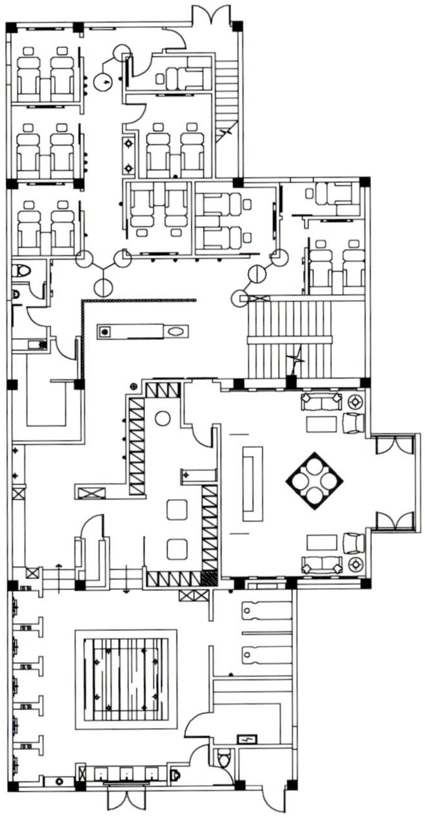

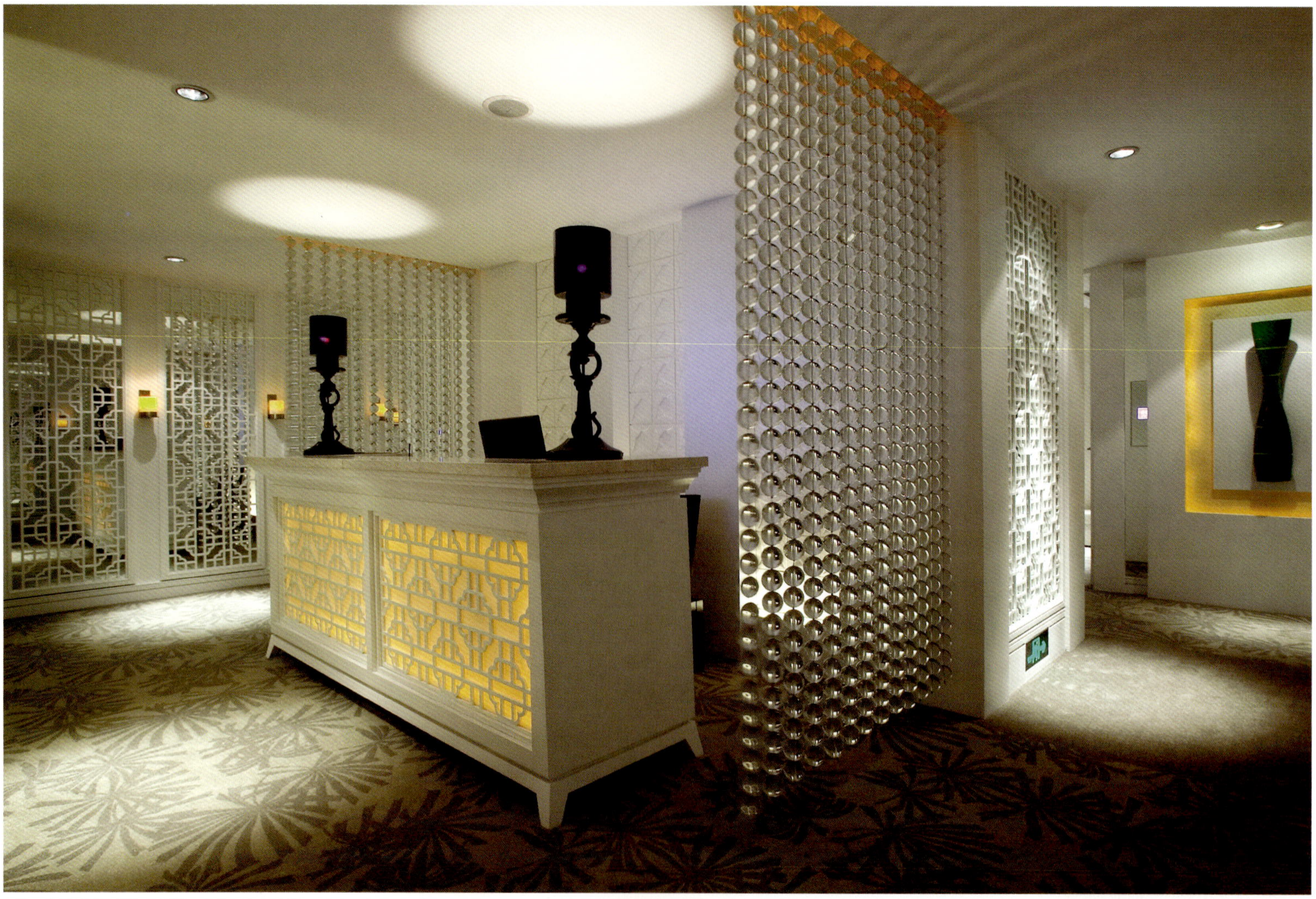

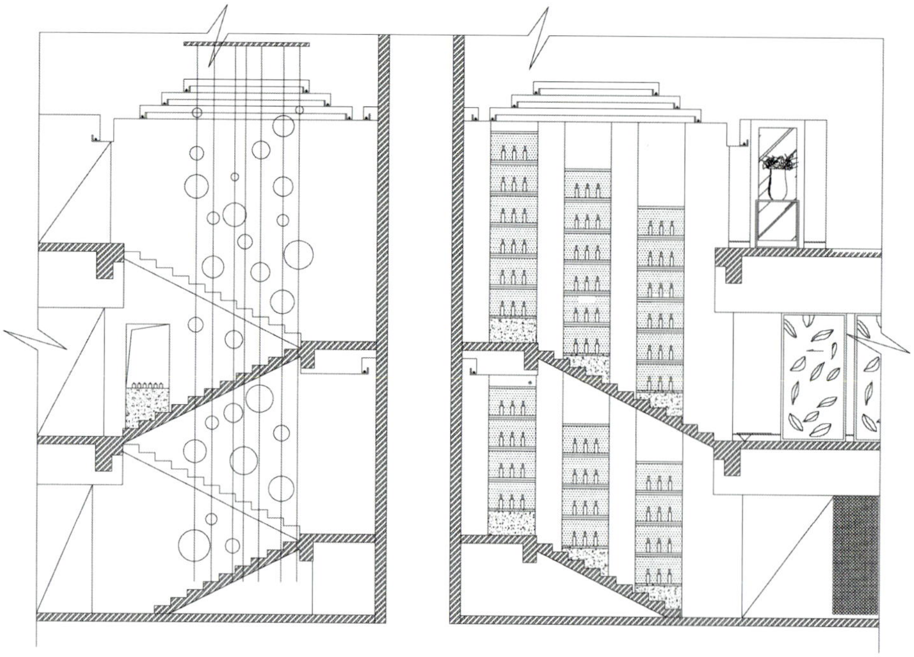

设计单位:VINYL GROUP / 设计:BEN LING / 参与设计:夏芹 / 软装设计:那格九号 朱岚

GREEN MASSAGE

面积:220m² / 坐落地点:上海市徐家汇 / 主要材料:砖、地毯、马赛克、清水泥、布、石材、曲面板 / 摄影:Don Yap

本案坐落在上海徐家汇一个小区的对面，不起眼的门口，很容易被忽略。一进门你就会被淡淡的薰香吸引住，整个神经便放松下来。休息间面积不大，时尚中透露出丝丝雅致，线条流畅的储物架既具备实用功能，又充满设计感，环境布置煞费心思。通过更衣室，随着“芝麻开门”似的一扇大门，才真正进入如会所般的大厅，暗暗的灯光，轻幽的音乐，身临其境，闻着淡淡的茶香，思绪顿时宁静下来。

It is located opposite to a community in Xujiahui District, Shanghai. Its doorway is very unobvious and can be easily neglected. Once entering it, you will be attracted by the faint incenses and feel relaxed completely. The area in the rest room is not very big with the traces of elegance revealed from the fashion. The smooth lined storage rack has both the practical functions and also is full of design senses. The arrangement is made with great efforts. Passing through the changing room to a big door like "open the door of fortune", you will really come to the club-like hall with dark lights and light music. You will feel calm down and peaceful after being involved in it and smelling the light tea fragrances.

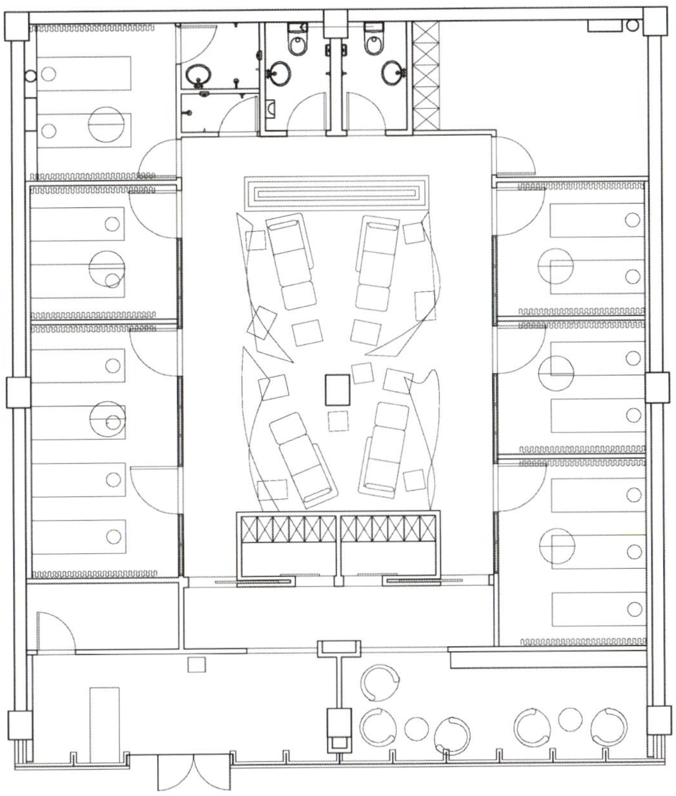

设计单位:大展·孙志刚空间设计工作室 / 设计:孙志刚、徐慧

希尔斯池典
Health SPA

面积:18000m² / 坐落地点:沈阳市铁西区建设中路 / 主要材料:雅士白理石、金属马赛克、铜板腐蚀雕刻、曲柳木饰面、玫瑰金色铜箔、西施红理石马赛克、黑镜、镜片马赛克 / 摄影:司徒颖

希尔斯为“health”的音译，健康、休闲是业主的初衷，整体设计融入了异域风格、东南亚的热带风情及中东的奢华宫廷风格，将“奢华的休闲”通过材料、造型、符号、整理、融合得以体现。

Xi Er Si is the transliteration of “health”. Heath and recreation are the original intention of the owner. The whole design has integrated with the foreign style, tropic style of Southeast Asia as well as the luxurious royal style in Middle East. “Luxurious recreation” is demonstrated through the materials, models, symbols, arrangements and integration.

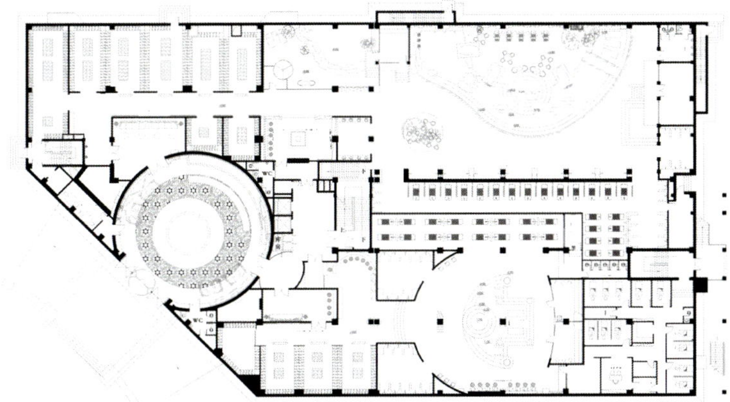

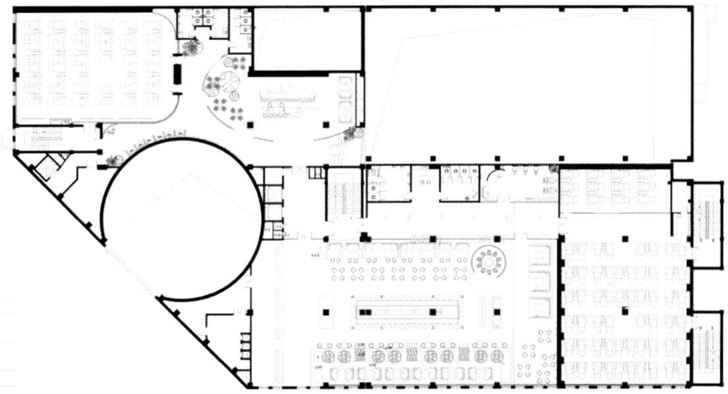

设计单位:C+S南京筑内空间设计顾问有限公司 / 设计:陈卫新 / 参与设计:陈妍妍

今生有约
Appointment Forever

面积:925m² / 坐落地点:南京 / 主要材料:米黄大理石、爵士白大理石、镜面不锈钢、实木复合地板、镜面磨砂纹 / 摄影:文宗博

色彩是感官的源泉，是喜怒哀乐的灵魂，而白色能记录一个瞬间也能表达一个女性的梦境人生。在这个略显沉闷的街道上，要想使这个婚纱摄影公主馆从林林总总的各类店铺中凸显出来，使用一个高调近白的基色是可行的，也是必要的。在这个长条形的三层店铺里，为了使空间动线更具合理，在平面布局上做了调整，重新确定了入口位置，在横向方向上大面积镜面的使用，使房子的进深得以扩展，也使空间内的展示目标更具趣味性。浅色木饰面、白色地砖、镜面不锈钢、磨砂纹镜面玻璃的使用，都在试图淡化空间的边界感，使整体空间在客人的行为参与中浑然一体。白色有多少可能性，或者说白色能为这个空间带来怎样的意义，是设计者深思的问题，作为一个展示女子最美状态的婚纱展馆，要表现的可能更多是客人的自我表达与自我实现，而环境是一个实实在在的托。

The colors are the sources of the sensory organs and the souls of pleasure, anger, sorrow and joy while white not only can record an instant but also can express a female' s dreaming life. In this a slightly dull street, to make this wedding photographing princess studio prominent in the various types of shops, it is both practicable and necessary to apply a distinguished off-white as the background color. In this long-strip three-story shop, to make the active lines in the space more reasonable, the arrangement on the plane also has been adjusted: the location of the entrance is re-decided and the large mirror is applied in the horizontal direction so that the depth of the house is expanded and the objects exhibited in the space become more interesting. The application of the light colored wood veneers, white floor tiles, and mirror stainless steel as well as frosted mirror glass are attempting to weaken the sense of boundaries of the space so that the integral space becomes integrated together through the actions and participations of the guests. How many kinds of possibilities white has, or what kind of meaning can white bring to this space? They are the questions the designer has to think deeply about. As a wedding dress exhibition shop to display the most beautiful status of a woman, what it wants to present may be more likely the self-expression and self-realization of the guests while the environment actually is a contrast.

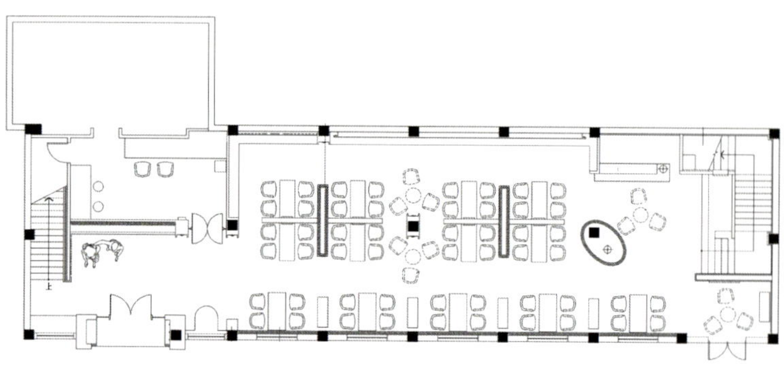

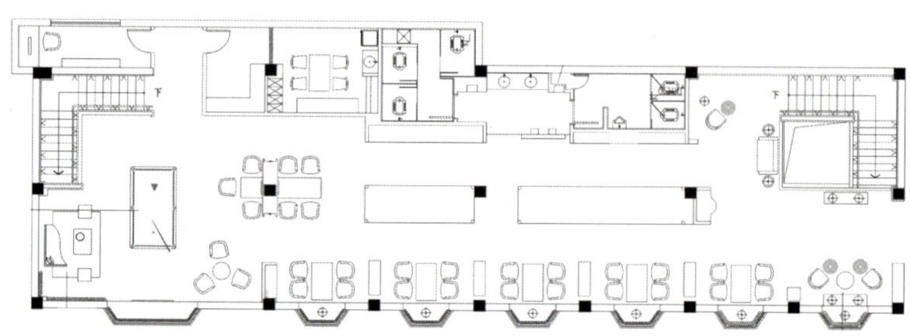

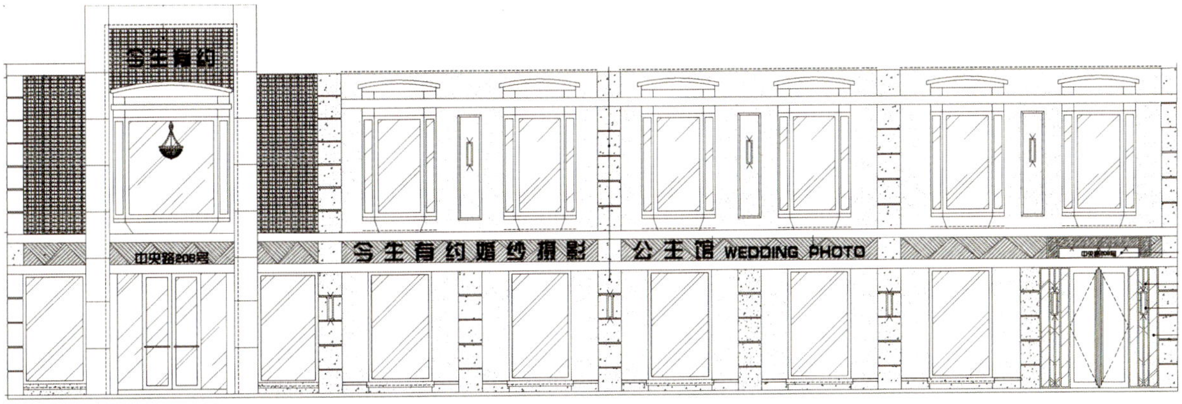
今生有约
中央路208号
今生有约婚纱摄影 公主馆 WEDDING PHOTO

设计单位：上瑞元筑

班雅SPA会馆
Banya SPA Club

面积：1750m² / 坐落地点：无锡 / 主要材料：实木板、藤、天然石材 / 摄影：文宗博

班雅SPA会馆在冰冷都市丛林当中开辟出一处静壤，让人们的身体得到恢复，使人们的精神和心智回归到本原的和谐状态。因此会馆将回归自然的主题贯穿始终，大量采用了木、藤等贴近自然的物料，旨在营造一种脱离尘世的静谧氛围。

一道定制藤编长案，将人们引入主题，大厅当中流水潺潺，暖香扑鼻，温暖而放松，让人们在短短的逗留当中得到视觉、听觉、触觉、味觉和直觉的洗礼，达到内心的平静，忘却尘嚣。

房间内秉承了回归自然的主题，将东南亚建筑的元素引入其中，同样使用了大量木料，营造如同置身自然的氛围，在沁人心脾的灵性音乐中糅入各种特色疗法，让人们的心灵充分放松和回归。

Banya SPA club has developed a quiet place among the cold city forest to restore people's health and restore their spirits and minds to the original harmonious status. Therefore, the theme of returning to the nature will run through the whole club. The materials to go with the nature such as the woods and vines are applied in a large quantity in order to create a quiet and peaceful atmosphere disengaged from this mortal life.

A customized long vine-woven table has introduced people to the theme. The water is purling in the lobby with warm fragrance greeting your smell, warm and relaxing so that the people can receive visual, hearing, touching, tasting and the sixth sense baptism in the short stay to reach the inner calmness and forget the worldly affairs.

The rooms have continued the theme of returning to the nature and have introduced the Southeast Asian architectural elements in. The wood materials are also applied in a large quantity to create an atmosphere as if you were staying in the nature. The various special therapies are integrated into the refreshing spiritual music in order to fully relax and restore the people's spirits.

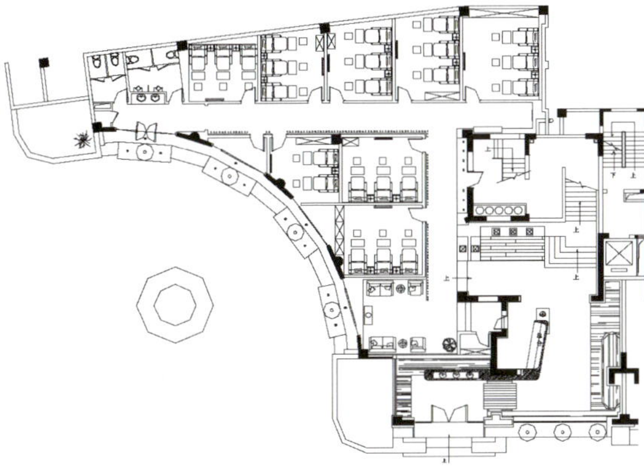

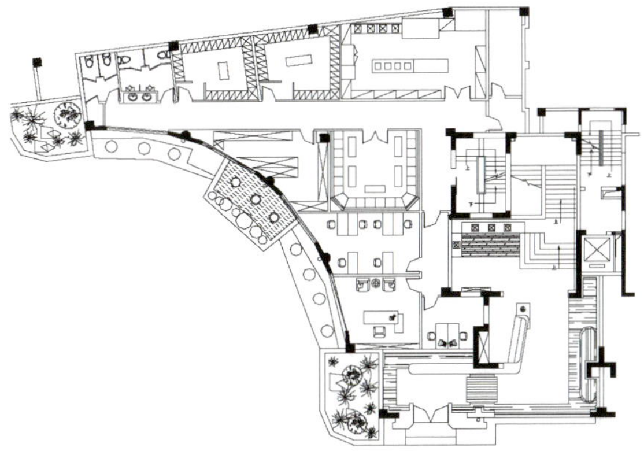

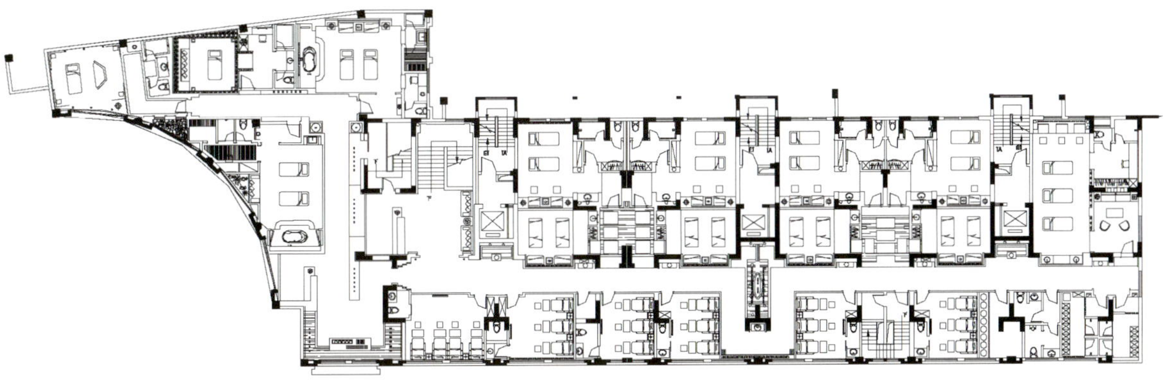

BEYOU
HEALTH CARE
CLUB

设计单位:陆文星室内设计事务所 / 设计:陆文星

王春美容院——龙江风尚馆
Wangchun Beauty Salon – Longjiang Fashion

面积:1500㎡ / 坐落地点:南京 / 主要材料:胡桃木、进口石材、个性马赛克、仿古砖

业主要求整个空间体现出浓郁的东南亚风格。设计师运用不同材质的质感与性能，诠释出健康、向上的美容生活新风尚。灯光与饰品的巧妙搭配，恰到好处的空间布局，彰显着空间的灵性。

金属质感的仿古地砖呈九十度交错铺贴，赋予空间灵动质感，再配上门厅的端景，使整个空间既包含风水意蕴，又趣意盎然。各个美疗间的色调不尽相同，但是红绿搭配的主色调给人眼前一亮的感觉，大片的红色铺底缀花纹，配以深浅的绿色细碎纹理交错，再加上木线条、方格框这些元素，平衡着整个空间的美感。用直线珠帘来分割不同的功能区域，感觉简洁利落，且隔而不断。此外，射灯投下来的暖黄光晕，可以营造出曼妙的光影效果，梦幻而让人欣喜。层次感极强的吊顶，配以深红色线条和用作软隔的草绿色帘子，无不曳动会所健康活力的主旋律。

The owner requires that the whole space shall demonstrate a strong Southeastern Asian style. The designer has applied the realties and performances of different materials to interpret a new healthy and positive SPA fashion. The artful combination of the lights with the ornaments as well as the appropriate space arrangement has revealed the aptitude of the space conspicuously.

The archaic ground tiles in metallic sense are paved in a 900 crossed way, which has endowed the space with the bright senses. In arrangement with the display shelf at the hallway, the whole space not only has the geomantic implication, but also looks interesting and vivid. The color in each SPA room is not completely same, but the main color in the arrangement of red and green will catch your eyes at once. The large area of red background decorated with the patterns, together with the interlaced light and dark green fine green grains as well as these elements such as wood lines and panes has balanced the aesthetics of the whole space. The straight bead curtains are used to divide the different functional areas, which looks simple and spiffy, separated but not blocked. Furthermore, the warm yellow halation projected by spot lights can create a lithe and elegant light-and-shadow effect, dreamlike and pleasing. The ceiling of strong hierarchic sense together with the scarlet lines and grass green curtain as the flexible separation has all without exception swayed the healthy and vigorous rhythm in the club.

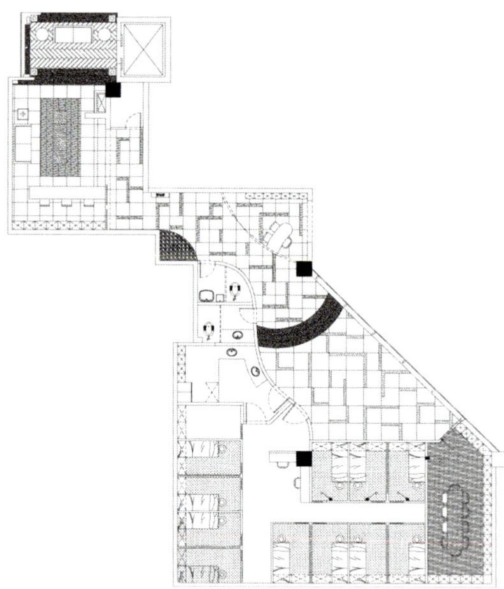

设计单位:南京目达配饰设计有限公司 / 设计:高铁、张苗苗

卡伊造型
Kayi Style

面积:220m² / 坐落地点:苏州市临顿路 / 主要材料:乳胶漆、清玻璃、壁纸 / 摄影:子常

作为一家连锁型的企业，卡伊每年都会有数家分店进行翻新，所以在可以达到设计效果的基础上，节约成本也是我们要替店主考虑的一个环节。每平方米1000多元的施工成本，我们觉得很合理。

为了既达到业主要求的另类、夸张，又不突破客户的心理接受防线，在设计时，我们从色彩及造型两方面进行了思考，顶部选用了桃红色与白色的乳胶漆，出彩而节约成本。吧台处选用亮眼的水晶灯使全店的顶部凸显。

楼梯方面，我们选用的是和镜台柜相同的木质结构，而且顶面、镜台柜、楼梯都选用了白色，形成一个整体。这就是设计空间时度的把握问题，如果颜色和材质运用得过多，反而会让人觉得杂乱无章。

As a chain enterprise, Kayi will have several branch stores renovated every year. Therefore, the cost saving is also a point we should consider for the owner while it reaches the design effects. We think it quite reasonable to estimate for with a construction cost of over 1000 Renminbi per square meter.

To reach the unusualness and exaggeration required by the owner but not to breach through the psychological defending line of the customers, during the design, we have considered both from the colors and modeling. The ceiling is applied with the pink read and white latex paints, outstanding and cost-saving. The brilliant crystal lamps at the bar counter has made the ceiling of the store distinguished.

As for the staircase, what we have selected is the same wooden structure as the mirror & cabinet combination. The top surface, mirror & cabinet combination as well as the staircase are all in white, which forms into an integration. This is the grasping issue of the designed space. The application of excessive colors and materials instead will make it look disorderly and unsystematic.

设计单位：无锡观点设计 / 设计：吕邵苍

名仕会SPA
Mingshihui SPA

面积：1300m² / 坐落地点：苏州桐泾路 / 主要材料：马赛克、乳胶漆、玻璃 / 摄影：文宗博

设计之初，我们重新梳理了对SPA的理解。SPA是一个关于水的娱乐项目，人在这个场景中应该是一次体验之旅。而什么东西能够象征性表达人的这一体验要求呢？我们想到了豪华游艇这一载体。豪华游艇作为与水有关的一个奢华的体验项目，可以很好地表达我们对SPA的理解。会所的概念即刻在我们的项目上脱颖而出，这次设计就有了这样一个主线，豪华游艇上的一场party，出海之行的体验之旅……

所有的设计素材、符号、灯光、色彩、形体，均是为这个精准的概念表达，所以就有了创造性的豪华游艇形状的泳池。干蒸房成为船的船舱架于水面之上，颜色只有白色和深蓝色，唯一跳跃的是深海里面五光十色的鱼。白乳胶漆墙面在LED蓝色光源的映衬下，显得尤为纯净。如水般的形体与动态十足的通道，恍如游艇在海上轻快地漂移。没有可以与不可以的东西在这个场景中出现，项目最终呈现的是一种从未有过的纯粹与干净。原创设计的本质是原创的思维体系与原创性的理解项目，从而塑造全新的设计语言，获得具国际化视界的可能性与突破性。

At the beginning of the design, we have reconsidered our understanding on SPA. SPA is a recreational object involved with water. People shall have a travel of experience in this scene. However, what kind of thing can symbolically deliver this kind of requirement for experience? We think of this carrier – the luxurious yacht. As a luxurious experiencing item involved with water, the luxurious yacht can well deliver our understanding on SPA. The conception of club becomes eminent in our project and thus this design gets such a main thread: a party in a luxurious yacht and a travel of experience in a sea outing……

All the design materials, symbols, lights, colors and forms are applied for this precise conceptual expression. Therefore, it comes to a creative swimming pool in a shape of a luxurious yacht. Dry sauna rooms become the cabin of the yacht and located over the water. There are only black and navy blue and the only jumping thing is the colorful fishes in the deep sea. The white latex-painted wall, in contrast by Blue LED lighting sources, looks extremely pure. The passage in the watery form and full of dynamics looks as if the yacht is drifting briskly in the sea. Nothing possible or impossible will appear in this scene and what the project will finally present is a purity and cleanness that never take place. The nature of the original design is the original thinking system and original understanding item to create a brand-new design language and obtain the possibilities and breakthrough in the internalized visual field.

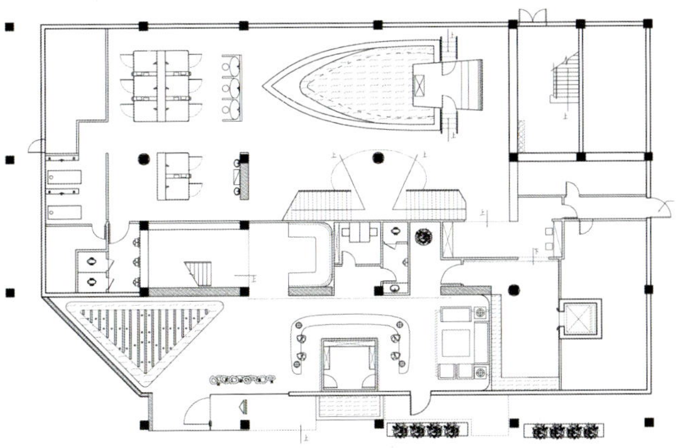

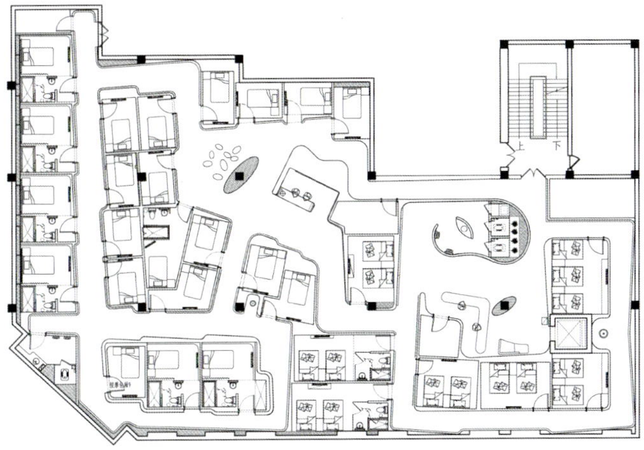

名仕会温泉会所
Welcome

A29

设计单位：哈尔滨唯美源装饰设计有限公司 / 设计：韩冠恒

Lea spa

面积：1000m² / 坐落地点：哈尔滨 / 主要材料：壁纸、仿木材地砖、亚麻地毯、石材马赛克、印花玻璃合镜片 / 摄影：张奇永

Lea-spa是隶属女王美容机构下的最新品牌，服务于五星级酒店。由于是多年的合作伙伴，每当有新项目的时候我们总要在一起讨论功能性的流程，这次也不例外。不过相互之间有多年合作的默契很快就达成了一致。我们的立意非常的本质，就是想让时间静止，让岁月停留不动。实施过程中最大的问题在于和施工单位之间强调质量问题，以及和酒店管理公司之间的多次会议。最后完成的效果还是让大家都很满意的。特别是客人，体验过的顾客都对环境给予了很高的评价，这也正是我们最在意的事情。设计中最大的亮点就在于每个房间的各种材质之间的质感强弱对比、色彩上的协调所营造出来的统一性。

项目位于酒店的三层，共9间房间3个类型，每个房间都有独立的更衣、淋浴、冲浪浴缸、桑拿房、卫生间及休息区美体床。房间的入口处都以半闭合的形态对内部进行遮挡，并且通过高度的变化保证了地面水管线的畅通。房间的地面以仿木材地砖铺装，结合亚麻地毯，既满足了要求又达到了很好的脚感。三个类型房间通过冲浪浴缸背景墙面的石材马赛克拼图的不同变化完成相互间的变化。有的是以年轮为背景，有的是以波浪形态的凹凸弧形面来体现一种意识流的想法。墙面多以壁纸为饰面，加以印花玻璃合镜片来完成多层次空间延展，整体的灯光都是用调光器形成微弱的光线，烘托慢节奏的气氛，再配以东南亚风格的家具，更能体现spa的特有魅力。

Lea-spa is the latest brand affiliated to Queen Beauty Organization, serving five-star hotels. As we have been the partners for many years, every time when there is a new project, we will always discuss the procedure of the functionalities together. It is not an exception this time. But we have reached the agreement very soon due to the years of cooperative privities. Our conception is very essential, that is, to make the time still and make years stationary. During the implementation, the biggest problem lies to emphasize the quality issues with the construction company as well as the multiple meetings with the hotel management company. The final completed effects are quite satisfactory, especially the customers. The customers who have taken the experience all have a very evaluation on the environment, which is also what we are mostly concerned about. The greatest shining point in the design is the real comparison between the various materials as well as the uniformity created by the harmony among the colors in every room.

The project is located at the third floor in the hotel with nine rooms in three types. Every room has an individual changing room, shower room, surfing bathtub, sauna room, toilet, as well as facial bed in the rest area. The entrance of the room is in a semi-enclosed type to hide the interior and ensures the water pipe lines on the floor clear through the change of the heights. The floor in the rooms is paved with the fake wood tiles. In combination with the linen carpets, it not only reaches the requirements, but also ensures the good foot senses. Three types of rooms have achieved the mutual changes through the different variations of the stone mosaic-assembled pictures on the background wall of the surfing bathtub. Some take the annual rings as the background while some take the concave-convex arc faces in a wave form to reveal a stream-of-consciousness idea. The wall surfaces are mostly applied with the wallpapers as the facing and realize an extension of the multi-hierarchic space together with the printed glasses and mirrors. The integral light is adjusted to the weak light by the light dimmers to set off a slow rhythmic atmosphere. Together with the furniture in the Southeastern Asian style, it can better reflect the special charms of SPA.

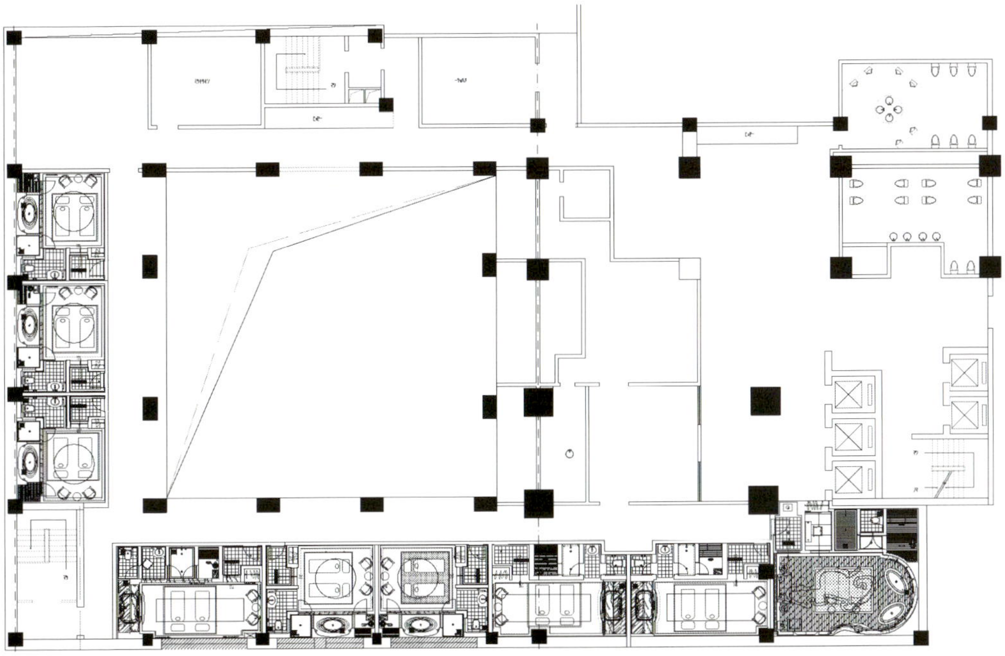

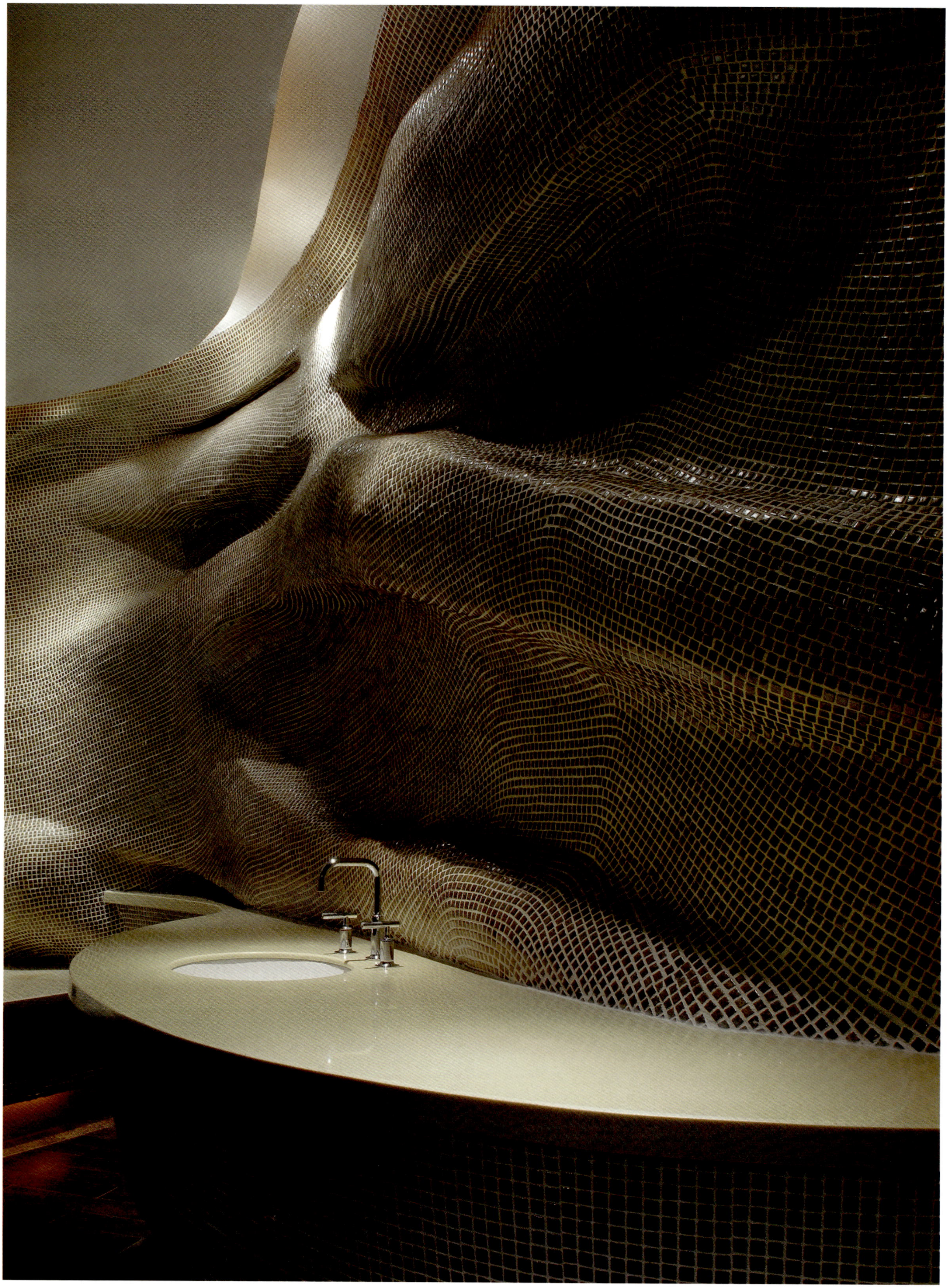

再次出发
Starting again

陈南 Chen nan

深圳市艺臣设计顾问有限公司执行总监

高级室内建筑师

李益中、范宜华	复地连城营销中心
张成喆	蓝海销售中心
张成喆	雁南园
王祎华	太湖温泉1858别墅VIP中心
张星	楼兰
D5组	正中集团科技大厦售楼部
万宏伟	高新科技广场售楼处
童武民、李信伟	中央公园二期售楼部
刘卫军	彼岸城售楼中心
李益中、范宜华	金众蓝钻风景花园营销中心
黄书恒	远雄建设上林苑接待中心
陈幼坚	万科中粮本岸售楼处
仇益国	中航楼吧
吴庆华、奚进、崔豪	南京世茂滨江新城会所展厅
内建筑	朗诗•国际街区售楼处
内建筑	南北西岸
吴政道	元一柏庄样板间
黄书恒	远雄•大未来

仅用文字的形式记述或议论一件事物并不见得就是对其有所了解，我愿怀着这样的陌生感，再次面对和重新审视售楼处空间的设计发展轨迹。

“售楼处”顾名思义，一个卖楼的地方，和房地产的关联不言而喻。众所周知，国内地产业从起步到现在30年的发展中，属于较少几个不被垄断的资本密集型行业。经济层面，是社会财富高度集中的代表；文化层面，更是时尚的潮手，立于时代之巅，引领生活潮流。

伴随国内地产业的蓬勃发展，售楼处的形象也由最初的简单沙盘加简陋桌椅的组合，升级到贴贴石材、做个造型之类的“豪华装修”，最后演变至今天需要聘请专业人员，结合总体策划、销售定位等要求，从建筑及环境设计的风格、展示方式、沟通形式等入手，到声、光、电以及家具、饰品的综合设计等，售楼处已然发展成一个复杂而庞大的系统工程，其面积规模和被重视程度也与口俱增且越演越烈。

如果样板房是设计界的时装秀，售楼处则是承托起时装秀的闪亮T台。随着创意时代的到来，这里更是头脑风暴的密集区，引领着多种消费的欲望。然而表面繁华的背后，作为一个真正的设计师不仅需要具有真诚的创作热情，以自己的认知来丰富设计的概念，更应直面设计的本质与内涵。记得曾在洛杉矶看过一个别墅的“售楼处”，只临时设在一栋已建好的200多平方米的别墅内，一个沙盘，简单资料，三两个妈妈级的营销员，亲切温馨。客人到后，闲问两句，或在起居厅随意小坐，气势虽远不如国内的售楼处豪华但却舒适贴心，温情而实在。这或许也是该有的方向。

随着国际经济放缓和国内新一轮的政策调控，2009年的地产业将会迎来新的考验。祸福相依，长远来看个人认为是件好事，至少可以增速行业优胜劣汰的发展进程，让鱼龙混杂，混乱无序的状态得以终结，促进地产业的发展迈上新的台阶。

在蝴蝶效应的扇动下，2009年设计界新的变化是不可避免的。我更期待能看到设计观及价值观的转变。希望可以在急功近利、浮夸奢侈的表皮背后，发现更多对人文的关怀，对内涵的追求及对气韵的求索。

温故而知新，这里记录的是曾经优秀的设计笔记，历经地产成熟十年的辛勤耕耘，在这片土壤上我们可以欣慰地看到设计者从生涩到成熟到再创作的发展历程，以及在部分作品里对精神领域的尝试性的探索。春华秋实，别人走过的路我们用几近奔跑的速度赶上了。现今我们更需要一批勇敢的开拓者，以充满自省的精神来深化和反思自身的设计意识，达到返璞归真的理性回归。这是设计的发展，更是我们的使命。

站在一个新的起点再次出发。在设计的道路上，让我们轻舞飞扬！

To record and narrate or discuss an object only in words is unnecessarily able to express the depth of understanding on it. With such kind of strange feelings, I would like to face again and review the development track of the space design for the show rooms in the sales pavilions.

“Sales Pavilion”, just as its name implies, is a place where the houses are sold and its relationship with the real estates is self-evident. As we all know, in the thirty-year development of the domestic real estate industry from the beginning to now, it belongs to one of several few capital-concentrated industries that are not monopolized. From the economic aspect, it is a representative of the high concentration of social wealth; from the cultural level, it is more a tend of the fashion, which stands on the top of the times and leads the life trend.

With the prosperous development of the domestic real estate industry, the image of the show rooms in the sales pavilions has been updated from the simple sand table with the simple tables and chairs at the very beginning to the “luxurious decoration” with laid stone materials and modeling. Today, it is finally evolved to hire the professional staff, begin with the styles, exhibition ways and communication forms of the buildings and environmental design to make a comprehensive design including sounds, light, electricity as well as furniture and ornaments. The show rooms of the sales pavilions have already developed into a complicated and giant systematic project. Its areas, scales and the attentions received also have increased steadily and become more and more intense

If the show room is a fashion show in the design industry, then the sales pavilion is a shiny T stage to undertake the fashion show. With the coming of the creativity times, it has more become an intensive area of the intellectual storm to lead the desires of various consumptions. However, behind this superficial prosperity, as a realer design, he/she not only needs to have the sincere enthusiastic on creation and enrich the design ideas through own cognitions, and shall further face the nature and connotation of the design directly. I remember that I saw a “sales pavilion” for single houses in Los Angles, which is only temporarily set in an already built single house with an area of over 200 square meters. There are only a sand table, simple documents and two or three mom-level sellers, friendly and warm. After the visitors come, they ask several questions or site freely in the living room. Although it looks far not luxurious from the domestic sales pavilions, but is comfortable and intimate as well as warm and practicable. This may be also the direction it shall have.

We can gain new insights through restudying old material. What is recorded here is the ever excellent design diary. After experiencing the ten-years of hard work during the maturing of the real estate industry, in this earth, we can gladly see the procedure in which the designers have develop from awkwardness to maturity to recreation. It flowers in spring and seeds in autumn. The road passed through has been caught up with by us in an almost running speed. Now, we more need brave explorers to deepen and think over their own design ideas with the spirits of full introspection in order to realize the rational return to its original nature. it is the development of the design and more our task.

Set out again on a new starting point. On the road of design, let’s feel dancing gracefully in high spirits!

设计单位:深圳市派尚环境艺术设计有限公司 / 设计:李益中、范宜华

复地连城营销中心
Fudi Liancheng Sales Center

面积:1900㎡ / 坐落地点:杭州 / 主要材料:木纹黄洞石、实木线条（造型）、铁刀木饰面、金属砖 / 摄影:吴永长

这个设计基本上只用了两种材料，一种是木纹洞石，另一种是木条，两种材料搭配建构成了一个干净整洁且富有空间趣味的销售中心。

这个销售中心空间最大的特点在一个中庭，空间构架很好，给销售中心提供了充足的天然采光。我们在做室内设计时，力图去表达这种空间感，并让它更加生动。具体表现在两个方面，一是用具有雕塑感的楼梯来表达块面体积，用木纹洞石表达界面的整洁干净；二是用垂直木条排列，形成美丽的韵律，在恰当位置做了一个异形的凹凸浮雕造型，阳光从透光天棚倾泻下来，投下优美的光影。以上两点都没有过多的去修饰空间界面，而是在整体之中去求变化，还原空间的构架感。

木线被广泛地重复，在洽谈区的靠外墙一面，由于柱子突出墙体以及过多的不规则的条形窗，使得空间界面极为凌乱，横向木条装饰则遮掩了这个缺陷，完整了界面，为平面道具（楼盘广告，平面信息）提供了整洁的背景。

在洽谈区的外侧，用倾斜的木条构筑了一个圆锥形的视听空间，有明显的空间界定。木条之间的缝隙又保证了空气的流通，也不会太过阻挡洽谈区的光线。

在两个签约区，同样用木条建构了方形的空间，有如一个盒子，为签约区提供了安定的气场。

Basically, the design only uses two kinds of materials, one is travertine with wood grain and the other is batten. These two kinds of materials build a clean and interesting sales center.

The most outstanding feature of the sales center is the atrium of good frame, which provides sufficient natural lighting to the sales center. When we design for interior space, we do our best to express the sense of space and make it livelier. It mainly manifests in two aspects. Firstly, the stairs with sense of sculpture are utilized to reflect cubage of block, while travertine with wood grain for cleanliness of the interfaces. Secondly, the vertical battens form a beautiful rhythm. An abnormal relief modeling is placed in the proper position so as to lay a beautiful shade with sunlight from translucent ceiling. The above two points don't adorn interfaces of space too much, but seek for change in the whole structure to restore sense of frame in the space.

Mouldings are widely repeated. On one side of external wall in the negotiation area, interfaces of space are rather cluttered because of pillar protruding wall and excessive irregular bar window. However, horizontal battens cover the defect, integrate interfaces and offer a neat background for the plane props (real estate advertising and plane information).

Outside the negotiation area, a conical audio-visual space is established with inclined battens with obvious interface for space. The gaps between battens ensure the air circulation and won't obstruct light in the negotiation area too much.

In the two signature areas, battens also form a square space like a box in order to provide stable atmosphere for signature area.

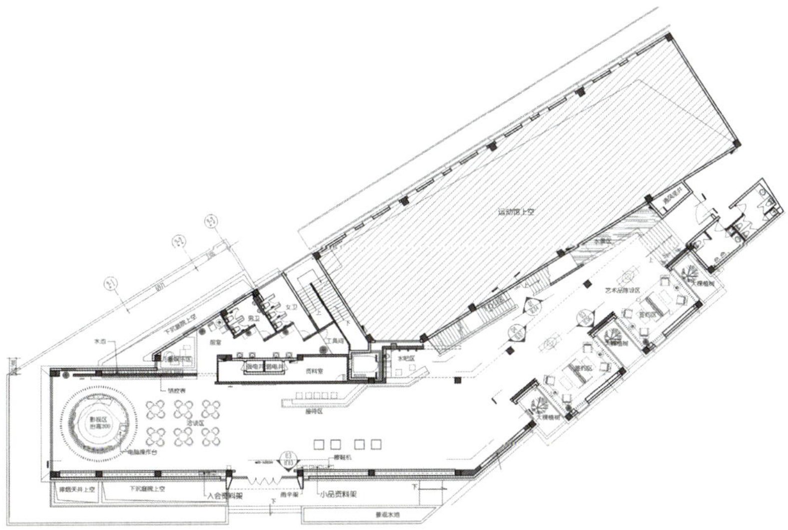
运动馆上空
男卫
女卫
资料室
水吧区
洽谈区
接待区
电脑操作台
下沉庭院上空
入会资料架
雨伞架
小品资料架

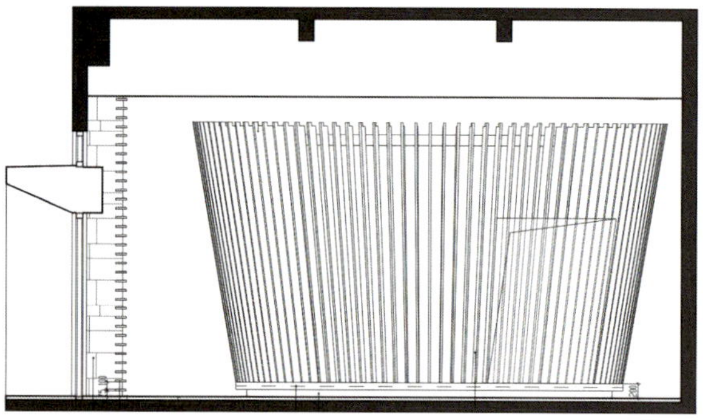

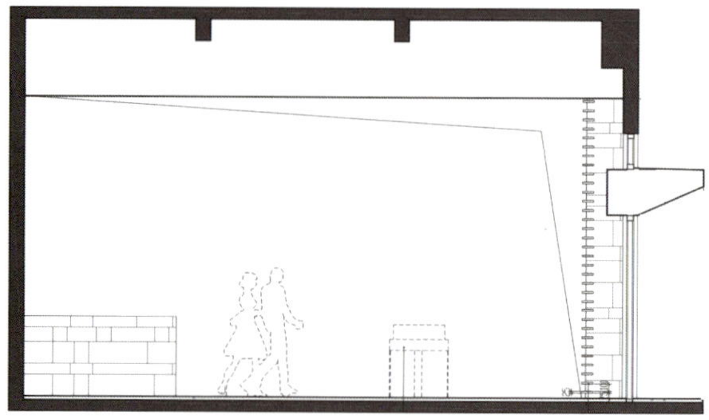

设计单位:IADC涞澳设计公司 / 设计:张成喆

蓝海销售中心 Blue Ocean

面积:1200㎡ / 坐落地点:上海 / 主要材料:环氧树脂、不锈钢、氧化铝、木、玻璃 / 摄影:蔡峰

设计师如同一位纯粹的雕塑家，摈弃过多表面的装饰，迷恋于最单纯的白色，将一切的形体元素还原为最单纯的极富逻辑感的直线，采用切割的手法在三维向度上对空间形体进行整体的塑造，让阳光来显现空间的优雅和形体的丰富变化，让质感的对比来自然地完成深度的细节刻画。依据室内中心的承重柱和天窗的位置，设计师设置了两片墙体直插向建筑顶部，墙体在平面上曲折行进，与建筑墙体对应分割出不同的功能区域。东面是洽谈区，墙体之间是前台和室内交通要道。西面则是模型沙盘展示区，而两片室内墙体在立面上也被“切割”，成为若干个不同角度的转折面，于是，单薄的墙体具备了厚实的建筑感，抽象的形体在阳光的沐浴下展现出如素描一般的细腻层次感和微妙变化，此刻，设计师也初步完成了建筑的室内延伸。

整座建筑物只具备两片混凝土实体墙面，于是，设计师选择远离主入口且最长的北墙来依次安排售楼中心的辅助功能区域，办公室、茶水间、更衣间、洗手间等。建筑墙体被“拉出”，同样以曲折的轨迹插入室内，细细考究室内每个部分的功能、尺度以及上下层空间的关系，同时，这一面墙体也成为重要的蓝海形象的展示界面。墙体由两层表皮组成，内层采用扭曲反光面的镜面玻璃，外层则是丝网印刷文字的透明玻璃。文字表达了蓝海的理念，而其排版方式也经过设计师的反复推敲。两层表皮之间装置白色日光灯，通过镜面玻璃的反射，文字成为变形与现实交叠的双重影像，非常具有数码时代的科幻感。在一片纯净的白色实体之间，这片展示墙似乎开拓了一片晶莹剔透的未知世界，引人遐想，而关于“蓝海”的表述也在无意间深入人心。

The designer, just like a pure sculptor, abandons the excessive exterior decorations and indulged in the simplest white. He restores all physical elements to the simplest lines rich with logic senses and applies the cutting technique to make the whole molding on the space from the three-dimensional approach so that the sun lights can display the elegance of the space and the abundant changes of the forms and the textural contrasts naturally realize the portrayal of the deep details. According to the locations of the bearing columns in the interior center and the skylights, the designer has set up two walls to be directly inserted to the top of the building while the wall is built in a curved way to be corresponding to the building wall and divide into different functional areas. The east is the discussion area while the space between the walls is the reception desk and the main interior path. The east side is the model exhibition area while the two interior walls are "cut" in the facade into various turning surfaces with different angles. Thus, the thin wall receives the thick architectural sense and the abstract form displays a fine hierarchic feeling and fine changes like a charcoal drawing in the shower of the sun lights. At this moment, the designer has also preliminarily realized the interior extension of the building.

The whole building only has two solid concrete walls. Thus, the designer has chosen the longest north wall far away from the main entrance to arrange the supporting functional areas, offices, pantry, change room and toilets for the sales center. The architectural wall is "stretched out" to be inserted into the interior area in a same curved trace to carefully study the function, dimension and space of every interior part and the space relationship between different levels. At the same time, this wall also becomes an important interface to display the image of Lanhai. The wall consists of two layers of coverings: the inner layer is the mirror glass with twisted glistening faces while the outer layer is the transparent glass with screen printed words. The words have expressed the idea of Lanhai while its set type has been considered over and over by the designer. The white fluorescent lamps are installed between the two coverings. After being reflected by the mirror glass, the words become a dual image overlapped by the deformed and actual pictures with strong sense of science fiction in the digital era. Among the pure white solid wall, this exhibition wall seems to have explored a bright and glistening unknown world, very fanciful while the expression about "Lanhai" is also deeply rooted in the people' s heart unconsciously.

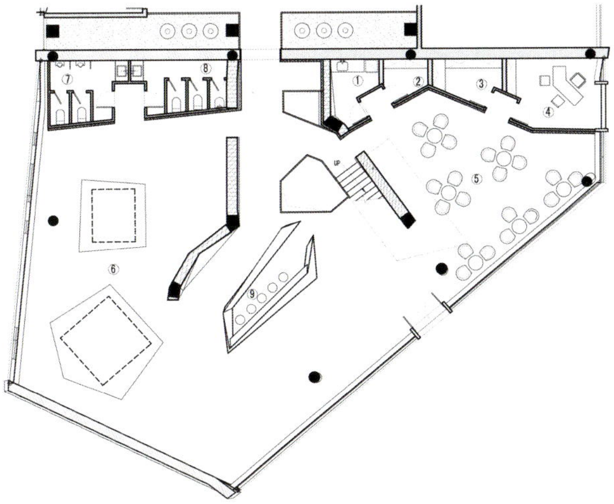

绿地
蓝海
科技岛广场 BLUE OCEAN
时尚休闲商业街 · 5A甲级写字楼 · 创意LOFT办公

WORKING
LIVING

设计单位：IADC涞澳设计公司 / 设计：张成喆

雁南园
Yannan Garden

面积：1510m² / 坐落地点：上海 / 主要材料：清水混凝土、钢、玻璃、木、石材 / 摄影：蔡峰

建筑立面主要由大面积的开窗和整齐划一的清水混凝土结构组成，有着严谨的秩序感。设计师意在如何使空间有着简洁而丰富的内容，同时充满想象的空间。经反复推敲，在一二层回廊立面筑起了分隔界面，用以区分入口与走廊的空间，同时抵消一部分建筑两侧开窗的一览无余。立面中开窗的形式借鉴了传统民居的造型。通往二层的楼梯挡板受到中国古琴弦排列方式的启发，纵横交错，充满韵律感，形成空间中有趣的构图。底层销售中心空间的设想，与二层其他机能空间在诸多细节中都做了有趣的尝试。此次会所设计，是设计师在东西方设计元素中寻找内在的契合点做的一次实践。

The elevation of the building is concrete structure with large-area windows. Designer' s target is to make the space become simple, abundant and imagination. The designer set up a boundary surface between 1st corridor and 2nd one to distinct the entrance from the corridor space. The style of opening windows in elevation space is come from traditional residents. The stairs board is connected to 2nd floor make the whole space interesting, the designer is lightened up by Chinese music equipment named Qin. There are a lot of interesting taste in the ground sales center and 2nd function space. This design is a new beginning of looking for some mixture with Chinese culture and western culture.

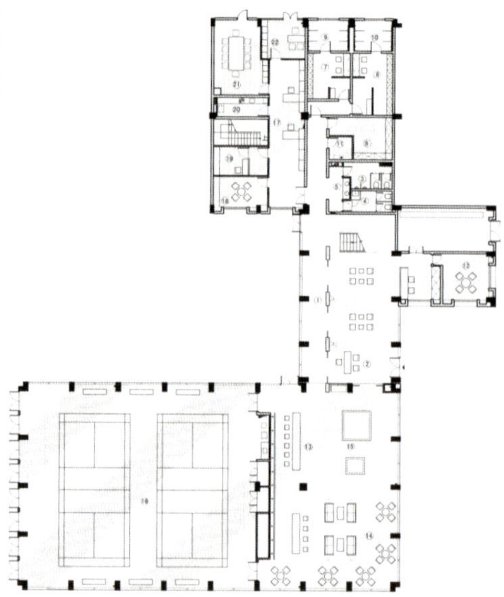

燕
城南 万科别墅生活
市与自然的可能

设计单位：金螳螂建筑装饰股份有限公司第一设计分院一所 / 设计：王祎华 / 参与设计：钱文宇、糜佳、刘岩、王启霖、孙铭

太湖温泉1858别墅VIP中心
Taihu Lake Spring 1858 Villas VIP Center

面积：600m² / 坐落地点：苏州吴中区太湖国家旅游度假区 / 主要材料：白色人造石、咖啡色烤漆玻璃、黑色烤漆玻璃、不锈钢 / 摄影：王祎华

该项目地处苏州八百里太湖之滨，别墅每户均有取自地下1858m深处的温泉，故而得名。

VIP中心集观赏、体验、销售、签约功能于一体，设计定位引太湖湖景为主体，牵温泉水文化于全线；浩浩荡荡，心存高远，人观水，水润周遭。

平面变轴以合流线，座位应湖景而立。以深色接待台背景反衬蓝光水纹图形；洗手间天顶打开，四壁深灰镜面马赛克使内外浑然一体，引阳光全入而生辉。取材上室外拓展至室内，均以天然木色、灰、黑白等石材、咖啡色烤漆玻璃、人造革为元素，不张扬，褪浮躁，除却大多售楼处的金银粉气，静瑟瑟沉其间，以现代简约之低调华丽娓娓讲述江南太湖情致往事。

It is located beside Eight-Hundred-Li Taihu Lake in Suzhou. Each villa has a hot spring taken from a 1858m depth underground and thus its name comes from this.

VIP center is one for viewing, experiencing, selling and contract signing all in one. The design is orientated to introduce the lake scene of Taihu Lake as the main part and lead the culture of spring throughout the line: vast and mighty, mind far away. The people view and admire the water while the water moistens the surroundings.

The axis in the plane is changed to meet the streamlines and the seats are arranged according to the lake scenes. The dark reception desk is applied as the background to set off the blue ripply patterns by contrast; the ceilings in the toilets are opened and the dark gray mosaic mirror in the surrounding walls makes the interior and exterior integrated together. The sun lights are introduced inside to brighten everything. The application of the materials has extended from the exterior to the interior, which all take the natural wooden color, gray, black and white stone materials, the coffee baked glass as well as artificial leathers as the elements. They are low-profiled and not blundering. The golden and silvery atmosphere in most sales centers is removed, quiet and peaceful. The Taihu Lake' s feelings and stories are tirelessly told in a way of a lower-profiled luxury in a modern simplicity.

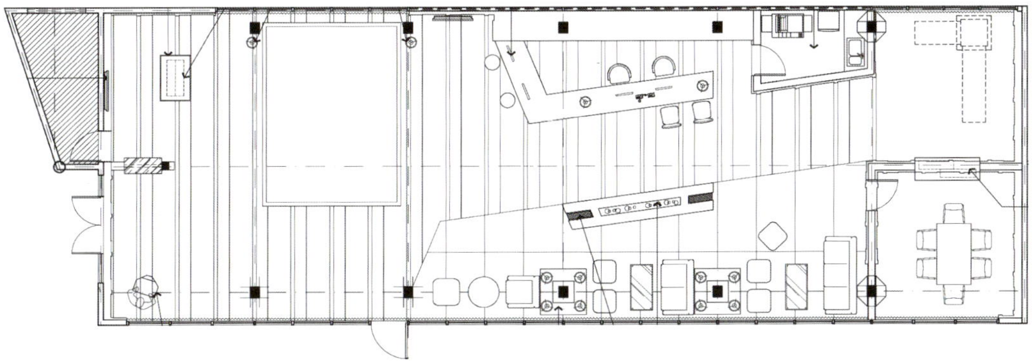

设计单位:香港东仓设计策划顾问有限公司 / 设计:张星

楼兰
LOLA

面积:1800m² / 坐落地点:佛山市 / 主要材料:砖 / 摄影:香港ATA空间摄影公司

我们试图在这个项目中用当代的技术手段和普遍审美观重新演绎东方的情绪。

中国的西北有一片被沙漠化的土地，历史上她像庞贝一样充满了神秘的气息和高度的文明，没有人知道她是什么样子，她叫做楼兰。

如今我们没有充分的证据来重现当年的楼兰景象，只有一些传说和间接信息告诉我们她的关键词：风沙 海市蜃楼 宗教 石屋……

所有的印象是需要实现一个空间的唯一依据，然而今天的审美和主流思维以及技术已经远离了那个年代，于是我们努力使用东方人特有的智慧和当代的工艺来再现我们心中的楼兰。

解构中国文明中的代表性元素，色彩从官袍中提炼黑和黄，形态从地貌中提炼出沙池，尺度从法则中提炼出方圆，精神从国画中提炼出“平静”。最终我们用寻找到的元素具象出一个新空间，精神需求大于视觉本质的空间。

We try to re-act the oriental emotions in the project through both contemporary technologies and general tastes.

There is a desertization land in the northwest of china, he has been filled with the tinge of mystery and the advanced civilization like Pompeii in history, nobody has known what he should be like, but was named LOLA.

Now we have not any evidences to repeat the vision of LOLA of those years,but some legends and indirect information can merely tell us the key words to describe him: windy and dusty, mirage, religions, and the houses made of stones……

All impressions are the only basis if we need to realize a space. however, the aesthetic of today and mainstream thinking and technologies have kept far away from those years, so we have to try our best to repeat the LOLA who have been living in our heart, using the wisdoms that are specially owned by the oriental as well as the contemporary technics.

Analyzing the representative element of civilizations in china,we can abstract the colors,black and yellow,from the robes that were worn by the officeholders in china,can abstract the shapes,the sand pools,from the physiognomy,can abstract the scales,the regulations,from the rules,and abstract the calmness,from the traditional chinese painting.Finally,we are able to sketch a new space with the found elements,in which the spirits are needed more greatly than the essence of the vision.

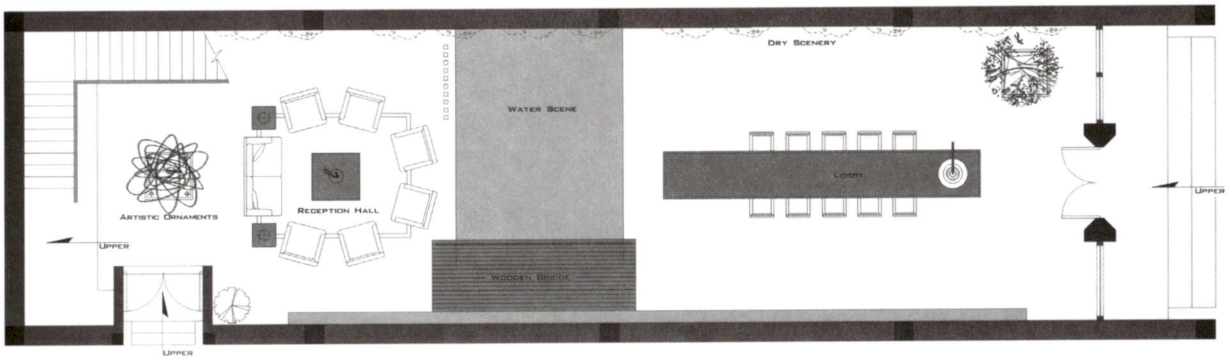
Dry Scenery
Water Scene
Lobby
Artistic Ornaments
Reception Hall
Wooden Bridge
Upper
Upper
Upper

设计单位:香港东仓设计策划顾问有限公司 / 设计:D5组

正中集团科技大厦售楼部
Sales Department for Technologic Building of Genzon Group

面积:1000㎡ / 坐落地点:深圳市 / 摄影:香港ATA空间摄影公司

当面与线在空间中产生碰撞时，气氛热烈而融洽。

作品延续现代而简约的设计风格，在界面处理手法上整体统一，既延伸空间同时又将材质的天然质感表现出来。

大空间大块面小细节的设计，不仅使空间视觉效果具有强烈的张力，而且满足商业空间需求，营造舒适轻松的环境氛围。

When the line collide with the surface in the space, the atmosphere was warm and cordial.

The works continue the modern and simple design style, the disposal on the interface is overall and unitive, which extend the space at the same time show the natural texture of the material.

Larege space, large surface and littele dedail not only make room for visual effects, which have a strong sense of tension, but also meet the demand of commercial space, create a comfortable an relaxed environment.

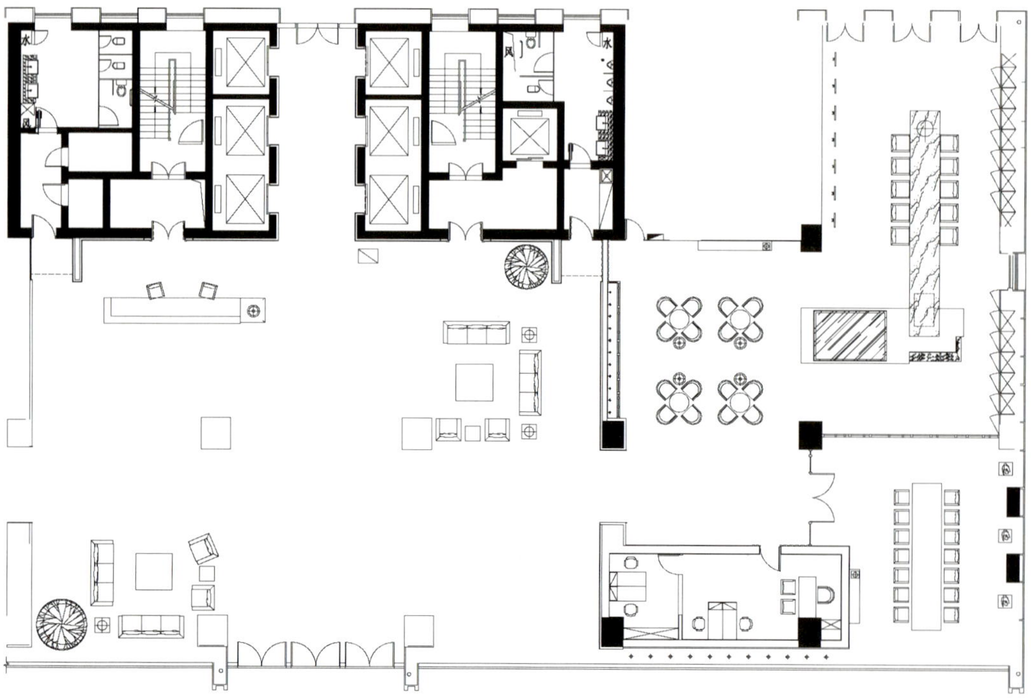

消火栓
会议室
中科大厦
科技办公

设计单位:汉文设计 / 设计:万宏伟

高新科技广场售楼处
Sales Pavilion of Hi-tech Plaza

面积:400㎡ / 坐落地点:宁波 / 主要材料:黑、白、灰微晶玉、爵士白石材、柳桉木染色、黑钛金饰面 / 摄影:杨旭

项目位于宁波国家高新区,自然环境优美。楼盘由四幢高层与一幢精装公寓围合组成。

我们从建筑外立面及销售功能的定位去细读,更好的找到一个建筑同室内设计的最融合与整合的合理完善的设计切入点。室外建筑与室内建筑的延续性与互通性，现代、硬朗，科技感、时尚感等词条的串联，运用条形码的意象，黑白灰基调的调控、细节的追求，完善了我们最终的设计构想。

从售楼处的外型结构体积的商讨到整个售楼中心传达的楼盘建筑及未来大楼大堂及公共空间室内设计信息的清晰表述，使得楼盘开盘热销，设计的切入点与设计的生产力得到更广泛的确认。

The project is located in the national hi-tech zone of Ningbo with a beautiful natural environment and the locational advantages.

We have carefully studied the exterior facade of the buildings and the orientation of the sales functions to better find a reasonable and perfect starting point that the architecutre can be best integrated with the interior design. The contiunity and interconnection of the exterior and interior architetures, the series connection of the words such as modernity, hardness, technology and fashion, the application of graphic symobls of the bar codes, the adjustment and control of black, white and gray colors as well as the pursuit to details have perfected our final desgin vision.

From the discussion on the exterior structural volume of the sales pavilion to the real estate delivered by the whole sales center to the clear expression on the interior design informations of the future building lobby and the public spaces, all these have realized that the real estate is sold well at the begignning and the starting point of the deisgn and the productivity of the desgin are widely recognized.

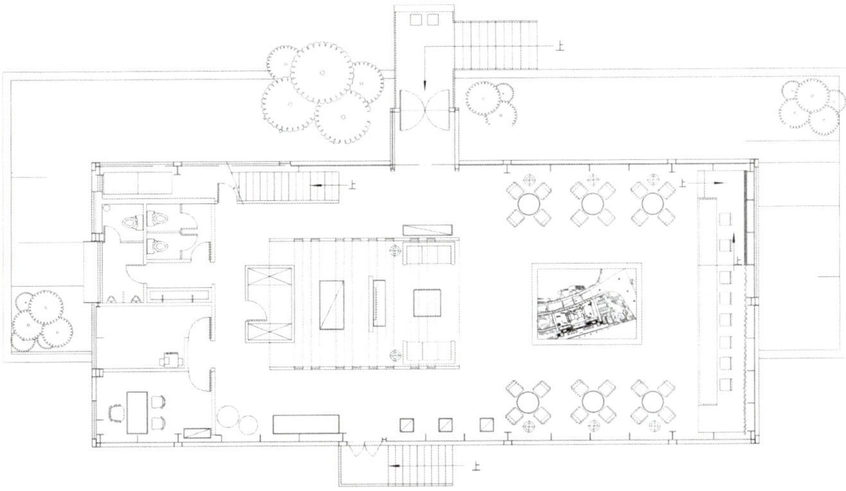

宁波高新科技广场
NINGBO HIGH-TECH
DEVELOPMENT PLAZA

设计单位:江西联动建筑装饰设计工程有限公司 / 设计:童武民、李信伟

中央公园二期售楼部
Sales Center for Central Park, Phase II

面积:800m^2 / 坐落地点:上饶市 / 主要材料:氟碳漆、羊皮灯、黄洞石、实木地板、电脑雕刻屏风、地毯、人造石 / 摄影:邓金泉

本案的设计延展了原有的建筑结构，整体功能空间强调了现代建筑艺术美感，功能分区导向合理，视觉效果和谐统一，设计师利用材质、色彩、灯光的简练要素充分表现出室内设计中的艺术效果，使原有钢结构空间不再寂寞，取而代之的则是浓重深厚的温情体验和售楼的时尚气息。

The design of this case has expanded the original architectural structure. The integral functional space has an emphasis on the modern architectural artistic aesthetics. The divisions of the functions are reasonable with the harmonious and uniform visual effects. The designer has applied the concise elements of the materials, colors and lights to sufficiently demonstrate the artistic effects in the interior design so that the existing steel structural space is not lonely any more and it is replaced by an experience of the thick and profound warms as well as a fashionable atmosphere of the sales center.

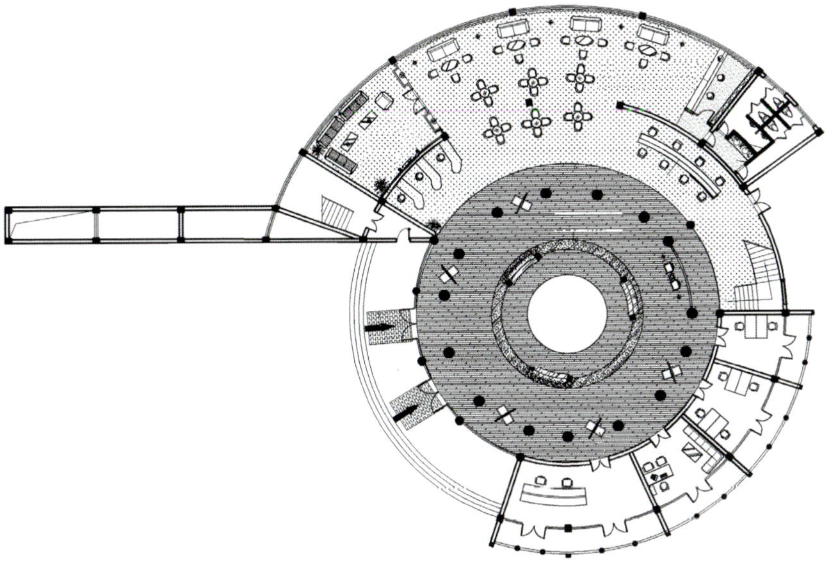

2008，只住中央公园

INTERNATIONAL HOUSE

设计单位:品伊创意机构、美国IARI刘卫军设计师事务所 / 设计:刘卫军 / 参与设计:黄昌浪

彼岸城售楼中心
Sales Center of Bi'an Cheng

面积:2100m² / 坐落地点:福州 / 主要材料:发光膜、艺术激光投影、人造石、实木地板、丝网玻璃、玉晶石 / 摄影:吴友长

设计师以简约的建筑艺术手法塑造了整体空间，强调现代建筑的艺术美感。在设计中既延伸了建筑艺术本身的风格，又增添了空间的互动性与适时的隐蔽性，如封闭或开放的洽谈区适应了不同客户的需求。

在功能分区上注重合理性、呼应性、延续性，使视觉效果富有节奏韵律感。VIP洽谈区中的不同灯光颜色营造不同的意境氛围，不但提升了空间的进深层次，回归空间的和谐与统一，给人明朗、动感、优雅的视觉空间效果，同时光、声、影又营造出温馨舒畅的环境氛围。设计师还特别注重材料的运用与设计造型相辅相成，如发光膜、吊链、艺术激光投影、人造石、实木地板、丝网玻璃、光纤水晶吊灯、玉晶石等，配以合理比例尺度构成的室内建筑形态，具有简约、大气的现代艺术美感，从而营造出一种超现实的时尚动感空间。设计师通过这一表现手法，启发并挖掘身处空间中的人们对高层次生活的憧憬与向往。

The designer has developed the whole space in a concise architectural artistic technique, emphasizing on the artistic aesthetics of the modern architectures. In the design, it not only extends the style of the architectural arts itself, but also increases the interactions between the spaces and the appropriate concealment. For example, the closed or open discussion areas have met the requirements of the different customers.

It lays emphasis on the rationality, echoing and continuity in the division of the functions so that the visual effects are enriched with rhythmic senses. The different light colors in VIP discussion areas have created a different artistic atmosphere. It not only upgrades the deep levels and returns to the harmony and unification of the space to provide a bright, active and elegant visual space effects, but also creates a warm and comfortable environmental atmosphere with the lights, sounds and shadows. The designer also pays special attention to the application of the materials as well as the supplementation of the designed models, such as luminescent films, hanging chains, artistic laser projections, artificial stones, solid wood floor, screen glass, fiberglass crystal pendant lamps, jade crystal tiles etc. in combination with the interior architectural forms in proper scale and dimension, it posses a simple and grand modern artistic aesthetics so that a surrealistic fashionable and active space is created. By this expressive technique, the designer has inspired and dug out the longing and yearning for high-level life of the people involved in the space.

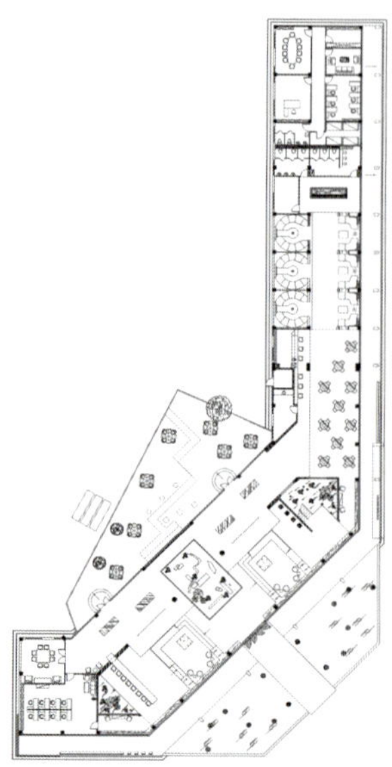

SHOW

设计单位:深圳市派尚环境艺术设计有限公司 / 设计:李益中、范宜华

金众蓝钻风景花园营销中心
Sales Center of Jinzong Lanzuan Fengjin Garden

面积:688m² / 坐落地点:深圳 / 主要材料:仿马毛条纹墙纸、白色及红色仿皮墙纸、光纤灯、黑镜、黑木纹大理石、灰蒂马拉白理石 / 摄影:吴永长

售楼处位于商业裙楼的一楼,平面纵深很长,在功能布置上有一些困难,曾经出过几次方案都觉得不满意。最后抓住了核心问题，于是设计迎刃而解。

这个核心问题就是如何解决售楼流程的问题，怎么迎接人流，又怎么引导？模型区该放在何处？洽谈区又在哪里？经过研究发现问题的关键在于接待台位置的设计。我们把接待台做成空间的转轴支点，把形象墙、接待台和水吧台一起设置，所有功能围绕这个支点展开，从而将平面长而纵深的问题化解，空间也有了起承转合。

接待台背景设计成发光的灯箱，展示的是“金众蓝钻”的楼盘LOGO，透过蓝色图案打出来的蓝光，把空间渲染得特别神秘且具有现代感。同时用灯箱界面围合的空间，刚好又成为一个贮藏间，实用和美观两者兼得。在造型语言上适应空间的形态，加强了纵向的线条表达，体现了一种现代的速度感，也在心理上将人流引向纵深处的洽谈区和签约区。这种纵向的线条组织表现在天花和地面上，宽窄不一，如条码状，有着强烈的视觉引导。

在空间的纵深尽端，用了黑色玻璃作为收束，让视觉停留，同时黑色玻璃上的人像剪影，有着青春跃动的意像。签约区的地面是升起的，也同时起到了空间收束的作用。

由于整个空间的层高比较高，所以我们设计了一个夹层，夹层上是办公用房，夹层下则是洗手间等配套用房。这样，既充分利用了资源，又极大地丰富了空间层次。

The sales pavilion is located on the first floor of the annex business building. As its plane is quite long in depth, it has some difficulties on the arrangement of the functions. The several plans that have been made are not satisfactory. Finally the key problem is seized and thus the design is also readily resolved.

This key problem is how to solve the procedure of selling the apartments, how to receive the streams of people, how to guide, where the modeling area shall be located? And where the discussion area is located? After the study, the key on the problem lies on the design of the reception desk. We make the reception desk the supporting point of the revolving spindle for the space. The image wall, reception desk and water bar counter are set together and all functions are carried out around this supporting point so that the problem of long plane and depth is solved and the space is also got properly connected and interacted.

The background of the reception disk is designed to be a luminescent light box, displaying the logo of the real estate “Jinzong Lanzuan”. The blue lights through the blue pattern have exaggerated the space especially mysterious and modern. At the same time, the space enclosed by the light boxes happens to form a storage room, practical and pleasing.

In the sculptural language, the form appropriate to the space has strengthened the longitudinal line expression, reflected a sense of modern speed and also in mind introduced the streams of people to the discussion area and contract signing area in the deep direction. This kind of longitudinal line arrangement is reflected on the ceiling and ground with variable widths just like the barcodes, which has a strong visual introduction.

At the end of the space in the deep direction, the black glass is applied as the stop point for the vision. The sketches of the portraits on black glass have a sense of jumpy youth. The ground in the contract-signing area is raised up to achieve the effect of space termination.

As the level height of whole space is quite high, we designed a mezzanine, above which are the office rooms and below which are the supporting rooms such as toilets etc. In this case, the resources are fully utilized and the space arrangement is also greatly enriched.

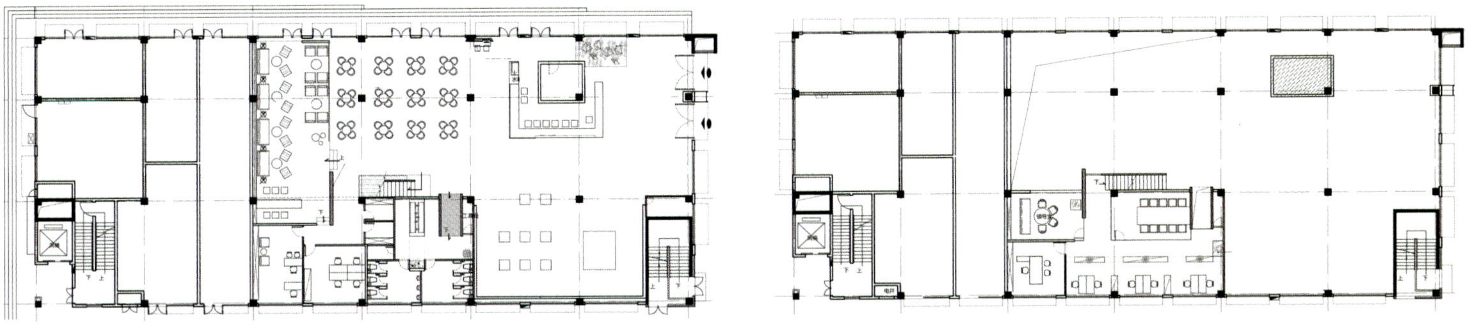

设计单位：玄武设计 / 设计：黄书恒 / 参与设计：欧阳毅、许棕宣、詹皓婷、胡春惠

远雄建设上林苑接待中心
Reception Center for Sky Mansion of Farglory Land Development

面积：2480m² / 坐落地点：台北市 / 主要材料：玻璃钢构、木作钢琴烤漆、纸筒、地毯、木板拼花、壁纸、抛光砖 / 摄影：王基守

设计师在构思整个接待中心时，借由拉大空间的比例尺度，强化戏剧性效果，凸显豪宅气势。踏入室内后，迎面为高达10m的雕塑墙，以微妙角度和高度的比例变化，耸立如教堂中雄伟的管风琴，白昼时透过光影变化展现不同律动感，夜间宛若由琴键缝里透出LED光芒，显现昼夜迥异的效果。

设计师透过此一夸张的尺度对应于人类的渺小比例，试图透过元素及语汇来捕捉空间之美与神圣性使人们产生微妙的心理变化。之后刻意180度反转的动线，让人们在路径上换置情绪，迎来柳暗花明的惊喜。一楼廊道呼应豪宅的宽阔尺度，经由导览动线的移动，视野不断穿透落地玻璃与外界绿荫交织。在企业文化介绍区里，以透明过桥衔接由清玻璃建构的环形剧场，利用平板玻璃部分交叠形成支撑，犹如悬浮于水面上一般。夜间时，整体上又创造出独特的渐变式光影效果。利用电动帷幔，围绕出无光的暗房，播送动态影像，传递着一种独特的视觉体验。在一楼导览动线尽头，以一组双螺旋楼梯衔接二楼的样品屋与洽谈区。

卫浴间则打破将其安置于幽暗角落的传统方式，刻意将其转为整个导览动线中的一景点，大器玄关，以极为时尚的手法，装点其空间。正中央以缆线悬吊起一透明玻璃的梳理台，将水龙头等皆嵌入如不锈钢镜面里，透过感应式的自动流泻水瀑，将机能与造景结合，展现了时尚与高科技巧妙连结的趣味，也再次呼应豪宅的不凡气势。往前则是最精华的洽谈区，面对整面预售案的建筑基地，一旁以桁架支撑整个玻璃帷幕的缓降斜坡道，作为整个导览动线的尾声。

整个空间不断探寻一种音符的律动与跳跃，企图让空间元素不断对话，映衬出灵动鲜活的空间。同时也引导着人们在暂时建构的接待空间里，重拾内心的渴求以及勾勒未来的豪宅模样。

When the designer is plotting the whole reception center, he has enlarged the scale of the space by this reason to strengthen the dramatic effects and show the momentum of the luxurious houses. After stepping in the building, you will be confronted with a 10m high sculptured wall, which varies at a delicate angle and height scale and is upright like a majestic pipe organ in the church. It has displayed with different rhythmic senses through the changes of the lights and shadows at daytime while it looks like LED light penetrated from the gaps of the organ buttons at nighttime and has shown an effect completely different from that at daytime.

By means of the comparison between the form in an exaggerated dimension and the tiny scale of human beings, the designer intends to catch the beauty and sacredness of the space through the elements and vocabulary to create a delicate psychological change in people’ s mind. Afterwards, the active lines are reversed deliberately at 180° so that the people can change their moods in the passage and have a suddenly enlightened surprise. The corridor at the first floor echoes with the wide dimension of the luxurious house while the visual fields, under the guidance of a moving line, have been interwoven with the exterior green shades constantly through the down-to-ground glass. In the introduction area of the enterprise’ s cultures, the transparent crossover bridge is applied to connect the circular theatre built by clear glass while the flat glass panels are overlapped partially to form a support, just like an object floating on the water. At night, the integral building also creates a unique light-and-shadow effect that changes gradually. The electric curtain is applied to form a circular dark room without lights to send the active images and deliver a unique visual experience. At the end of the active guiding line at the first floor, a pair of double -spiral staircases is connecting the show room and discussion area.

The convention that the bathroom used to be arranged in a dim corner has been broken through. It is deliberately turned to be a scenic spot in the whole active guiding line. The space is decorated in an extremely fashion way. A transparent glass counter is suspended by a cable in the right center while the faucets etc. are embedded in the stainless steel mirror so that the functions and the creation of scenery are combined together through the induction-type automatic flowing water to show an interest of the delicate connection between the fashion and high-tech as well as echo with the unusual atmosphere of the luxurious houses again. The gradually descending slope to support the whole glass curtain by the trusses at one side is designed as the end of the whole active guiding line.

The whole space is constantly seeking the rhythm and jump of a music note in order to have a constant dialogue between the space elements and set off a bright and vivid space. It also guides the people in this temporarily built reception space to re-pick up their inner desires and outline the appearance of the future luxurious house.

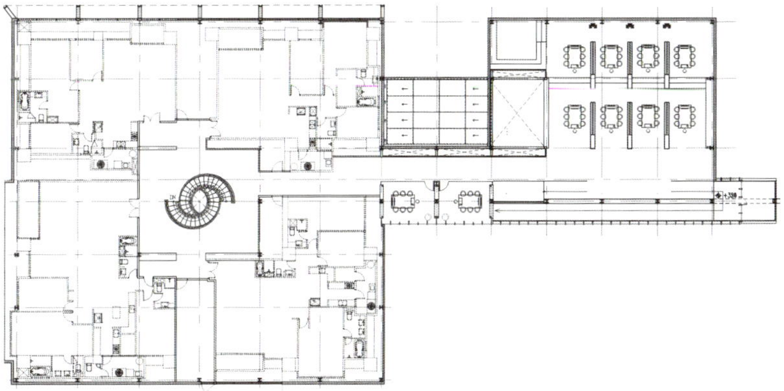

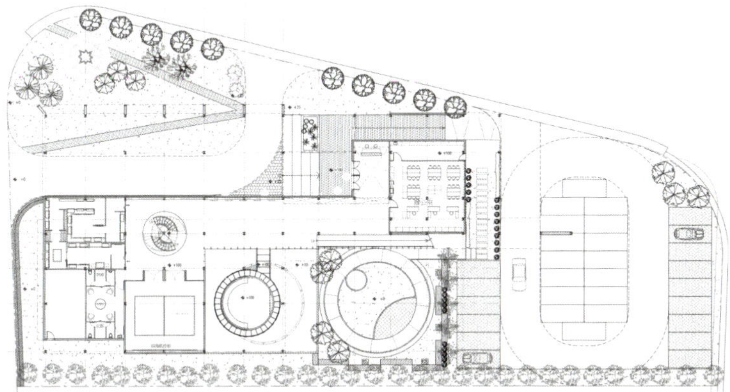

设计:陈幼坚

万科中粮本岸售楼处
Sales Pavilion of Wanke Zhongliang Ben'an

坐落地点:苏州 / 摄影:文宗博

万科中粮本岸，位于苏州工业园区、金鸡湖东侧，占据行政文化中心、科技文化艺术中心、中央商务中心三大中心。总体规划遵循“自然和谐、人本便利、弘扬文化”的原则，以传统村落鱼骨状自然为原形，以苏州民居院落为特色，以白墙黑瓦的中式元素为设计风格，优雅舒展于斜塘河景观一带。

Located in Suzhou Industrial Park and to the east side of Jinji Lake, Wenke Zhongliang Ben' an has occupied three centers including administrative and cultural center, sci-tech cultural art center and the central business center. Its overall planning follows the principle of "natural and harmonious, people-oriented and convenient as well as promotion of cultures". With the fishbone nature of traditional village as the original shape, Suzhou civilian courtyard as the characteristics as well as the Chinese element of white walls and gray tiles as the design style, it elegantly stretches along Xietang stream.

设计：仇益国

中航楼吧
CATIC Real Estate Club

面积：1000m² / 坐落地点：深圳市

在外观上，我们把整个建筑物看作是一个完整的形象。如果单一地把建筑的两个角作造型或广告较难形成视觉中心，无论是在体量上还是形式上。所以我们建议把整个建筑物看作一个完整的装置、一个完整的雕塑，形成鲜明的广告形象，同时突出阳光下和在夜晚灯光造型的效果。

楼吧作为长期存在同时为多个项目服务、展示企业文化的中心，其室内设计更多的应带给人们以想象而非功能。这是一个新概念主义室内设计作品，放弃直接的细节功能，一切只为瞬间抓住观众，深深地刻在人们的脑海里，忘却不了，却又无穷无尽。

具体功能由四个空间相连的展示区构成。进入楼吧一眼就能看完四个区，清晰醒目，展示感及空间感极强。展示长廊是整个空间的背景，服务总台镶嵌其中，会议室和办公区是悬挂其中的白色发光体。最有趣的是下沉的交流广场，它把商务展示区和休闲交流区作了明确的区分，产生了两种不同环境的文化介面并且有室内和室外两种气氛。空间的区分使功能区域得到了最大的统一，同时也拓展了功能内容，如下沉广场可举办媒体见面会，客户交流会，甚至举办高尔夫、音乐等虚拟会所交流。

作为一个长时间多项目的房地产交流中心，它与招商的餐厅形成良好气氛的文化长廊，加强了品牌宣传，同时也起到领导潮流的品牌效应。提供一个概念化空间才是这一设计的目的，而不是一个功能化的装修。

From the appearance, we take the whole building as a complete image. If only two angles of the building are used as the model or advertisement, it is quite difficult to become a visual center, regardless of the size or form. Therefore, we suggest that the whole building shall be deemed to be a complete device and a complete sculpture to form a vivid advertising image. The sculptural effects are also very prominent under the sunshine and the lights at night.

Real estate club is a long-standing center serving several projects and exhibiting the enterprise' s cultures, so its interior design shall bring with people with more impression instead of its functions. It is a neo-conceptualism interior design work, which directly gives up the detailed functions. All what it does is to attract the audience instantly and leaves deep impression on people, unforgettable and infinite.

The specific functions are consisted by four exhibition areas with the spaces connected with each other. Once entering the real estate club, you can see through the four areas at one sight, clear and distinguished with strong sense of exhibition and space. The exhibition corridor is the background of the whole space with the service & reception desk among it. The conference room and office area are the white luminescent objects suspended among it. The most interesting is the sunken communication square, which has made a clear division between the commercial exhibition area and recreational communication area so that the two different environmental cultural interfaces are created, together with the two interior and exterior atmospheres. The space division has unified the functional areas to the greatest extent and also expanded the contents of the functions, for example: the afternoon communication meetings, media meetings, client communication meetings and even the virtual club communications such as golf and music can be held in the sunken square.

As a long-standing real estate communication center for many projects, it has formed into a cultural corridor of good atmosphere together with the merchant' s canteen, which not only strengthens the brand promotion, but also becomes a brand effect of leading the trend. Therefore, its design purpose is to provide a conceptualized space instead of a functionalized decoration.

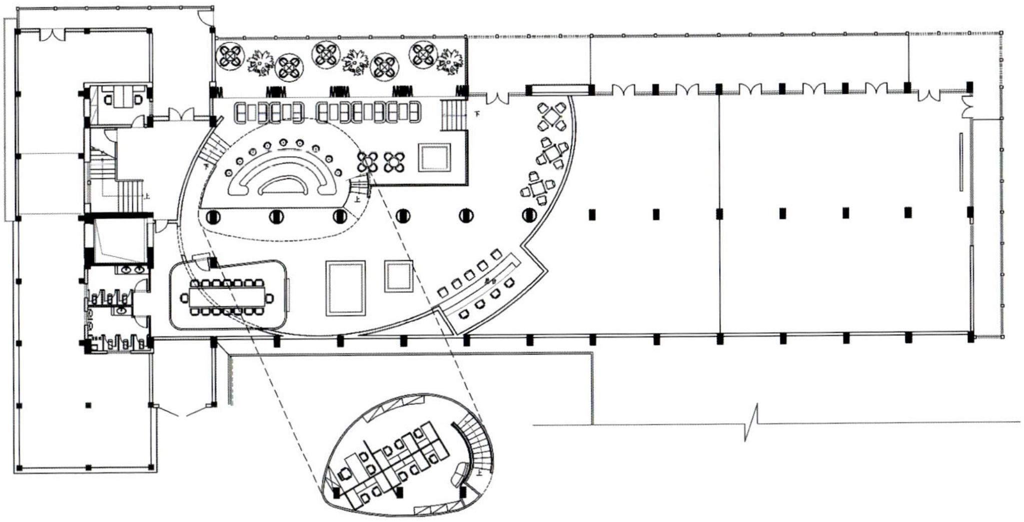

鼎誠國際
DYNAMIC WORLD
中航地产

CATIC

鼎誠國際
DYNAMIC WORLD
中航地产
红酒·霓裳
HAPPY PARTY
嘉年华

设计单位:上海火翼装饰工程有限公司 / 设计:吴庆华、奚进、崔豪

南京世茂滨江新城会所展厅
Exhibition Hall of Nanjing Shimao Riviera New City Club

面积:550㎡ / 坐落地点:南京 / 主要材料:进口塑胶地板、地毯、水纹玻璃、装饰镀膜镜面玻璃、钢化玻璃、镜面不锈钢、铝板、大花白大理石、冷轧钢板/
摄影:胡文杰

设计师意欲通过“解构”的手法来营造一个梦幻般的环境。“梦境”的序曲由入口开始，看似平淡的米黄大理石和橙黄色的灯光在丝丝细节中透露出不同。入口右侧楼梯用铝塑板饰面，扶手全由清玻璃组成，形成了一种软、硬不同的质感对比。楼梯下，地面上是一泓清水，倒映出周遭环境——两面铝塑板斜包饰面的墙壁上开出道道倾斜的灯槽，投射出迷人的红色光晕。墙壁的对面便是一根用镜面不锈钢包面的柱子，把周围的环境色彩、质感毫无保留地进行传递，给人神秘的臆想。前方是多媒体演示区，整个空间被笼罩在充满幻想的蓝色调中，但是大面积的发光地面，镜面处理的展架和背景墙以及背景墙上所悬挂的多台等离子电视机都在提示人们：这是一个展示现代建筑观念和企业文化的场所。设计师在顶部创造了一个巨大流线形的灯具，弯弯探出头，梦幻般照亮整个空间。翠绿色的夹丝玻璃前摆着大红色的三人沙发，这是色彩与质感的“梦境”。几十片造型玻璃被插入蓝色柱子中，顶上圆形灯槽中泛出暖光，光、色、材质之间一片迷离。而这种反射又与面前的镜面背景墙之反射形成一种微妙的契合与差异。

在楼盘模型展示区中，圆形玻璃展架的底座是一朵泛出金属光泽的巨型花瓣。在它的花蕊上陈列的便是楼盘的模型。底座金属花瓣被设计师运用“解构”手法，使其质感、尺度、造型与人们的固有观念形成一种奇妙的错位，加上背后夹丝玻璃的映衬和顶面满天星灯的渲染，营造出奇妙的“梦境”。

业绩展示区，是满得快要溢出的红色。顶面的满天星灯和螺旋形的灯槽一起烘托了这个红色的“梦境”。地面上倾斜的方块分割，墙壁上开出道道倾斜的灯槽。倾斜的灯槽投映在木龙骨的玻璃地面上，把整个室内环境推向了反复循环的天籁。

The designer tends to create a dreamlike environment through the technique of “disorganization”. The prelude of “the dreamland” starts from the entrance. The seemingly ordinary off-white marbles and orange lamp light have revealed with the differences through the details. The staircase on the left of the entrance is veneered with aluminum-plastic composite panels while the handrails are all made of clear glass, which have formed a contrast between soft and hard materials. Under the staircase, there is a pool of clear water, in which there is some inverted image of the surrounding environment —— the wall veneered slantwise by aluminum-plastic composite panels on both sides is dug with rows of gradient light channels. The charming red halation is projected from them. Opposite to the wall is a column covered by mirror stainless steel, which transmits the colors and sense of realities of the surrounding environment without reservation and creates a mysterious illusion. In the front there is a multi-media presentation area. The whole space is overwhelmed in a dreamy blue tone. But the large luminescent floor, the exhibition shelves with mirror arrangement, background wall as well as several plasma televisions hung on the background wall are all reminding the people: it is an area to display the modern architectural conceptions and enterprise’s culture. The designer has installed I huge streamlined lighting fixture on the ceiling with the bent heads to lighten the whole space up in a dreamy way. In front of the jade green wired glass is installed with a red sofa. It is a “dreamland” of colors and sense of realities. Tens of pieces of modeling glass are inserted in the blue column while the round light channel on top are floated with warm lights, misted among lights, colors and materials. This kind of reflection together with the reflection of the front mirror background wall has formed a subtle integration and difference.

In the exhibition area for the model of the building, the base of the round glass exhibition shelf is a huge petal floated with metal gloss. On its pistil is displayed with the model of the real estate. The designer has applied the technique of “disorganization” to the metal petal on the base so that its sense of reality, dimension and modeling have formed a marvelous misalignment with people’s inherent opinions and have created a wonderful “dreamland” together with the setoff of the wired glass in the back and exaggeration of starry lights all over the ceiling.

The exhibition area for the outstanding achievements is nearly overflowed with red. The starry lights all over the ceiling and the spiral light channels are together setting off this read "dreamland" by contrast. The gradient square blocks are divided on the floor while the rows of gradient light channels are dug on the wall. The gradient light channels are projected on the glass ground with wood joists to put the entire interior environment to the repeatedly circulating sounds of nature.

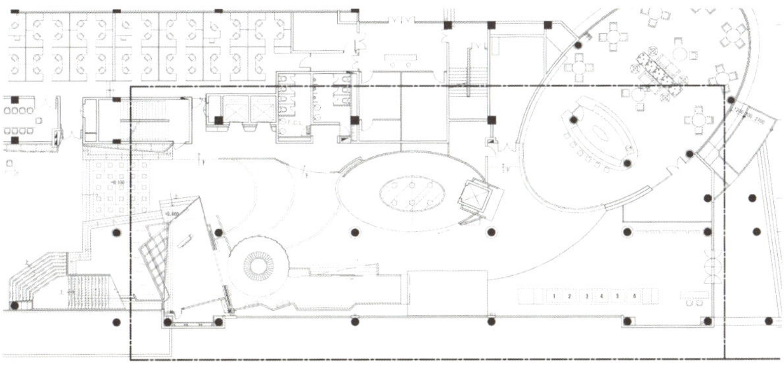

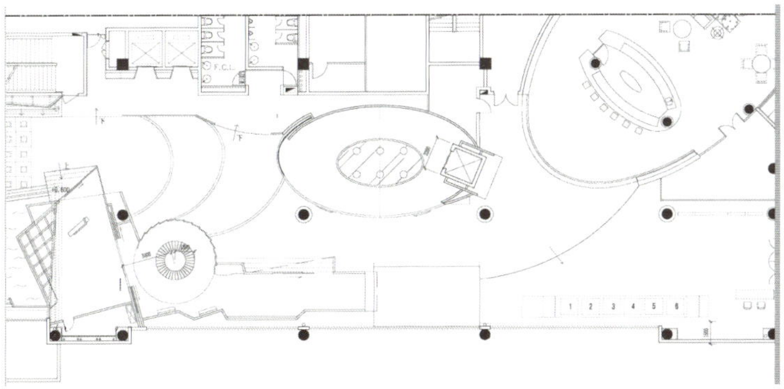

设计单位：内建筑

朗诗·国际街区售楼处
Sales Pavilion for Landsea·International Block

面积：800㎡ / 坐落地点：杭州 / 主要材料：清水混凝土、橡木饰面、白色烤漆

呈黛色的点阵外立面既迎合了南方独有的内敛气质，又成就了空间的秩序感。开敞的一层大厅内，白桌白椅，简洁而清新。由于层高和面积的限制，设计师放弃使用吊灯，于顶上规置几排轨道射灯进行基本照明，每张桌子上安排一盏台灯，用"光"在有形的大空间中划分出无形的小天地。大面积落地玻璃毫不吝啬地将窗外的景色揽收室内，让客人忘却时间，变等待为享受。

The exterior facade with black lattice has both met the unique introverted temperaments of the south and achieved the orderly sense of the space. In the open lobby at the first floor, white tables and white chairs look simple and freshen. Due to the limitation of the level height and area, the designer has given up the pendant lamps and arranged several rows of track lamps for the basic lighting. A table lamp is put on each table and "light" is applied to divide the visible space into the invisible small worlds. The large ground-to-ceiling glass has generously included the scenery outside the windows into the interior so that the visitors have forgotten the time while the waiting becomes an enjoyment.

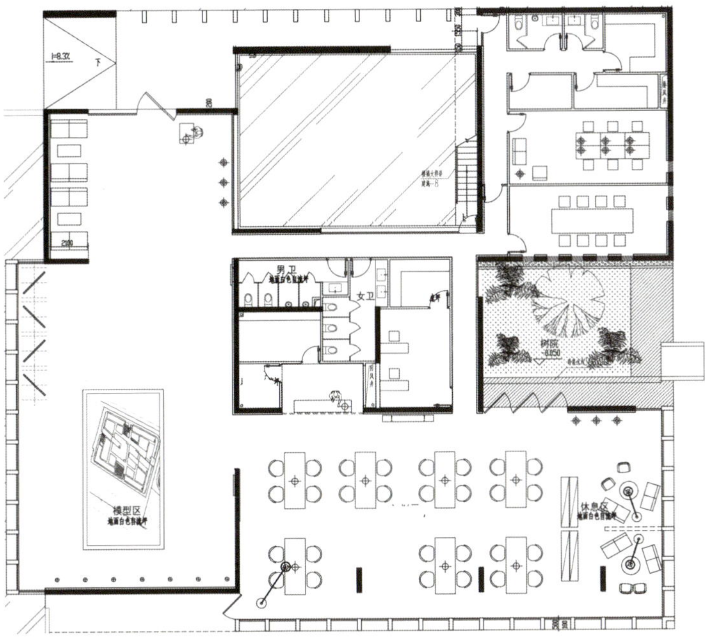
男卫
女卫
模型区
休息区

设计单位：内建筑

南北西岸
West Bank

面积：1000m² / 坐落地点：杭州 / 主要材料：黑金沙石材、镜面玻璃、亚克力成品灯、皮革饰面

建筑追求传统与现代风格相结合，充分体现运河文化，让西岸的每一栋建筑都有幸成为运河景观的有机部分。立面追求光影效果，有机分段处理，整体形象简洁而又不失细部。简练到极致的建筑元素，将浪漫情怀、人居理想通通融入西岸的空间模式之中。样板设计在内部空间上，层次错落、通透舒畅。整体气质现代、简洁，布局设计精巧，内部空间灵活，赋予变化。大开间、小进深的标准设计，空间方正实用，布局合理，通风对流。

The buildings are in pursuit of the combination between the traditional and modern styles to fully reflect the culture of the Grand Canal and make every building of West Bank fortunately become an organic part of the canal scenery. The facade is seeking for the effects of the lights and shadows, which are arranged properly by sections so that the entire image is simple but detailed. The architectural elements which have been concise to the extremity have integrated the romantic feelings and the ideal of human residences into the space mode of West Bank. The show room is designed to the interior space split-level, clear and comfortable. The entirety looks modern and simple with the delicate arrangement and design. The interior space is flexible and full of changes. The standard design of big width and small depth makes the space square and practical. The space is arranged properly with ventilation and convection.

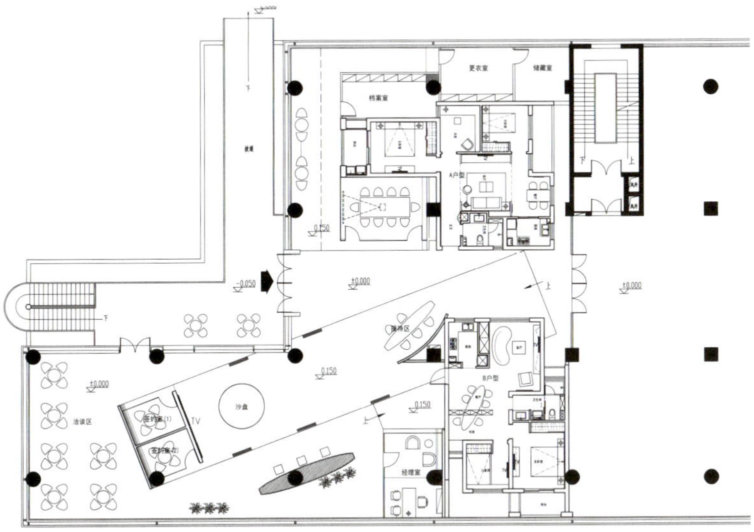

WEST BANK

设计单位：吴政道 / 参与设计：刘育葶、赵静

元一柏庄样板间
Show Room of Yuanyi Baizhuang

面积：105m² / 坐落地点：合肥 / 主要材料：烤漆玻璃、自然风壁纸、草皮、橡木地板、马赛克 / 摄影：周宇贤

此案设定28～35岁之间的年轻族群，男主人从事平面设计工作，女主人则是时尚杂志记者，因此对于色彩有相当的敏锐度。在建筑体构成既定的室内空间格局下，规划空间上并无敲改任何墙面，只在家具的配置上让空间动线更流畅。空间设计构思上，希望给人明亮、轻快、活力的感受。

客厅与餐厅处于同一长形空间内的左右两侧，天花板设计了连贯两空间的灯沟，客厅主墙采用色彩抢眼的芥末绿烤漆玻璃，墙体延伸连贯餐厅墙面，视觉上让两空间有延伸扩张的效果，也达到同一空间却不同属性机能的整合，另餐厅加入了大自然草皮的元素，增添此空间的丰富性及亲近大自然的感受。书房具备基本阅读空间机能，设计上采用色调明快的青苹果绿搭配使用具自然风的藤草压制成的壁纸，架构此空间轻松阅读、愉快沟通的调性。主卧室希望呈现加拿大枫叶给人的有氧气息，色调采用枫叶红温暖色系，卧室一角则选用一款具英式贵族的精致线条但以现代化的有机玻璃雕制而成的台灯，形构出温馨、雅致、浪漫的氛围。

This case is orientated at the young group of 28~35 years old. The host is engaged in plan design while the hostess is a journalist of a fashion magazine. Therefore, they are quite sensitive to the color. As the building has fixed interior space arrangement, no wall is removed or changed during the planning of the space. But the arrangement of the furniture makes the active lines of the space smoother. As for the conception of space design, it hopes to provide people with a bright, brisk and vigorous feeling.

The living room and dining room are located on the left and right sides of a long space. The ceiling is designed with a light channel connecting both spaces. The main wall of the living room is applied with distinguished wasabi green coated glass and the wall is extended to connect with the wall of the dining room to provide a visual effect that both spaces are extended and expanded and also realize an integration of different features and functions in the same space. Furthermore, the dining room is also added with the element of natural greensward to increase the wonderfulness of this space and an intimate feeling with the nature. The study has the basic functions for a reading space. As for the design, the vivid apple green is arranged with the wall paper made of natural liana to constitute the atmosphere of relaxing reading and pleasant communication in this space. The master bedroom is purposed to deliver an aerobatic sense that is brought by Canadian maple. The color is adopted with warm color of maple red while one corner of the bedroom is chosen with a table lamp of British-style noble delicate lines but made of modern organic glass to deliver a warm, elegant and romantic atmosphere.

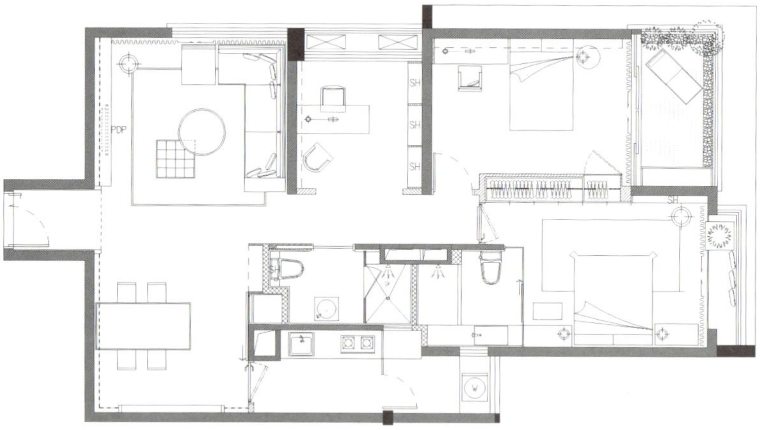

设计单位:玄武设计 / 设计:黄书恒 / 参与设计:欧阳毅、许棕宣、蔡明宪、许宜真、胡春惠、胡春梅

远雄·大未来
Farglory·future town

面积:320m² / 坐落地点:台北 / 主要材料:烤漆玻璃、木皮、金箔、钢琴烤漆、观音石材、马鞍皮 / 摄影:王基守

“居所”是用来包裹人与其相关活动的容器，这个容器是承载生命的另一种形态，而“家”是心灵得以安稳、身躯得以沉静的“居所”。

在设计的表现手法上，让居所潜藏着不对称的协调性、沉稳的空透性与气韵的流动性，借由“以简御繁”的哲学，来平衡空间中的适切度，让眼睛所视的多余能够消失，并伴随着生活中的喜怒哀乐情结而相互纠缠。从玄关开始，企图营造开阔空间的气度与对生命的感性诉求，从简练到内敛，所以，偏移的物件在视觉导引下能够感到自在，客厅、书房、餐厅、厨房是一整体的空间，可依机能分开使用，也可以串连成一处无主宾之分的场所，相间着无绝对的黑与白，如钢琴键般，自谱出温和、宁静的氛围，在视觉音符的律动中，穿透至每一个所能想到的位置。优雅、诗性，让脚步在慢中舒缓开来，让一切的呼吸是用享受来解释的，并借由石材的敦厚性、木材的质朴性与玻璃的洗练，彼此相互影响。而卧室的魔力，在于理所当然中多了些许的闲情逸致。让“家”提供生活的每一天，都有着诉说不完的角落。

陶渊明的诗中写道：“此中有真意， 欲辩已忘言”。设计的初衷，便是希望让空间的暗哑无声与欲语还休的气质，再次以另一种态度与生活碰撞，让生命中最真切的与最严肃的态度，都能在从容中慢慢开展。

“Residence” is a container applied to enclose human beings and their involved activities. This container is another form of carrying the lives while “home” is a “residence” where the souls can be calmed down and the bodies get quiet.

As for the demonstrative techniques of the design, the residence is hidden with the harmony of asymmetries, transparency of quietness and fluidity of mood to balance the appropriateness in the space and remove necessities seen by eyes by means of the philosophy of “control complicatedness by simplicity, which also get interlinked with each other accompanying with the pleasure, anger, sorrow and joy in life. Beginning from the front door, it has attempted to create an atmosphere of open space and a perceptual appeal to life from concision to introversion. Therefore, the deviated objects can feel easy with the visual introduction. The living room, study, dining room and kitchen are an integrated space, but it can be used separately according to the functions or connected in series into a field without clear hosts and guests distinction. It is alternated with non-absolute black and white, just like the piano buttons to generate a mild and quiet atmosphere. In the rhythm of visual notes, it has penetrated to every location that can be thought of, elegant and poetic, so that the steps get relaxed in the slowness and every breath is interpreted by enjoyment. They affect each other through the steadiness of white stone materials and simplicity of wood materials as well as the clearness of glass while the magic of the bedrooms of course appears more leisurely and carefree so that “home” provides every day of the life and always has a corner that can be talked about endlessly.

It is written as follows in the poem by Tao Yuanming: “there is genuine meaning in it and the talk has been forgotten when attempting to argue”. The initial purpose of the design is to make the muteness of the space and the shyness of the atmosphere collide with the life again in another attitude, and the most real and serious attitude in life get developed in the easiness.

编委（排名不分先后）
陈耀光、陈南、高蓓、胡文杰、黄玉枝、满登、张乘风
孙天文、沈雷、宋桥林、王琼、王兆明、吴海燕、叶铮

主编
陈卫新

国书在版编目（CIP）数据

2008中国室内设计年鉴. 1 /《中国室内设计年鉴》编委会编. —沈阳：辽宁科学技术出版社，2008.11
ISBN 978-7-5381-5585-3

Ⅰ.2… Ⅱ.中… Ⅲ.室内设计—中国—2008—年鉴
Ⅳ.TU238-54

中国版本图书馆CIP数据核字（2008）第151005号

出版发行：辽宁科学技术出版社
（沈阳市和平区十一纬路29号 邮编：110003）
印 刷 者：上海美雅延中印刷有限公司
经 销 者：各地新华书店
幅面尺寸：230mm × 300mm
印　　张：87
字　　数：300千字
插　　页：8
印　　数：1~2000
出版时间：2008年11月第1版
印刷时间：2008年11月第1次印刷
责任编辑：陈慈良
封面设计：赵宝伟
版式设计：赵宝伟
责任校对：周　文

书　　号：ISBN 978-7-5381-5585-3
定　　价：498.00元 (1) (2) 册

联系电话：024-23284360
邮购热线：024-23284502
E-mail:lkzzb@mail.lnpge.com.cn
http://www.lnkj.com.cn

2008 China Interior Design Annual

2008中国室内设计年鉴（2）

《中国室内设计年鉴》编委会 编

辽宁科学技术出版社

目录

餐厅 Restaurant

酒店 Hotel

CONTENTS

RESTAURANT

孙天文 Shun tianwen
上海黑泡泡建筑装饰设计工程有限公司总设计师
高级室内建筑师
中国建筑学会室内设计分会理事

斗艳
——当代餐厅设计现象的分析与思考
Beauty Competition
——Analysis and Thinking on the Design of Contemporary Restuarants

世界各地的饮食习惯都具有特殊性，与其地域和民族的文化密不可分。文化，统而言之，是给生活以意义和尊重的体系。作为分享固有文化的单位个体——民族，其多样的饮食和消费习惯成为世界各地餐厅设计和经营的基础。

在世界变得“越来越平”的今天，现代化及由此引起的产业化社会、消费社会的潮流，使许多设计发生变化。产业社会的幽灵仍在我们社会周边出现。靠巨大制度结构造成社会分工和分割的物质基础的产业社会也好、消费社会也好、情报社会也好，看来似乎完全变了样，但是很难说其秩序的根基已经崩溃。西方文化和生活方式的冲击，挑战着和甚至几乎湮没了本身已被革命过的中国传统文化和生活习惯。虽然如此，还是有越来越多的餐厅设计希望传达对自身文化渊源的敬意。在众多不同的案例中，对传统文化符号、空间手法、造型原型等等的不同尺度和程度的借鉴和创新，创造了很多丰富的别具一格的设计形式。比如华丽与温馨并存的“历家菜”，具有现代禅意的“来福楼火锅餐厅”等。

与国家的野心和企业的野心相比，室内设计更接近于个人行为。新登场的时尚由象征大众文化转为象征个人文化。崇尚个人固有价值的潮流，是靠肉眼看得见的方法——设计而开创出来的。室内设计更能体现这个潮流。而餐厅空间给人的最初印象，对于多数人来说，更能唤醒个人的情感。以个人思想为对象的设计的产生，能够通过餐厅准确地传达社会导向。于是，设计师（业主）对餐厅趣味的主导使得其个人化痕迹常常随处流现。诸如“半岛明珠酒家”中设计师对“光”的介于梦想和现实之间的表达，“星月湾时尚快捷酒店”中设计师钟爱的个人的语汇流露——设计师通过餐厅设计与集体文化相对抗，用它的形象作为未来的生活模型。

今天，餐厅设计已经对人们的生活产生了巨大的影响，综合性杂志常常编辑饮食专栏，饮食指南相对于介绍饮食特点则更加重视设计。各式各样的设计，不仅是个人化的，甚至挑战性的设计——如在“蛋蛋屋”中的夸张——已成为理所当然的现象，并通行无阻。所有的主义和主张、正相反的理论、设计与艺术的合体、莫名其妙的观念、乡愁、地方主义等等，在设计领域各踞一方，其结果是设计风格的多元化并存。这就是今天的现状。

限于篇幅，不能在这里详细说明何以出现这种情形。可以说在现代城市的无机状态乃至混乱状态中，“室内”创造了掩埋这种状态的独自的文化。但是，在社会与个人的夹缝中产生了认识上的分歧，而商业空间成为个人获得特性的场所，这也是事实。但是，比起以上两点更为重要的事实是：因为企业资本作为消费时代的战略而追求表面的差异所以产生了这种多样性。就是说，企业资本作为差异化战略选择了设计。这种现象使得以民族固有文化为基础的设计向中性的设计偏移。这种现象不能说是坏事，就是说，人们得以重新获得造型的趣味、彩色世界的魅力、装饰复杂的魔力，这在过去都被现代主义暗暗地视为禁忌。

随着中国社会和经济的迅猛发展，餐厅如雨后春笋般地发展起来，似乎大家已坦然接受饮食作宴的“朝秦暮楚”。但是，喧嚣过后，留下的仍是头脑冷静、脚踏实地的社会，那时，我们更可以等待和评价那些成功并富有魅力的餐厅——融本民族传统风格与当代艺术为一体的设计作品。

The drinking and eating habits around the world are usually special with regard to the other places. It is closely related to the regions, nationalities and the relevant cultures. However, the culture is a system to bring the significance and prestige to life. Any nationality that brings the system to human life is an individual unit to share the inherent culture. The restaurants around the world more or less are based on this kind of culture.

Although the contemporary design has varied slightly due to the changes of the design history and communications between different cultures, wherever it is, the traces of its inherent cultural origins are inerasable. (e.g.: the design of Lai fu Hot Pot Restaurant in the yearbook has integrated the modern vocabularies on basis of the tradition)

Today when intercourses of people of the world become more and more frequent, the impact of western cultures seem to have ruined Chinese traditional cultures and living habits. But, it is not the fact. The designs of many restaurants have absorbed the new intelligences from different cultures. In this point of view, the bosom of the traditional cultures is unfathomable. (e.g.: Li Jia Cai is designed to tactfully combine the western vocabularies with the eastern patterns to create a new space effect of the coexistence of new luxury and warmness).

Even though, the modernization and the trends of its consequent industrialized and consumption society have still made some changes on many designs. The spirits of the industrial society still appear around us. No matter either the industrial society with a material basis of social division of work and segmentation caused by the giant system structure, consumption society, or information society, it seems to have changed thoroughly. But it is difficult to say that the origin of its system has collapsed. (e.g.: For Xiang Pin Fang Restaurant, you still can dimly see the colors of Chinese red in a very modern environment).

The advocacy on the expression of individual values and the trend of “self-fulfillment” are the visual methods by eyes developed by the design. The fashion coming on stage at the moment is changed from symbolic popular culture to symbolic individual culture. In such a trend, interior design has played an important role. The present design of restaurants in China is achieved in this trend. (e.g.: the design of Xin Yue Wan Hotel has delivered the individual understanding of the designer on the design through a completely new vocabulary).

For majority of people, their first impression on the space of restaurant can more wake up the individual feelings. The generation of the design with the individuals as an object can precisely deliver the social orientation through the restaurants. The designer has fought with the collective culture through the design of the restaurants and taken its image as the model of future life. (e.g.: for Peninsula Pearl Restaurant, what the designer intends to express is light. He hopes to fix the light in the dream into reality).

The design of restaurants now has created great influences on the people’s life. The comprehensive magazines often edit the catering columns. Catering guides put more emphasis on the design compared with the introduction of catering characteristics. Today, various types of design and even the challenging designs have become a natural phenomenon without any obstruction. (e.g.: in the design of Dandan Wu, the designer has applied an extremely exaggerated technique to challenge the environment of the normal restaurants).

Due to the limited length of the article, I can not explain why this situation occurs in detail. It also can be said that “interior” has created an individual culture to bury this status in the nonorganic, even confused status of the modern cities. However, the divergence on the acknowledgement has been created in the gap between the society and individual while the business space has become a field where the individual can receive the peculiarities. But, the fact that is more important than the above two points is: as the strategic core in the consumption era, the enterprises’ capitals have been pursuing the superficial differences, thus this kind of variety is generated. That means the diversity “desired” by the enterprises’ capitals has selected the design. This kind of situation is deviating from the design based on the regional and inherent national cultures to the eclecticism.

With the rapid economical development, the restaurants are spring up like bamboo shoots to cater for the social requirements. But, a cool-headed and surefooted society is left behind after the festivals. Any region is still reserved with the traces of its unique cultures. If this view point is correct, we shall feel lucky here.

设计：卢迅、吕彪、黄君瑜

湘品房餐厅
Xiang Pin Fang Restaurant

面积：1416m² / 坐落地点：东莞市东城区莞长路 / 主要材料：地毯、玻璃、石膏板、乳胶漆、灰色麻石、水曲柳饰面、红色烤漆玻璃、红色双面铝塑板、拉丝不锈钢、仿古砖、欧松板

一种细腻含蓄、一种热情和张扬，两种不同时代的文化和情怀，却在同一空间得到融合。八角形的构建、“笼子”的区间渗透着阵阵新东方主义色彩，直线的魅力细腻而含蓄的延展整体空间，大厅水泥地板的质朴洒脱与“笼子”天花的直率及红白色彩融洽对应、天花的自然裸露与地面的整齐规矩一张一弛，在协调、简约中透现张弛之道。

One is exquisite and implicit while the other is passionate and aggressive. Two kinds of cultures and feelings of different eras get integrated in the same space. Octagonal construction and “cage” part have revealed gusts of neo-orientalism moods. The lines are extended in the entire space with exquisite charms in a controlled way: the simplicity and ease of the concrete floor in the hall and the frankness of the “cage” ceiling, the harmonious echoing of red and white colors, as well as the natural exposure of the ceiling and the orderliness of the ground, which have perfectly matched with each other. Hereby, the principle of perfect integration between tension and relaxation are well demonstrated in the coordination and concision.

设计单位:中山天艺空间设计制作有限公司 / 设计:蔡旭民、吴立成

王子私房菜
Prince Private Cuisine

面积:800m² / 坐落地点:中山市 / 主要材料:花岗石、梨木、裂纹玻璃

本案以其功能空间的细致划分，运用多种独特的材质及装饰摆设，造就出特有餐饮文化的主题空间。设计师巧妙利用原有建筑的高差错落，在有限的空间里将功能与装饰充分结合，使各功能区域划分得更合理、更具个性化。入口区开放式明档渲染热闹的中式餐饮气氛，感染着每一位食客。最高的挑空层为开放大厅，透过6m高的弧形玻璃墙，室外幽雅的园林环境尽收眼底。高身半圆的卡座设计，结合落地金属纱窗，围合成一组半私密空间。仅有的五间贵宾房各具特色，均配有独立备餐区和洗手间。

整个餐厅的色调以深色为主，普光源与点光源的完美演绎，营造出浪漫的用餐氛围。

This case is to create a subjected space of the unique catering culture by the delicate division of its functional space and utilization of various unique materials for decoration and ornaments. The designer has cleverly utilized the height differences of the original building and fully combined the function with the ornaments in the limited space to make the division of each functional area more rational and special. The active Chinese-style catering atmosphere rendered by the show kitchen at the entrance has touched every customer. The highest pick-air story is the open lobby and the elegant outdoor garden can be viewed panoramically through a 6m high arc glass curtain wall. The design of the half-round high backing seating in combination with the floor metal window screen is enclosed into a half-private space. The only five VIP rooms have their individual characteristics, which is equipped with the independent pantry countertop and toilet separately.

The dark color is the main mood of the whole restaurant. The perfection performance of general light sources and spot lights has developed a romantic dining atmosphere.

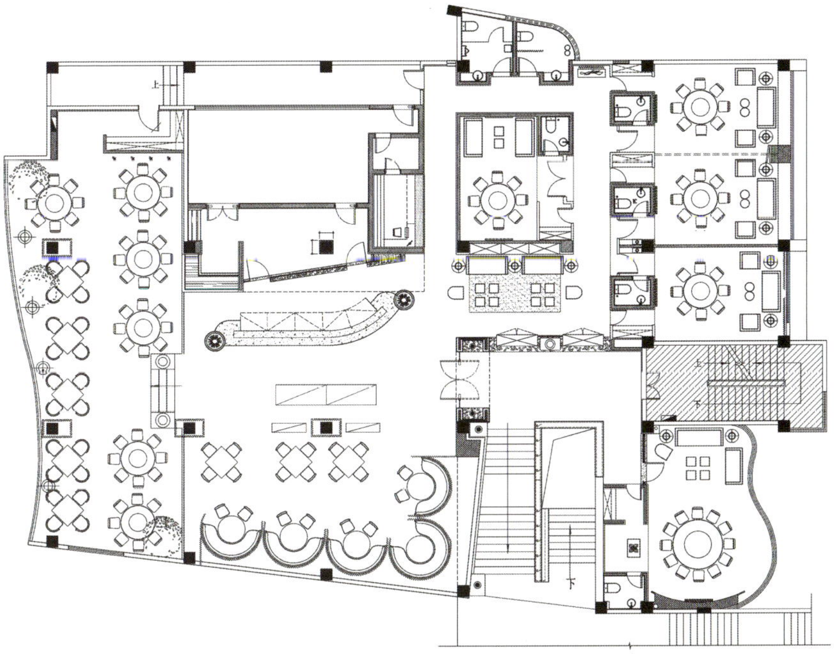

设计单位:上海泓叶室内装饰有限公司 / 设计:叶铮 / 参与设计:陈燕妮

锦江之星-北京大观园店
Jinjiang Inn – Beijing Daguanyuan Hotel

坐落地点:北京

设计以中国古典元素为主，运用了传统建筑的结构特点，营造出时尚与传统相融合的设计风格。

设计师在处理空间关系中采用了两种手法：一是空间的对称与秩序，二是空间的多层次。随着层高的不断变化，给人以一连串多层次的感官享受。在设计中多方面运用了中国古典特色的元素，如框架、拱门、花格栅、屏风等等。根据不同形式、比例和材质的搭配，产生一种古典、优雅、柔和之美。

空间色彩选用了艳丽的中国红，在强调主题的同时更给人视觉上的冲击。形与色的有机结合，使空间感觉更为协调统一。几组中式吊灯，无论从造型抑或材质上，均充分展示了中国传统文化的优雅气质，起到了画龙点睛的作用。

The design is focused on Chinese classic elements and applied with the structural features of traditional buildings to create a design style integrated with fashion and tradition.

The designer has applied two ways in the arrangement of the space relationship: one is the symmetry and order of space. The second is the multi-hierarchies of the space. With the constant changes of the level height, it has provided people with a series of multi-hierarchic organic enjoyment. In the design, the elements of Chinese classic features have been applied from many aspects, such as: frames, arches, patterned grids and screens etc. The arrangement of different forms, scales and materials has created a classic, elegant and gentle beauty.

The space colors are chosen with flamboyant Chinese red, which has made greater impact on the vision while emphasizing the subject. The perfect combination between the forms and colors make the space more harmonious and unified. No matter from the modeling or material, several sets of Chinese-style pendant lamps have sufficiently demonstrated the gracefulness of Chinese traditional cultures and become a shining point.

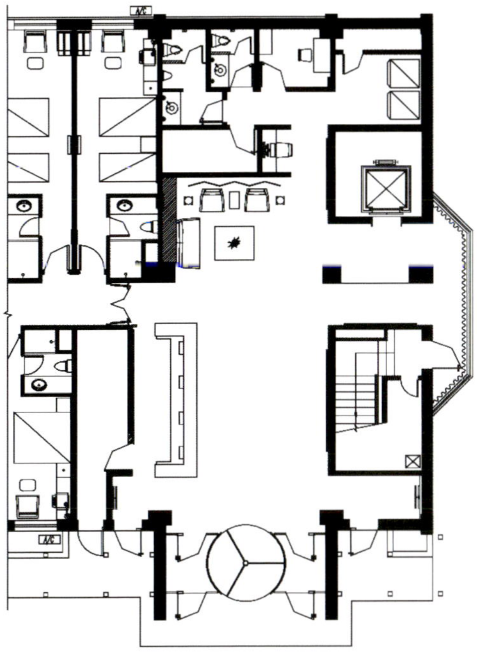

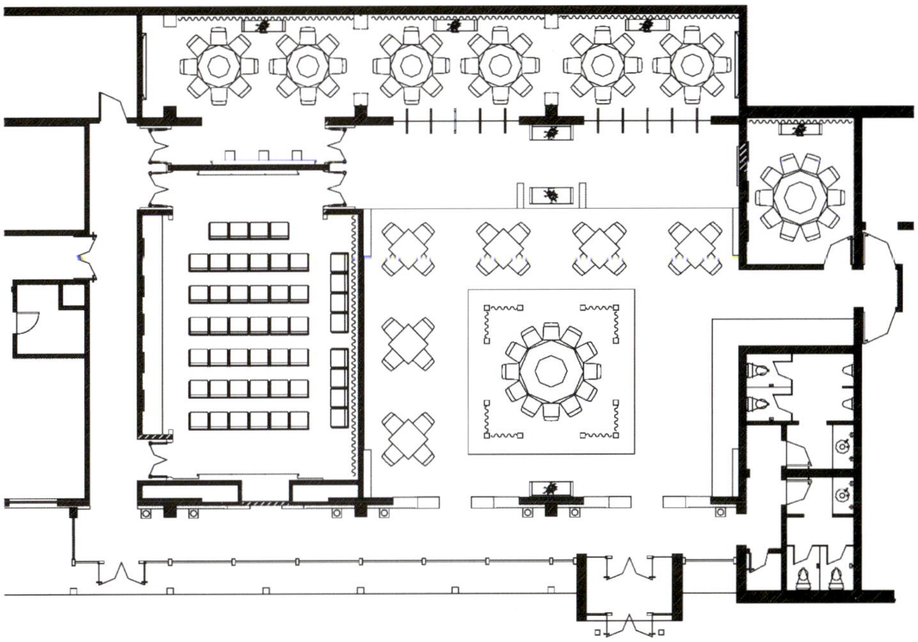

设计单位：上海泓叶室内装饰有限公司 / 设计：叶铮 / 参与设计：戴俪俪

锦江之星-济南解放店
Jinjiang Inn – Jinan Jiefang Hotel

面积：757m² / 坐落地点：济南

本案坐落于济南市中心解放路上，总建筑面积为757m²，客房数为112间，是锦江之星在济南的首家直营酒店。室内设计主要涉及大堂、餐厅、会议厅、电梯厅等公共区域。这是一家典型的小型酒店。在有限的空间限定下，运用平面构成的概念，将斜线拼板式的造型语言空间化，同时利用斜线不规则切割所形成的凹缝，联系着立界面与顶界面，使各界面的拼合转折相互贯通。以粉绿、墨绿、普蓝、中灰这四色为基调，构成冷艳的室内氛围。在形与色的高度和谐中构成新的空间秩序感。

This hotel is located in Jiefang Road in the downtown area of Jinan with a total floor area of 757m2 and 112 guest rooms. It is the first hotel directly operated by Jinjiang Inn in Jinan. The interior design is mainly related to the public areas such as the lobby, canteen, conference hall and elevator hall etc. This is a typical small hotel. Due to the limitation of the limited space, the conception constituted by the plan is applied to spatialize the diagonal-and-assembled-board-type modeling language. At the same time, the recessed joints which are formed by the irregular cutting of the diagonals are applied to connect the vertical interfaces and ceiling interfaces so that the assembly and transition of each interface can be interconnected. Pastel green, black green, Prussian blue and gray are the four basic colors to create a quiet elegant and magnificent] interior atmosphere. The new sense of space sequence is generated in the extreme harmony of forms and colors.

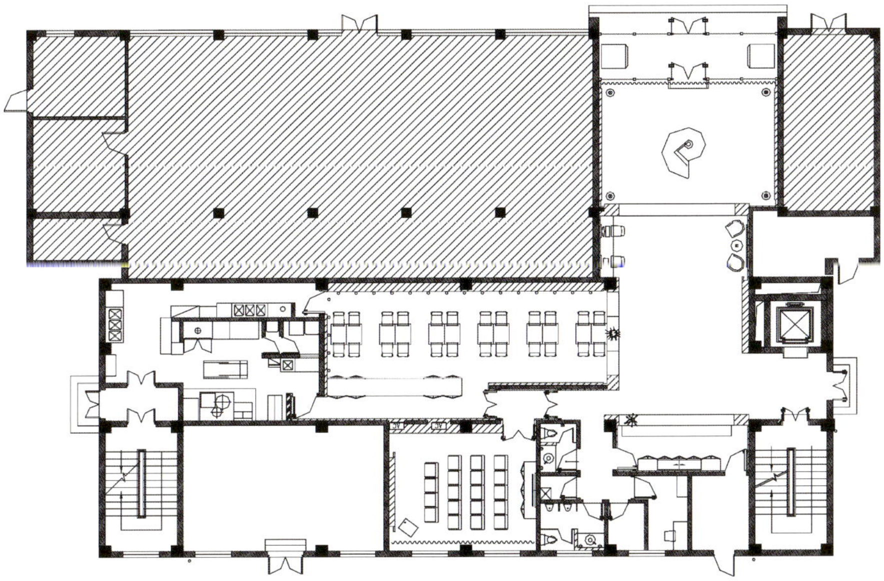

设计单位:上海泓叶室内装饰有限公司 / 设计:叶铮 / 参与设计:郑思南

锦江之星-南宁民族大道店
Jinjiang Inn – Nanjing Mingzu Avenue

面积:960m² / 坐落地点:广西南宁 / 主要材料:原木、毛石

锦江之星南宁民族大道店设计主题概念为“自然、原生态、亚热带”。

本案设计范围包括一层大堂、会议室、电梯厅、走道、四层餐厅、会议室兼包房、电梯厅等，总计设计精装修面积约960m²。

对界面造型上的处理采用排线的手法，以不同宽度、不同长度、不同间距及不同方向的排列方式，对不同材质和不同色彩的线条进行排列组合。在材料选择上配合主题概念，多采用原木、毛石等感觉天然的材料。整体设计风格清新、自然，悠闲中更见幽雅、和谐。

The design subject of Jinjiang Inn Nanning Mingzu Avenue is nature, original ecology and semitropical ……

This design scope includes: the lobby at the first floor, conference rooms, elevator hall, passageway restaurant at the fourth floor, conference room and VIP room as well as elevator hall etc. The total decoration area is designed about 960m2.

The technique of arraying the lines is applied to arrange the modeling of the interfaces. The arrangement methods in different widths, lengths, distances and directions are applied for permutation and combination of the lines in different materials and colors. The materials are selected to match the conception of the subject, so most of the materials are the timbers and rubbles which look and feel more natural. The style of the whole design is fresh, natural and leisure with elegance and harmony.

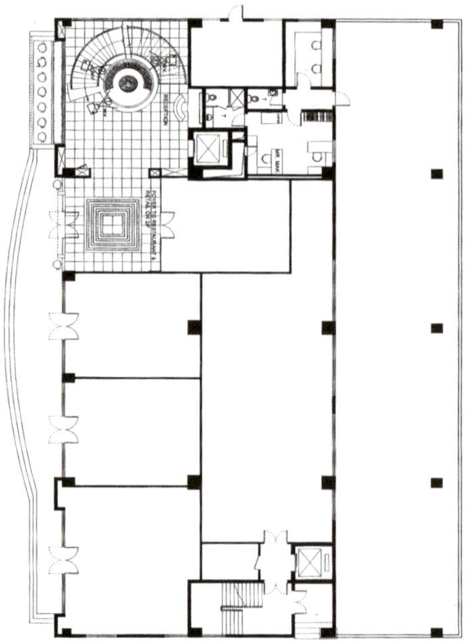

设计单位:上海泓叶室内装饰有限公司 / 设计:叶铮 / 参与设计:皱俊波

锦江之星-天津南开店
Jinjiang Inn – Tianjin Nankai

面积:8200㎡ / 坐落地点:天津

锦江之星天津南开店，虽说是一经济型酒店，却在室内设计中体现出强烈的风格。该项目总建筑面积约为8200㎡，拥有客房220间。

建筑平面成“V”字形。大堂位于两翼相交处。室内设计利用成角的“V”形视线来构成空间的动线感，并赋予空间以波谱感极强的“卷片”和“卷盒”为造型语言，以此来限定空间，重构秩序感。室内设计的重点主要位于大堂、餐厅、会议厅、电梯厅等公共区域。大堂顶部的卷片与地坪相呼应，同时与总台前的圆柱形成穿插关系。餐厅中卷盒的排列成为空间的主要特征，构成半开放的箱式空间。联系大堂与餐厅的长廊被处理成筒状的卷盒，配以蓝色的夹胶玻璃及灯光处理，宛如一条时空隧道。

Although Jinjiang Inn Tianjin Nankai is an economic hotel, it has been reflected with a strong designed hotel style in its interior design. It has a total floor area of about 8200m2 and possesses 220 guest rooms.

The plane of the building is in a V shape while the lobby is located at the intersection of the two wings. The interior design has applied the angular "V" shaped line of sight to form the sense of active lines for the space. The space is also endowed with the "wound pieces" and "wound boxes" of strong sense of spectrum as the modeling language to limit itself and reconstitute the sense of sequences. The interior design is mainly focused on the public areas such as the lobby, canteen, conference hall and elevator hall. The wound pieces on the top of the ceiling echoes with the floor and also forms into an alternate relationship with the round columns before the reception desk. The arrangement of the wound boxes in the canteen becomes the main feature of the space and constitutes the half open box-type space. The corridor connecting the lobby and canteen has been arranged in a tube-shaped wound box. In combination with the blue laminated glass as well as the light arrangement, it looks like a cosmic tunnel.

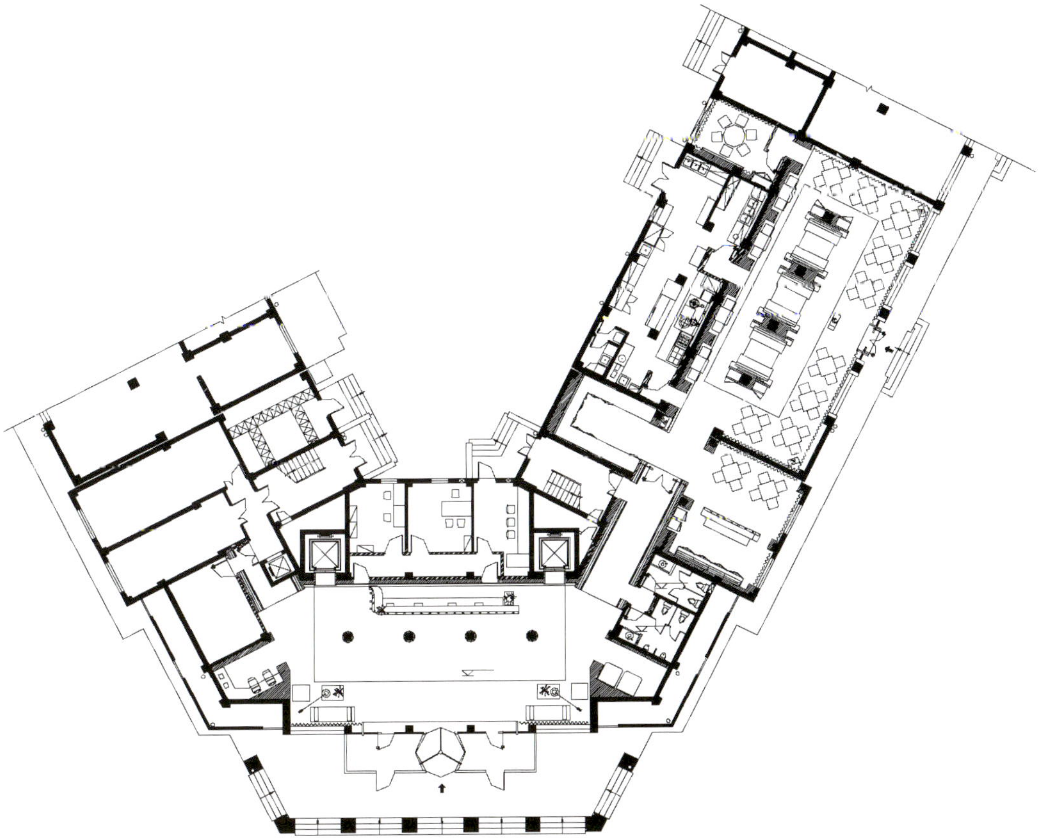

设计:洪约瑟

凯逸酒家
Kaiyi Restaurant

坐落地点:深圳 / 主要材料:大理石、玻璃

通道因为灯光的照射呈现出梦幻现代的感觉，条纹状的灯光镶嵌在墙上，地板上的灯光给顾客指明了主接待区的方向。接待区内蓝色与橘黄色的光形成对比，在此营造出令人舒适的气氛。22间包房被设计成三种颜色，红、黄、橘黄，给人一个良好的视觉效果。

The passage has demonstrated a dreamy and modern feeling due to the project of the lights. The striped lights are embedded on the wall while the lights in the floor have introduced the customers to the main reception area. The blue and orange lights in the reception area have formed a contrast to create a comfortable atmosphere. Twenty-two VIP rooms are designed into three colors including red, yellow and orange with a good visual effect for the people.

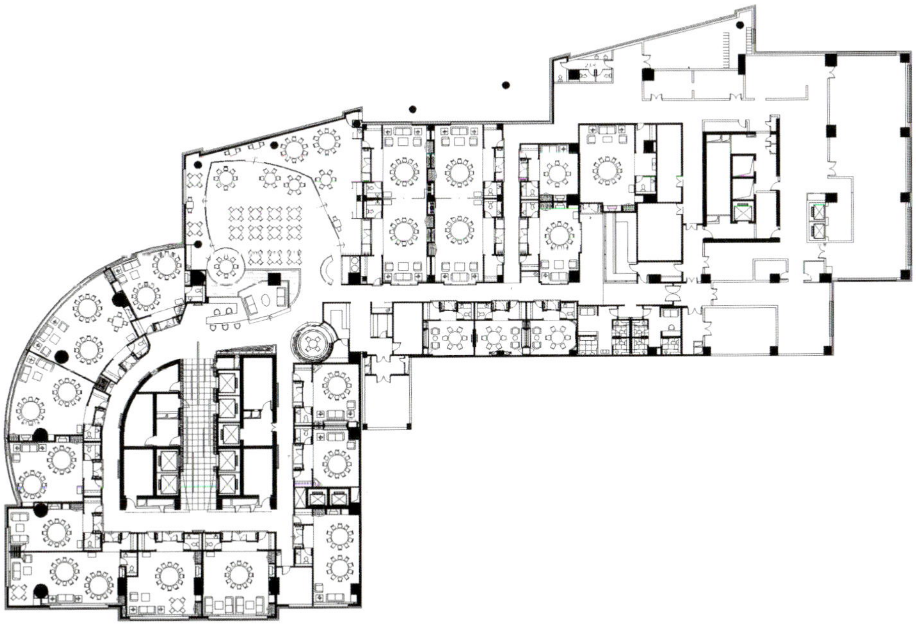

设计:洪约瑟

半岛明珠酒家
Peninsula Pearl Restaurant

坐落地点:广州 / 主要材料:磨砂玻璃、黑色大理石、木料

善于巧妙地利用照明来营造适宜的空间氛围是设计师的强项。观此设计如万花筒般热闹非凡，五彩缤纷的灯光效果烘托着建筑，吸引了路边及隔岸行人的视线，在其中就餐的人们饱览珠江的景色，经过的人们则陶醉在建筑带给人的无尚享受中。

内部一隅，螺旋形楼梯从地面一直延伸到顶层，灯光的色彩则由桃红变为蓝紫，极富视觉冲击力。走廊沿途的侧墙里布置了灯箱，仿佛走进现代科幻的新空间，这种新奇给人惊喜之余更不失为一种美的享受。

The designer is good at smartly applying the lights to create a proper space atmosphere. This design is very lively like a kaleidoscope. The colorful lights have set off the building to attract the eyes of the pedestrians on the road and opposite to the bank. The people dining in the restaurant can enjoy the Pearl River' s scene while the people passing by are toxicities by the supreme enjoyments brought by the building.

At one interior corner, the spiral staircase extends from the ground to the top level. The color of the lights is changed from pink to purple blue with a strong visual impact. The side walls along the corridor are arranged with the light boxes, as if you have entered the modern and science-fictional new space. This novelty is also an enjoyment of beauty while surprising the people.

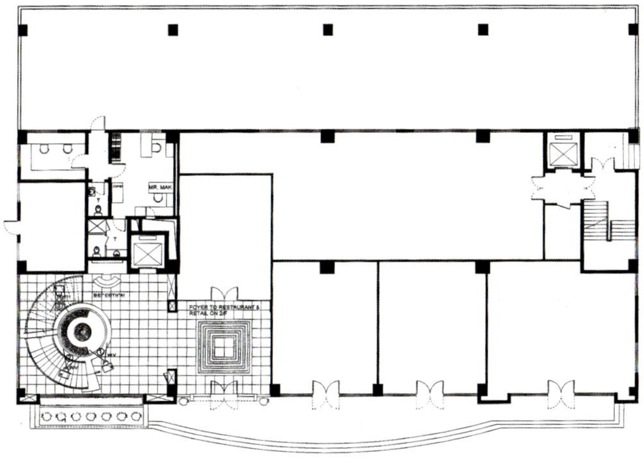

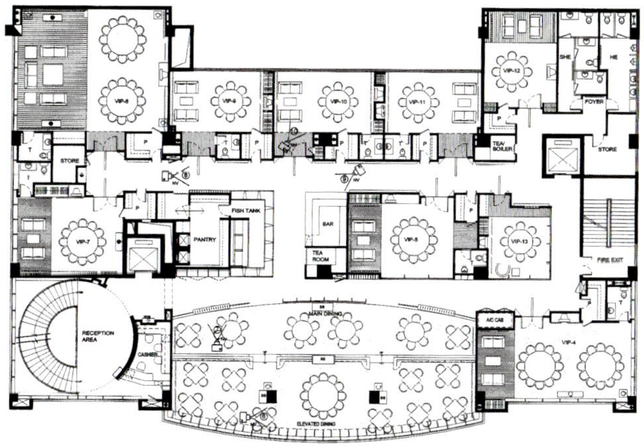

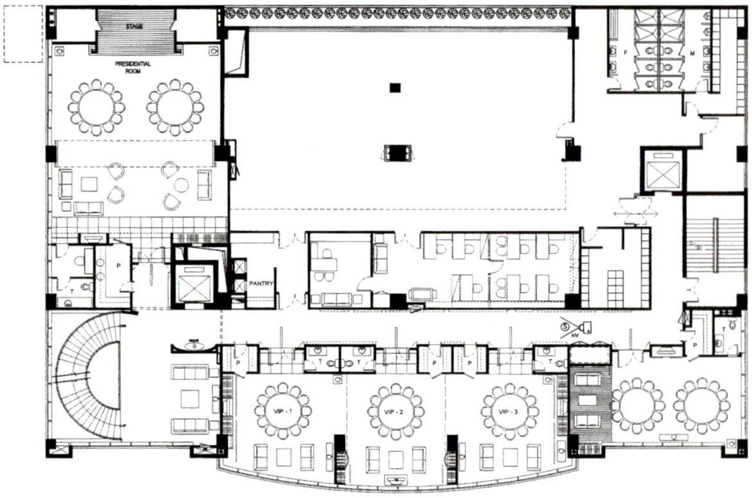
STAGE
PRESIDENTIAL ROOM
F
M
P
T
PANTRY
NV
VIP - 1
VIP - 2
VIP - 3

设计单位：中山天艺空间设计制作有限公司 / 设计：蔡旭民、吴立成

“长安一号”餐厅
“Chang’an No.1” Restaurant

面积：6000㎡ / 坐落地点：西安 / 主要材料：西班牙旧米、凿面黄麻石、橡木、玻璃砖、乳胶漆

“长安一号”是西安的一家高档中餐馆，地处西安市长安路一号，故餐厅命名为“长安一号”。

由于西安是一个拥有十三朝古都璀璨历史的都市，设计师在室内设计上不流于表面装饰，而是多以空间造型来营造气氛。风格尽量能体现汉唐长安的风韵、恢宏的帝王气概、醇厚华夏文明的内涵，实现古老文明与现代文明的珠联璧合。

餐厅的菜式上也汇集南北之长，萃取长安美食的雍容风情，汉唐主题宴会中历史典故和民间传说化入美食当中，让顾客在醉心之余，领略到最具古典文化气息的健康高品质。

“Chang’an No. 1” is a superior Chinese restaurant in Xi’an. As it is located in No. 1, Chang’an Road, Xi’an, it is named as “Chang’an No. 1”.

As Xi’an is a city with the resplendent history as the capital of thirteen dynasties, the designer just develops the atmosphere in design by the space models instead of the superficial decoration. Its style tries to reflect the charm of Chang’an in Han and Tang Dynasty, vast regal spirits and the connotation of pure and honest Chinese civilization in order to achieve the perfect combination of ancient and modern civilizations.

The foods styles in the restaurant also collect the excellences of the south and the north, and have extracted the elegant and graceful charms and integrated the historical stories and folklores into the Han-Tang banquets to indulge the customers while enabling them to appreciate the healthy high qualities of the most classic cultural temperaments.

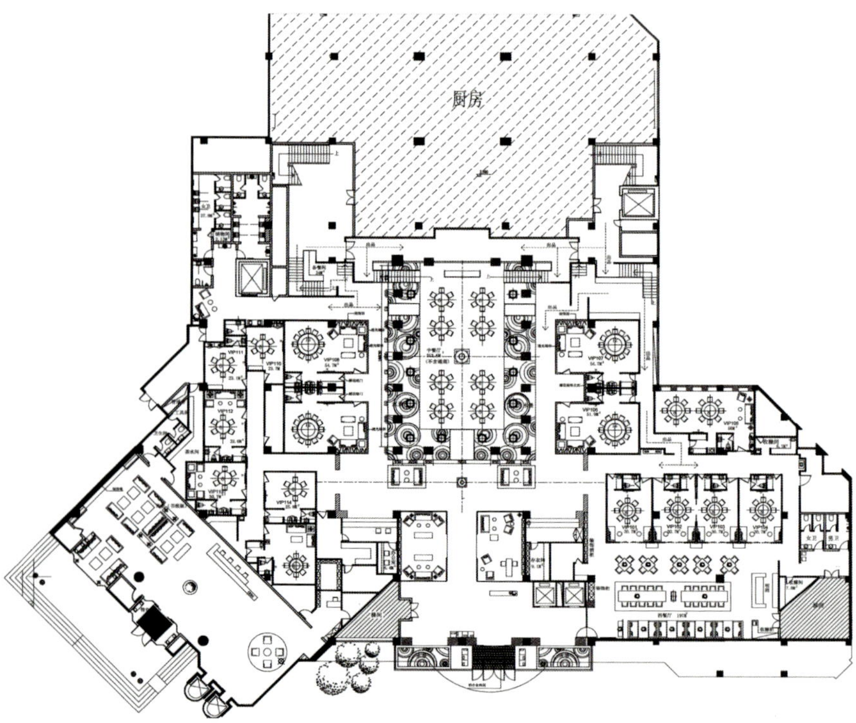

设计:王善祥 / 参与设计:管鹏

来福楼火锅餐厅
Laifulou Hot Pot Restuarant

面积:300m² / 坐落地点:上海 / 主要材料:旧青瓦、白水泥、铁刀木夹板、黑色板岩 / 摄影:申强

来福楼火锅餐厅是一家在上海颇有口碑的火锅店，此次设计的是第五家店。店址在一个综合商场的二楼，长方形，基本是一面落地窗，一面是墙。靠窗一侧是一个建筑工地，至少一年半内不会有什么景观。

这是个小面积又是小成本的项目，场地细长，设计师便将焦点放在了立面与顶面的氛围营造上，图形、色调、肌理成为主角。餐厅面积较为狭长单调，没有阻隔，空间可以稍作分隔。项目根据自身的空间特征做出来的一些意趣，满足了业主的要求，故而自己戏称为：小“本”生“意”。

为了延续来福楼一贯的既传统又现代、各分店又都不完全相同的环境特征（这是与业主几年来达成的共识），每次都会设计出一种符合这一要求的气氛。此家店由于场地较为狭长，面积也不大，就没有在空间上做多少分割，仅划出了两间包房和吧台，剩余的面积就是一个大空间，以显示业主所需要的一定的热闹。这样的空间实际有些单调，于是就把主要焦点放在了立面和顶面的塑造上来了。墙面用旧青瓦切成约20mm宽的片状，在墙面上摆满类似于传统园林的漏窗图案，然后用白水泥将其糊满，等干后以磨光机打磨，直到其由内而外露出青瓦所摆出的图案。每面墙均摆出略有不同的图形，以产生变化。顶面则是满吊石膏板，但在上面做了几条相互交叉的弧形灯槽，与墙面青瓦的弧形图案相类似。一种从传统符号中抽离出来的分子被重新组合，数量增多，从传统建筑中配角的地位一跃而成为了主角。这样，黑白灰的基调形成，再配以深咖啡色的铁刀木饰面，浅咖啡色的座椅，半透的屏风，于是，典型的江南古建筑色调被再一次渲染。

Laifulou Hot Pot Restaurant is a quite well-known hot pot restaurant in Shanghai. This design is for the fifth restaurant, located at the second floor of a comprehensive shopping mall. It is rectangular, mainly consisting of one French window and one wall. The side along the window is a construction site, where there will be not any scenery within one and a half years.

It is a small-area project with low cost. The field is long and thin. The designer has focused on the creation of the atmosphere on the faade and top surface. The patterns, colors and textures play the main role. The area in the restaurant is quite long, thin and simple without any block, thus the space can be slightly separated. The project also has made some changes according to the characteristics of the space to meet the requirements of the client. Therefore, it also call itself: to do business in a small way.

To continue the persistent environmental characteristics of Laifulou that is both traditional and modern as well as not slightly different from each branch (this is the common understanding with the client since years), the atmosphere that meets this requirement will be designed every time. As this restaurant is quite thin and long with small area, the space is not separated greatly and only two VIP rooms and bar counter are arranged. The remaining area is a big space to show the certain liveliness required by the client. Such a space in fact is a little bit simple. Therefore, it is focused on the buildup of the faade and top surface. The wall is by about 20mm wide slice forms with the old gray tiles. The wall is filled with the patterns similar to the ornamental perforated windows in the traditional gardens, then pasted fully with the white cement and ground by the grinding machine after being dried until the patterns formed by the gray tiles are revealed from the inside to the outside. Every wall has some slightly different pattern as the changes. The top is suspended fully with the gypsum boards, on which several intercrossed arc light channels are made, similar to the arc patterns of the gray tiles on the wall. The components taken out of the traditional symbols are reassembled and increased in quantity sot that they suddenly become the major role from the minor role in the traditional architecture. Consequently, the basic colors of black, white and gray come into being. Together with the dark coffee wood veneer of the iron knife, light coffee chairs, semi-transparent screens, a typical ancient architectural atmosphere in the south of Yangtze River is exaggerated again.

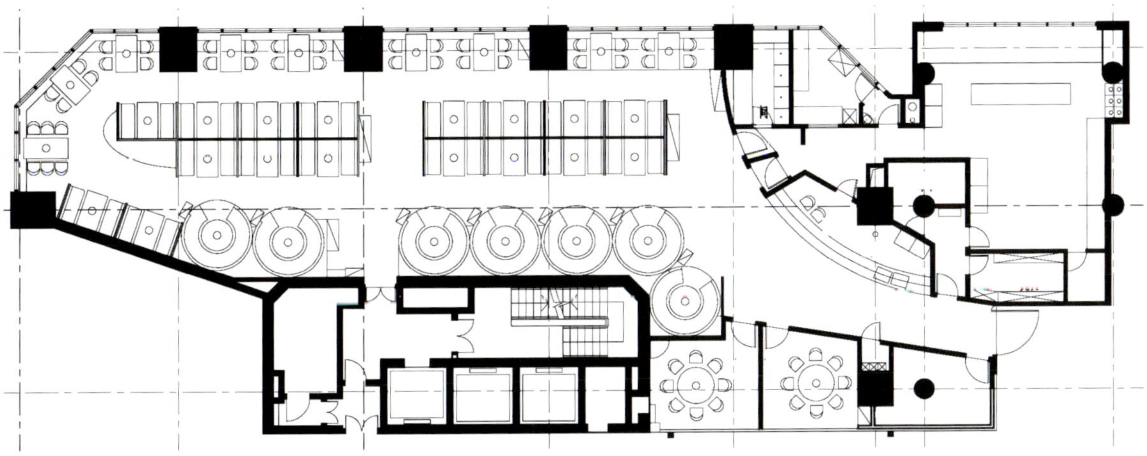

火锅餐厅
HOT POT KING

设计单位:南京智点设计顾问有限公司 / 设计:潘开富 / 参与设计:徐高林、孙传忠、王宇

鸿兴隆酒店
Hongxingrong Restaurant

面积:2000㎡ / 坐落地点:南京 / 主要材料:樱桃木、茶镜、不锈钢 / 摄影:贾方

本案共三层。一层为接待区，二层为敞开式就餐区及私密餐区，三层只有私密餐区。以一种从容、淡定、含蓄、低调的贵气，诠释极雅极致的气质。进门首先映入眼帘的是一条长廊，一旁酒架上的红酒将人带入一种渐入佳境前的微醺。电梯出口处施华洛世奇的水晶吊灯如宝石般璀璨，优雅而华丽地登场，光和影朴实自然的含蓄、新概念镜面木格屏风似隔非隔的含蓄……整个空间浪漫柔美，令人遐想。

This restaurant has three stories: the first floor is the reception area, the second floor has an open dining area and private dining areas and the third floor only has the private dining areas. It has applied a leisurely, simple, implicit and low-profiled nobility to depict the extremely elegant and delicate atmosphere. At the entrance, what first greets you is a long corridor, beside which the red wines on the wine racks make people slightly drunk before in the good mood. The Swarovski crystal pendant lamps at the exit of the elevators, as brilliant as the gemstones, are presented elegantly and garishly. The simple and natural implication of the lights and shadows, the constringency of the wood grid screens with new conceptual mirrors that separate dimly…… the whole space is romantic and mellow, full of fancies.

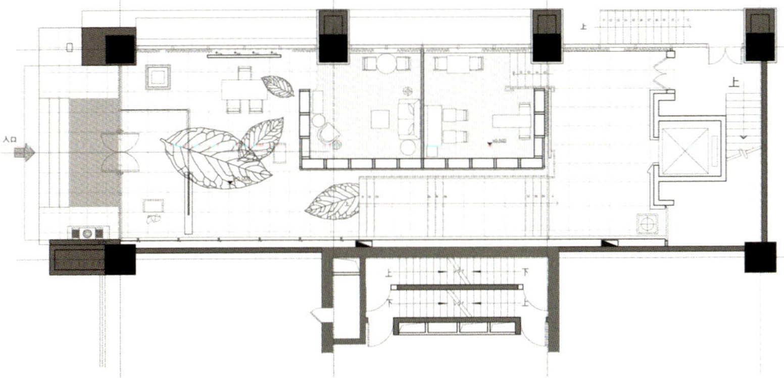

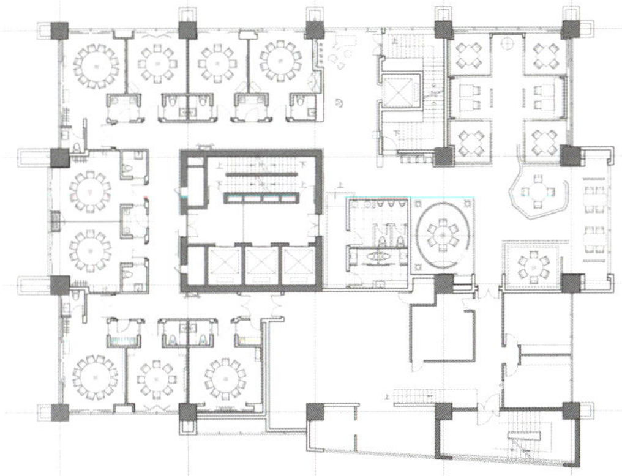

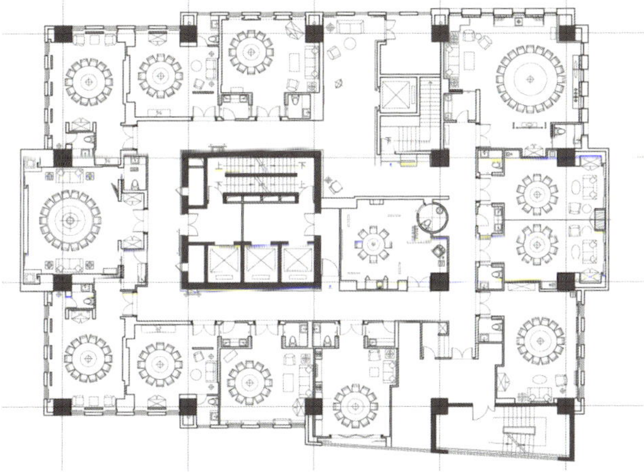

设计单位:MoHen Design International / 设计:赵牧桓 / 参与设计:方凯、陈君正、罗玉龙、魏芝云

蛋蛋屋
Danbo Fun

面积:180m² / 坐落地点:上海 / 主要材料:塑料地板、不锈钢、防火板、亚克力、FRP / 摄影:周宇贤、莫尚勤

蛋蛋屋，从中文的片面字义上来解释，就是蛋包饭，但是因为重复字的关系，感觉上比较童言童语，更可爱而且更平易近人。餐厅的消费阶层设定在15岁到35岁之间；这是一群年轻的消费群体，特性是富有活力与朝气，勇于创新并喜欢尝试新鲜的事物，也是一群可爱但是又以女性居多的客户层。针对这样的消费群，决定用比较激进的设计手段来吸引他们的关注。

基地本身的地理位置是在上海中山公园旁的一个高层百货公司的地下二楼，直接挑高到10楼的挑高空间，两侧有手扶梯进出整个百货商场，整体气氛条件偏暗而且并不显眼，严格来说，基地的条件并不是太优越。

经过几次的研讨，最终决定用一种夸张的表达方式来诠释这个个案，目的是要能够非常抢眼地吸引进出人群的目光。平面构图基本顺着客户已有的清楚定位：以蛋作为原始设计概念，并试图表达并加强蛋的能量、活泼、自由的原型意义。位于平面正中间的大型的有机形体取自于蛋白流动的动能，自由且张力比较大。地板用视觉艺术的概念把塑料地砖做成复杂拼花并赋予对比较强的颜色，这样可以进而让家具能够更明显地跳跃出来，从楼上各个楼层往下看的时候也能有更强烈旋转视觉的活跃感。

最后，当我们再次在现场检测实体空间构图时，虽然还是有些形体无法按照原始意念完整地表达出来，在家具的选择上也显得有些被动且不尽人意，但这个案子还是给了我们对于设计逻辑的根本改变和冲击，我们也因此对以后的设计方式有了新的认识和方向，这也应该算是件成长并值得欣慰的事情吧。

Danbo Fun, literally speaking, means omelet in terms of Mandarin, which is kind of cuisine made of rice covered with eggs and other recipes. It is regarded as a western and oriental mixed food. The design concept of Danbo Fun comes from the original Mandarin meaning:, which is vivid and energetic power of eggs.

The site is located on the basement atrium of a department store close to the Zhong Shan Park, which is one of the major city park in Shanghai. The department store itself is about 10 stories in height with 2 escalators access to all stories at the sides. We decided to give a dramatic visual impact from the top view of each floor. The composition of the floor plan consists of the shape of egg cracking with the protein jumping out. The major access to the restaurant is from the bottom tunnel looking toward the major free form of the protein. The vinyl floor tile with optical arts patterns reinforce the visual impact of the yolk and further popping out the forms of the loose furniture.

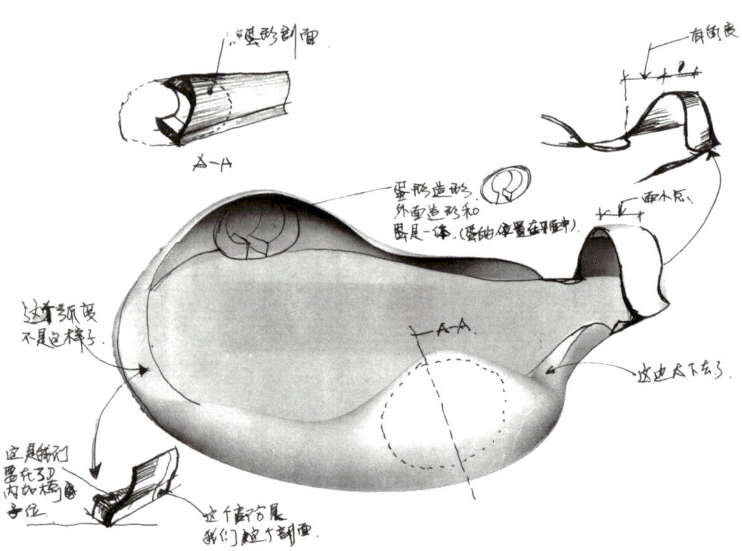

Danbo fun
蛋蛋屋

设计：殷艳明、胡崴

富京美膳酒楼
Fujing Meishan Restaurant

面积：2500m² / 坐落地点：西安 / 主要材料：黑镜、云石、大理石、陶砖、水晶琉璃 / 摄影：臧兵

该空间设计要表达的主题是带着大地气息的“亲切质朴”和王者官府的“富贵典雅”。

接待大堂的空间不大，为了更好地拓展视觉空间，大量运用了黑镜和琉璃相结合的设计手法。云石灯片环绕出的牡丹花吊灯、黑色云石内透光的接待台、地面上纵横交错的金色圆圈、深木色“锁子甲”的折扇门……优雅婉转地述说着富贵堂皇的气息，营造出大堂的视觉中心，也向来者表达出设计文化理念的第一感觉。椭圆的空间造型把大堂和中餐大厅连接在一起，晶莹剔透的珠帘围合出就餐大厅的中心区域，大厅两个角落精巧的旋转楼梯很好地把功能性和展示性结合在一起。旋转而上，进入包房区。整个包房空间主要运用了红色、黑色、金色、紫色来传达丰富的情感，同时也穿插了蓝色、灰绿色来打破前述主要色彩可能带来的压抑，让整个空间活跃起来。

整个设计营造出真正的“王者”风范：它不强调材料的奢华和金银的堆砌，而重在通过有意味的形式表达出设计师对餐饮文化内涵的深刻理解和把握。我们端出了一道地道的“美膳空间”。

The subject expressed by this space design is “kindness and simplicity” with the breath of earth and “wealth and elegance” of royals and governments.

The space of the reception lobby is not very big. To better develop the visual space, the combination of black mirrors and colored glazes is applied greatly in the design. The peony pendant lamps circled by marble tabs, the luminescent black marble reception desk, the crossed golden circles, dark wood “hauberk” folding door…… which have elegantly and tactfully told a rich and gallant mood and created a visual center in the lobby as well as expressed the first conceptual feeling of the design culture to the visitors. The elliptical space has connected the lobby with the hall of Chinese restaurant while the brilliant and bright bead curtain has enclosed out a central area for the dining hall. The quaint spiral staircases located in the two corners of t he lobby have properly combined the functionality and exhibition together. Turning upward, you will enter VIP rooms. The whole VIP space is mainly applied with red, black, gold and purple to deliver the rich feelings. At the same time, the alternated blue and gray-green have broken the oppression which may be brought in by the afore-mentioned main colors and animated the whole space.

The whole design has created a real “royal” style: it emphasizes on neither the luxury of the materials nor the pile of gold and silver; instead, it lays emphasis on the expression of deep understanding and control over the connotation of the catering culture by the designers through a meaningful form. Thus, we have put on one authentic “Delicious Food Space”.

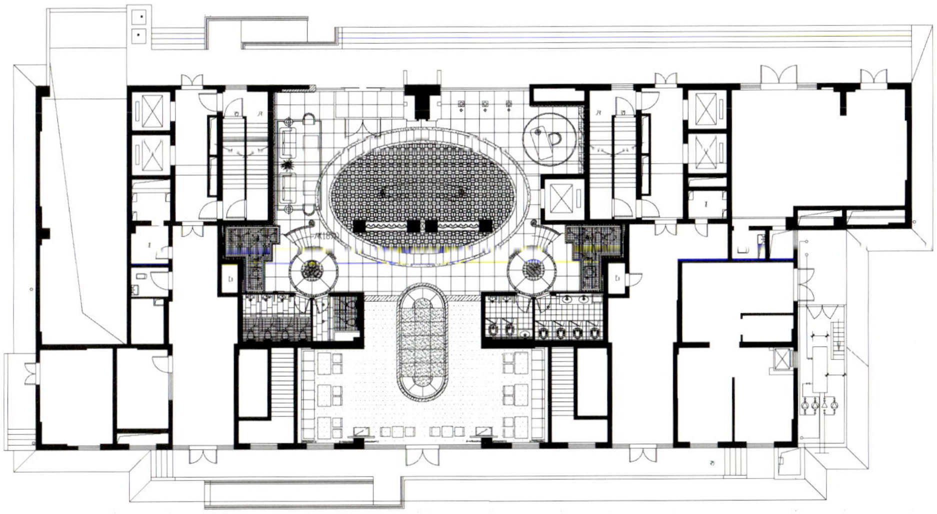

设计单位:七观国际有限公司 / 设计:潘礼旺、周哲毅、游田安、王振宇、李雅琇

豪鼎饭店
How Dine Restaurant

面积:992m² / 坐落地点:台北 / 主要材料:樱桃红大理石、烤漆玻璃、喷砂玻璃、金丝玻璃、明镜、不锈钢、亚克力、地毯、半反射玻璃 / 摄影:吕荣德

这是一首听不腻的爱情故事主题曲,一段属于两情相悦者真爱旅程的起点。

婚宴在现代世纪,是一个将新人幸福与所有亲友嘉宾分享的场所。本案在规划婚宴空间时,采取的是复合多元的形态——包含婚礼与纪念会的概念。形象包装上,则塑造话题,打造白金级的价值服务内容及新美食空间体验。空间的风格上采用古典简约的色彩表现,如璀璨金、儒雅金、威尼斯晨金、缎带红等。白金飨宴期待给予宾客一种尊贵、专属的礼遇。在这个为相约到老而承诺的宴会中,让新人接受众人无限的祝福,留下永恒的回忆。

It is a theme song about love stories that can never be tired of listening to and a starting point of a journey belonging to the two who truly love each other.

In the modern century, a wedding dinner is a place where to share the happiness of a new couple with all relatives, friends and honorable guests. When the space for the wedding dinners is planned for this case, a composite and multiple form is applied —— including the conceptions of the weddings and commemorations. The image package is to create a topic on making the contents of the platinum-level value services and the space experience on the new delicious foods. The space style is demonstrated by the application of the classic simplicity, such as resplendent gold, elegant gold, Venice morning gold and ribbon red. The platinum banquet expects to provide the guests with a noble and special courteous reception. in this banquet for the commitment to make an appointment till death, the new couples can receive the infinite good wishes and be left with the eternal memories.

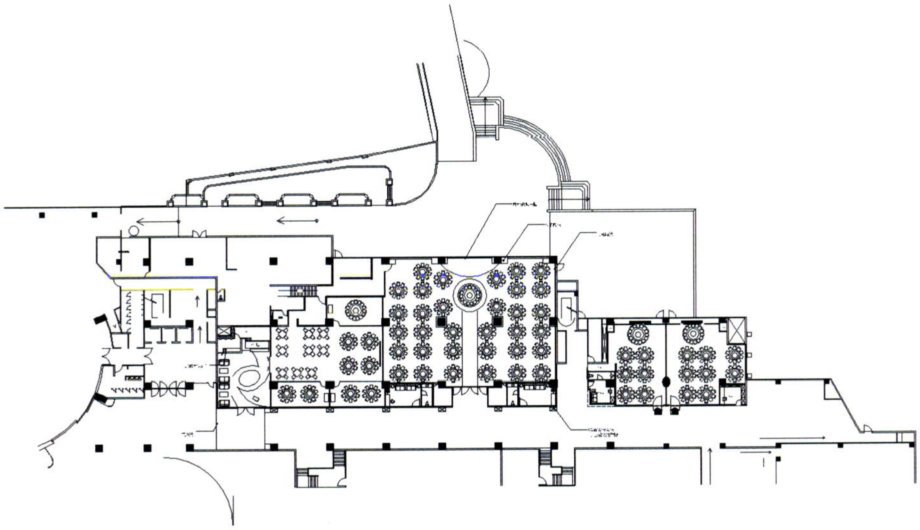

设计单位:如恩设计研究室 / 设计:郭锡恩、胡如珊

黄浦会高级餐厅
BEIJING WHAMPOA CLUB

面积:2080m² / 坐落地点:北京 / 主要材料:石英石、南非胡桃木、不锈钢装饰天花 / 摄影:Derryck Mene

黄浦会高级餐厅位于北京金融街一家经过改造的传统四合院中，周围高楼大厦林立。设计的关键问题在于如何将传统的中国四合院模式与餐厅巧妙地结合起来，其目的在于将现代的餐饮体验与中国的传统烹饪技艺形成对照。为此，NHDRO探索了传统四合院不同的利用方式，提供了新式、老式，新颖、陈旧，明亮、昏暗，敞开、封闭的固定不变的并列式布局，在建筑体验中来说明这种烹饪概念。

进入餐厅通过走廊，就会进入一个完全洁白的空间，庄严肃穆。空间的纯净将大家的注意力吸引到建筑的细节上来，而不是通过华丽的形象将其模糊。白色的走廊在过渡到各个精致目的地之前让人们的眼睛得以休息。

与白色走廊形成强烈对比的是黑色的酒吧。这里传统的中式屏风已更换成一种定制的图案—— 交互重叠的水圈，将水从天井中引入，并重新进行了诠释。包间的颜色各不相同，采用定制的清漆屏风进行分割。中国文化的典雅在最北的房间表现得尤为突出，从历史角度来说是最醒目最精致的。

经过重新改建的四合院，下面开凿出一个停车场，这种结构被视为“半历史的”。正如一个对最近的拆迁潮流和对于历史保护漠视现象的评述中所说：第一个庭院已经挖空并通向地下，从而导致了地势下降，接着安装上玻璃，形成一个注满水的反射池，宛如静静地沉思着消失已久的四合院。

灯光穿过上方“损坏的”天井并反射出公共就餐空间中成千上万的不锈钢圆盘。不锈钢圆盘同时反射了人们各自的影像并一瞥上方空无一物的天井。这种下部天井由单一的水流构成，与上方的水池形成对应的情景，并用黑檀木屏风来重塑在传统上环绕天井布置的大厅和房屋。屏风外面的空间提供了半私密就餐空间以及回忆四合院的机会。

Beijing Whampoa Club is situated in a reconfigured traditional Chinese courtyard house amidst a cluster of modern curtain wall high rises on Beijing’s financial street. From the onset, the overriding question concerning the design was how to “properly” engage the traditional Chinese courtyard typology with a restaurant whose aim is to contrast a modern dining experience with traditional Chinese cooking techniques. To address this, NHDRO explored different ways of occupying the traditional courtyard house, providing a constant juxtaposition of new/old, fresh/stale, light/dark, open/enclosed, exemplifying the culinary concept in an architectural experience.

Upon entering the restaurant and traveling through the corridors, one is surrounded by a completely white space. Solemn and serene, the purity of the space draws attention to the Chinese construction details rather than obscuring it with colorful imagery, as in history. The white corridors provide rest for the eyes before their transition to the various decadent destinations.

In contrast with the white corridors, the Bar is all black. Here, the traditional Chinese screen has been replaced and re-interpreted with a custom-made pattern – rings of water overlapping one another, drawing in the water from the courtyard.The Private Dining Rooms are each a different color, framed by lacquered custom screens. The decadence of Chinese culture is evident with the northernmost room – historically the most auspicious – the most elaborate.

The project site is actually a courtyard house re-built where one used to stand, with a parking garage trenched below. For this reason the structures are considered “semi-historical”. As commentary on this recent demolition trend and disregard for historic preservation, the first courtyard was hollowed out and opened to the level below, resulting in a physical lowering. Glass was then installed to create a reflection pool filled with water, a peaceful setting in memory of courtyard houses long gone.

Light streams through the 'broken' courtyard above and reverberates across thousands of stainless steel discs in the Public Dining Room. The steel discs reflect one's own image as well as allowing glimpses of the gutted courtyard above: a reminder. This lower courtyard is framed by a single stream of water creating a dialogue with the pool above – as well as ebonized wood screens which function to recreate the halls and house that traditionally surround a courtyard. Spaces beyond the screen allow for semi-private dining and the opportunity to look back into the courtyard.

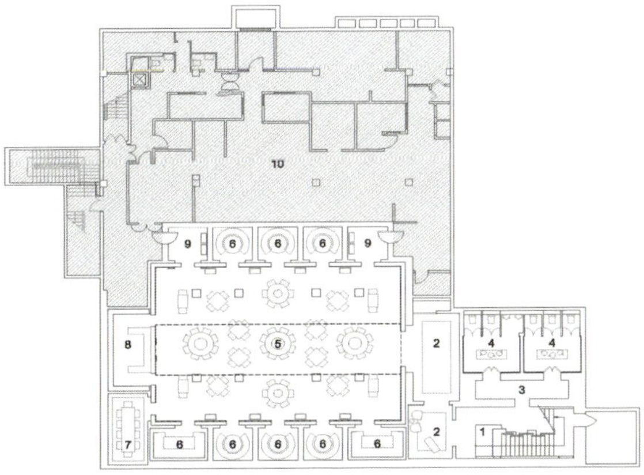

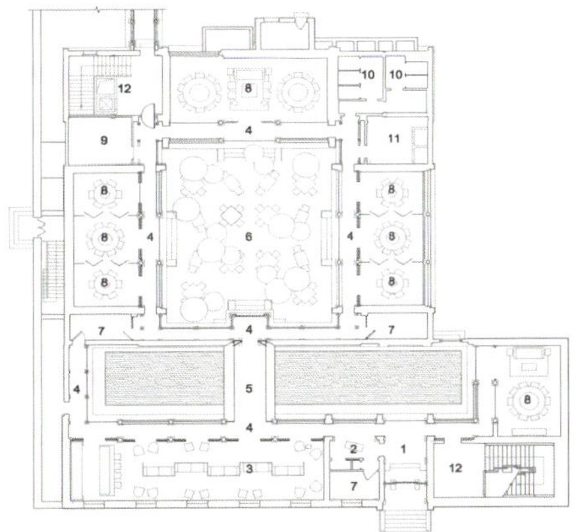

设计:陈旭东

星月湾时尚快捷酒店
Xing Yue Wan Hotel

面积:3000m² / 坐落地点:长春 / 主要材料:爵士白理石、玻璃钢、地毯、壁纸、玻璃、银镜

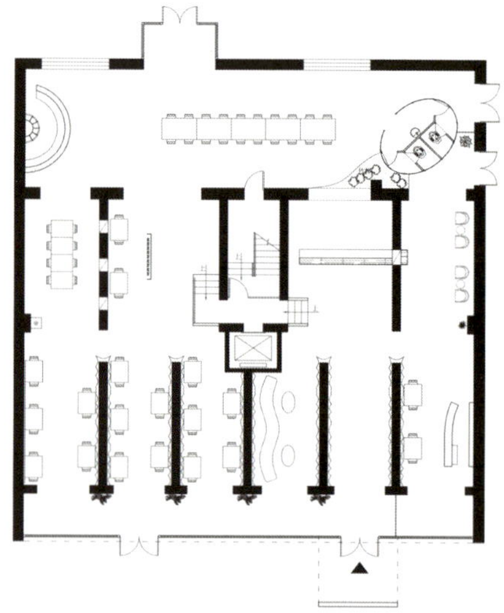

在最初构思时，设计师首先考虑了周围的环境因素，星月湾酒店位于长春市内的最佳风景区——南湖的旁边。南湖内水波荡漾，昼与夜的风景迥然不同。设计师结合星月酒店一贯的经营理念，以及注重带给入住者舒适睡眠的倡导，最后决定用“星、水、月”做为设计主题，这也正好契合酒店“星月湾”的店名。

在实际设计中，最制约设计师的是原来旧车库中的隔墙。在这里用“水”的元素与原结构中的实墙结合，不仅给规矩的室内空间带来动感，也消除了普通砖墙呆板的感觉。原有的车库开间正好用做餐厅隔间，创造小范围的用餐气氛。墙面上的开洞顺应墙面“水流”的动势，是“水中月”在空间的具体体现。洞与洞之间，环环相扣，在正对时可以一望到底，扩展了空间在视觉上的延伸感，给用餐环境增添了不少趣味。

在色彩使用上，整体色调以白色与暗红色的泥土色为主。白色让人联想到皎洁的月光，配合柔和的灯光，能产生直达人心安宁、静谧的力量。砖红色代表东北地区特有的“高天厚土”的环境，给人温暖与踏实。月光与泥土之间，天与地之间，晶莹的水晶灯点缀其中，给安适的空间带来些许跳跃、活泼的感觉。

In the initial conception, the designer first takes the surrounding environmental factors into consideration. Xing Yue Wan Hotel is located beside Nanhua Lake, the best scenic spot inside Changcun City. The water waves in Nanhu Lake undulate with the widely different scenery between the daytime and nighttime. In combination of the persistent business idea of Xing Yue Hotel and the advocation on the emphasis on bringing the visitors with comfortable sleep, the designer finally decides to apply "star, water and moon" as the design subject, which right corresponds to the hotel name "Xing Yue Wan".

In the actual design, the most condition that restricts the designer is the existing partition wall in the old garage. Here, the combination of the element of "water" and the solid wall in the original structure not only brings the innervations to the regular interior space, but also eliminates the stiff sense of the normal brick wall. The bay of the existing garage is just used as the partition of the restaurant to create a dining atmosphere in a small area. The penetrations on the wall just conform to the movement of "streamline" on the wall, which is a specific embodiment of "the moon in the water" in the space. The holes are cross linked. It can be seen through completely at one sight when directly confronted with it, which expands the extendibility of the space in vision and also adds more fun to the dining environment.

In the application of the colors, the whole tone gives priority to the earth color including white and dark red. White can associate the people with the clear and bright moonlight. In cooperation with the soft lights, it can create a peaceful and quiet strength that can directly touch the heart. The brick red represents the environment of "high sky and thick earth" that be unique to the northeast of China and brings people with warmness and firmness. The brilliant crystal lamps are decorated among the moonlight and earth as well as the sky and the ground to bring a certain jumping and vivid mood to the quiet and comfortable space.

设计单位:海岳建筑装饰工程设计有限公司 / 设计:姜湘岳、吴海燕 / 参与设计:纪红、徐云春、钟铮、潘婷

国品燕鲍翅馆
Guopin Swallow's Nestles and Shark Fin Restaurant

面积:2000m² / 坐落地点:南京中山陵 / 主要材料:大西洋灰石材、黑色钢琴漆、日本特色墙纸、酸枝木 / 摄影:莫尚勤

设计之初的立意，希望营造一个既高贵又轻松的感觉。高贵的气氛往往会和庄重的深色联系在一起。于是我们运用了一个极色——“黑”。黑色和温暖的木色作对比，更能体现木色的纯粹。渐渐地“黑”成为了背景，木色开始唱主角，轻松的气氛随之慢慢体现。

采用类似木地板铺设的方式，将夹板错落并追求反差的拼贴，使得流水般的纹理、暖暖的咖啡色，在黑色背景前特别具有戏剧的味道。以梅花、鸟、树这三个要素作为装饰要素，运用在了墙面、地面及顶面上，力求在室内的气质、氛围上贴近中山陵的自然气息。

在灯的做法上，基本已抛弃原来灯具加工的方式，自己设计加工，自己组合，让艺术品和照明结合起来，尤其水晶树枝和鸟都飞上了顶部，给来客带来了自然的惊喜。墙面上的画，都是让摄影师到中山陵寻找可爱的树拍下来的，它们极为细密的交叉树枝，让人深深吸引。在一楼有一堵墙，上面停满了色彩斑斓的鸟，那是我的最爱，这次终于有机会将它们凝固在我的作品里。

室内作完之后，我们又对室外进行了调整。夜晚将远处的水杉树用灯光照亮，有了光，可爱的树在夜晚也有了展现的舞台，让来此的客人能充分享受到中山陵树影婆娑的姿态。

我们的设计，一切都围绕这可爱自然的精灵。

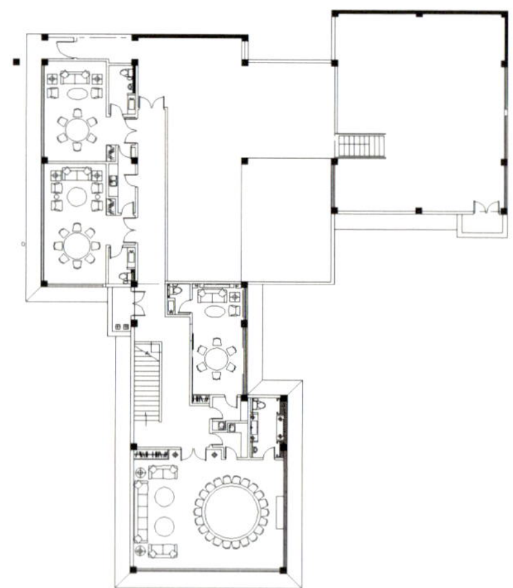

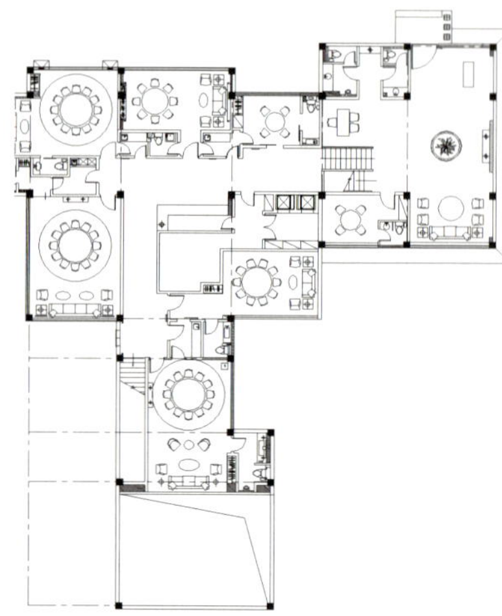

The initial purpose of the design is to create a nobler and comfortable feeling. The noble atmosphere is always connected with the grand dark colors. Therefore, we have applied an extreme color "black". The contrast between black and the warm wood color can better reflect the purity of the wood color. Gradually, "black" becomes the background and the wood color starts to play the major role. Consequently, the comfortable atmosphere is gradually reflected.

We have applied the method similar to the layout of wooden floor to alternate the plywood for contrasted assembly so that the stream-like texture and the warm coffee colour have a special dramatic atmosphere in front of the black background. The three elements including plum blossom, birds and trees are taken as the decorative elements which have been applied to the wall, ground and the ceiling to make the interior atmosphere as close to the natural style of Sun Yat-sen Mausoleum as possible.

The fabrication of the lamps has mostly given up the original fabrication way of the lighting fixtures. The self design, fabrication and assembly have integrated the arts with lighting; especially the crystal tree branches and birds have flied to the ceiling to bring the visitors with the natural surprises. The pictures of the lovely trees on the wall are taken by the photographers in Sun Yat-sen Mausoleum, in which the closely crossed branches are very impressing and attractive. There is a wall full of the colorful birds, which is my favorite. I have liked them since I was a kid. This time, I finally find an opportunity to freeze them in my work.

After the completion of the interior work, we also have made an adjustment on the exterior work. The distant metasequoia trees are lightened up by the lights at night. With the lights, the lovely trees find their exhibition stage at night and the customers who come here at night also can fully enjoy the dancing trees and their shadows of Sun Yat-sen Mausoleum.

Our design is fully focused on this lovely natural goblin.

设计单位：上瑞元筑

外婆人家
Grandma's Restaurant

面积：2905m² / 坐落地点：无锡 / 主要材料：乳胶漆、仿古砖、天然石材、实木 / 摄影：文宗博

外婆人家引用了江南水乡的街景画面，描绘出一幅粉墙黛瓦，竹林深处有人家的幽美画卷。

空间是以简洁而单纯的白色为主调，用现代、精练的体块关系营造出一处处房屋，一道道回廊院墙，形成整个场景的大布景。将整个大厅变成一个表演的舞台，邀朋唤友，不亦乐乎。设计师独具匠心地把两处明档置于其中，将炊烟袅袅，香气扑鼻的街景氛围烘托得更为生动，让人不禁唤起儿时的记忆。

意象的竹林，潺潺的流水，形成一道天然的屏风，几步之遥，却悠远僻静。走入深深小巷，闹中取静，别有一番情趣。

Grandma' s Restaurant has introduced the vista scenes in the water town of the South of the Yangtze River to show a beautiful picture where there are white walls and black tiles and families in the deep of bamboo forest.

The space is taken with the concise and simple white as the main color. The modern and polished block relationship is applied to create the houses here and there and one-after-another winding corridors and enclosures and form the big scene of the whole setting. The whole lobby is changed to a big performance stage. It will be a big pleasure to invite the friends here. The designer has specially put two show kitchens among it so that the vista atmosphere with the smoke from kitchen chimneys curling upwards and sweet smells greeting are set off more vivid so that people cannot help calling out the memories in the childhood.

The imaginary bamboo forest and purling water has formed a natural screen. Although there are only several steps away, it is quite remote and quiet. Walking into a small alley and feeling the quietness in the nosiness, you will get another kind of enjoyment.

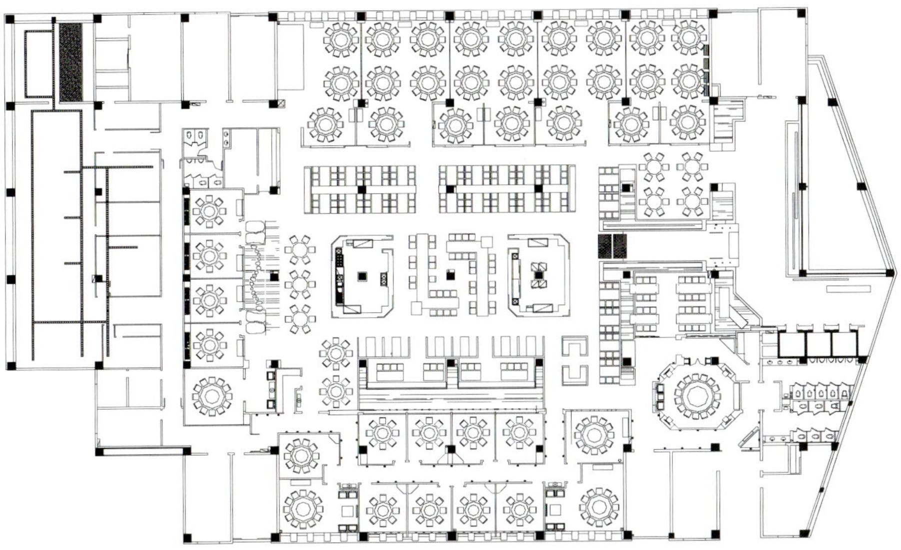

小妻巷

设计单位:BEIJING NDC STUDIO, INC / 设计:迫 庆一郎

蜂巢印象中餐
Epicurean Element

坐落地点:深圳市 / 摄影: Misae Hiromatsu

餐厅主要以粤菜、杭帮菜、湘菜为主，由日本专业店面形象设计师设计，红色的墙面与白色的“蜂巢”相间，每个蜂巢里面就是一个小小的包厢，可以坐六七人。风格独特之余依旧不忘人体美学、餐饮美学的要求，既饱了眼球和味觉的享受又给人典雅舒适的享受。

The restaurant is mainly involved in Cantonese food, Hangzhou food and Hunan food. It is designed by a professional Japanese shop image designer. Red wall and white "honeycomb" are alternated. Each honeycomb has a small VIP room inside with a hold of six to seven people. Besides the uniqueness of its style, it still remembers the requirements of the human and catering aesthetics. It not only provides a visual and tasting pleasure, but also a elegant and comfortable enjoyment.

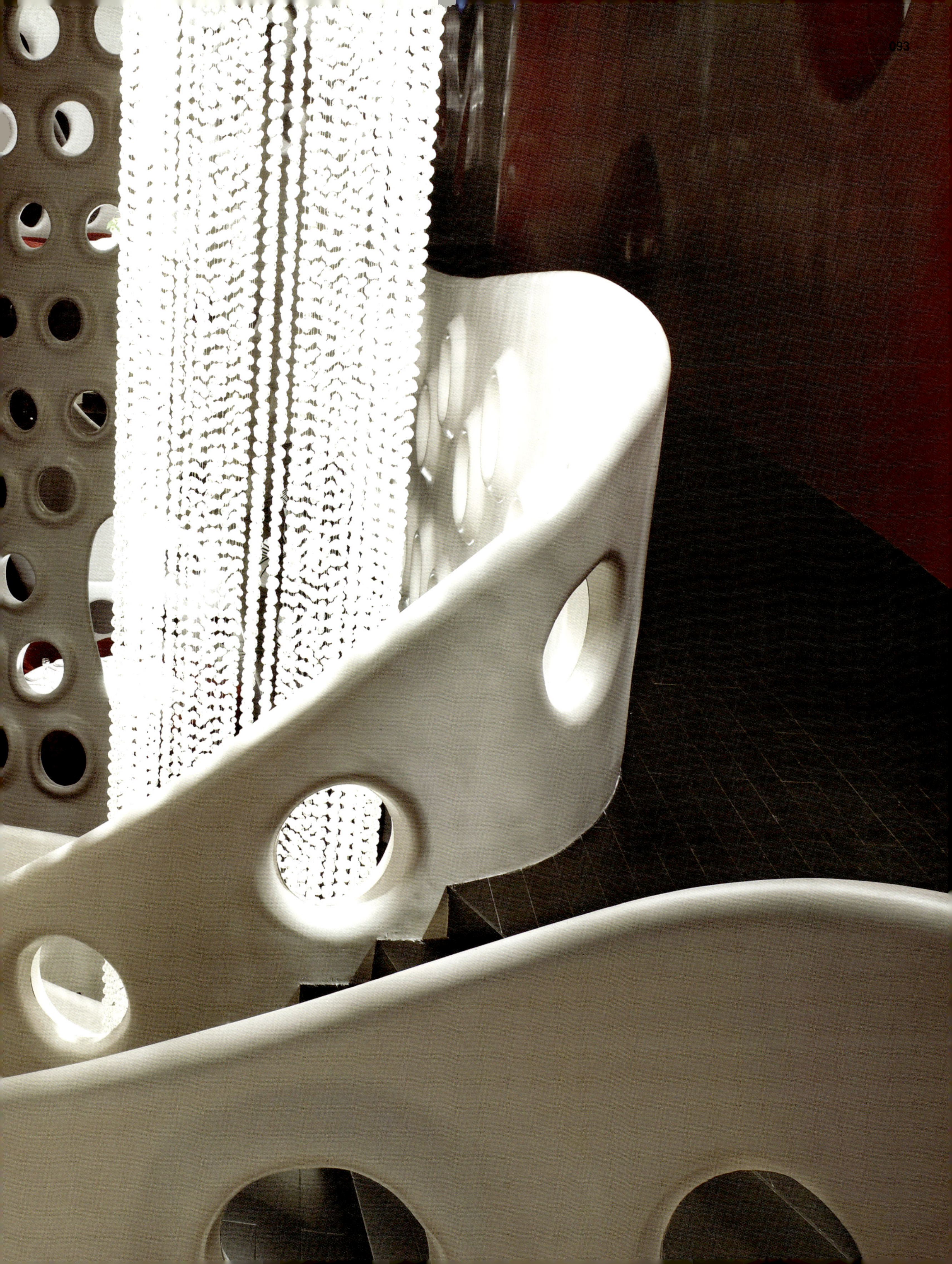

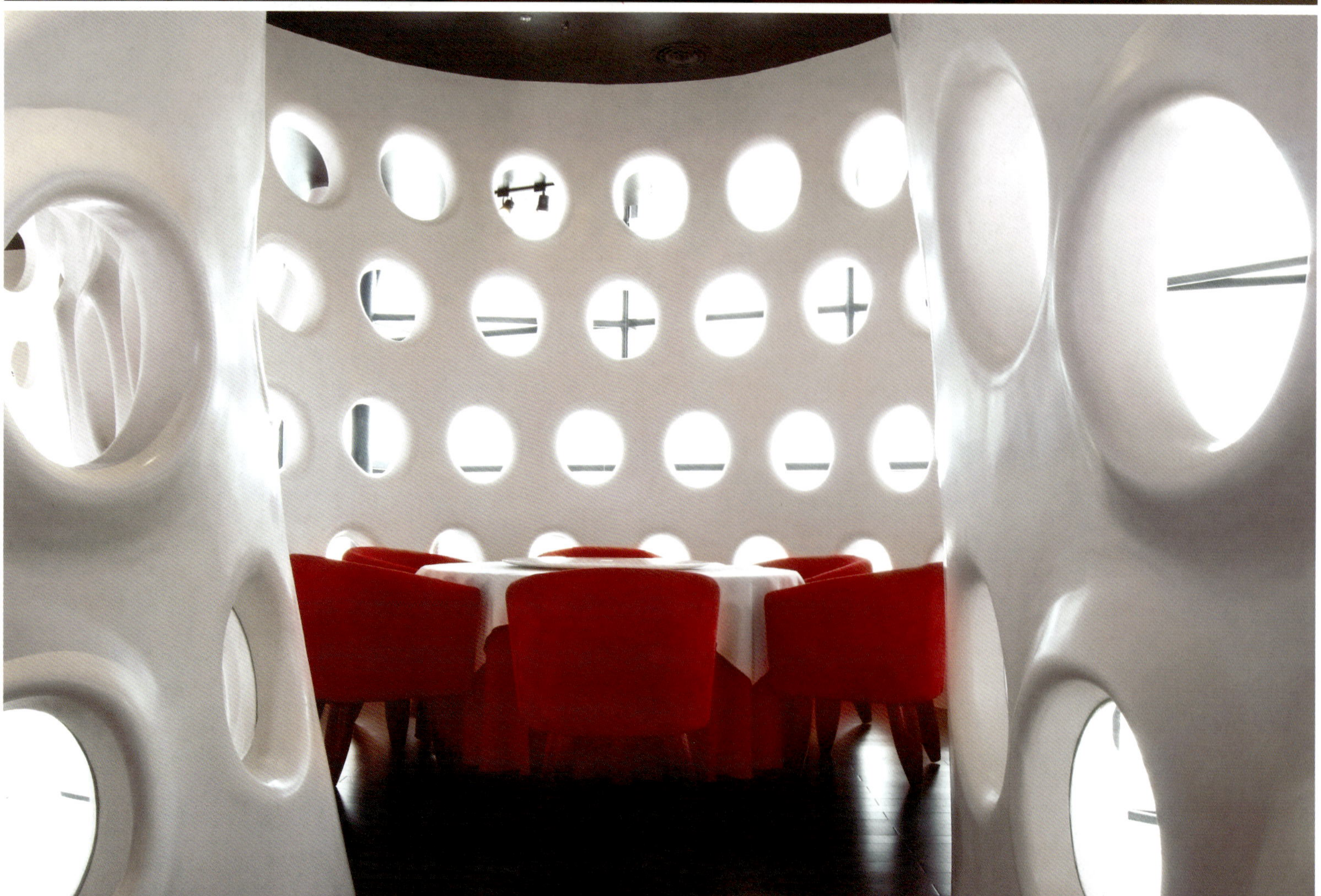

设计单位:哈尔滨唯美源装饰设计有限公司 / 设计:张奇永

非尝食间
Fei Chang Shi Jian

面积:700m² / 坐落地点:哈尔滨上游街 / 主要材料:地砖、柳条、水泥 / 摄影:张奇永

子非尝，焉知食间之美，故得名“非尝食间”也。非尝食间为一家专营粤菜与花式饮品的休闲美食空间，位于哈尔滨著名风景旅游名街中央大街与上游街交汇处。就其地理位置的特有性与经营内容的专属为此项目的构思主线，来进行对空间外在形式的塑造与内在灵魂的赋予。

在岁月的长河里，无论是谁都会对时间有所记忆，它可有形亦可无形，那么在此空间中留下这片记忆的便是用柳条编织的“非常空间”，以便共证岁月的记忆。

设计的灵感主要来自音同字不同的“非常时间”一词。“非常”二字，寓意着空间的专属性与不寻常之处。而“时间”二字则有趣的被分为此时与彼时。那么对于这个空间来说也就有了所谓的“此一时彼一时”的内在灵魂与“非常”的外在形式。

因我们来自不同的境遇，有着不同的价值观与世界观。所以我们就像墙面上的一片片大小不一、位置不同的镜片和充满无极的时空数字一样，相互反射着，彼此穿梭着，想必每一个人都会在此找到适合自己的那一片镜片与数字，尽在此时彼时间。

How can know the deliciousness of the foods if you do not try them? Therefore, it acquires its name as “Fei Chang Shi Jian”. Fei Chang Shi Jian is a space for leisure and delicious foods mainly engaged in Cantonese foods and fancy drinks. It is located at the crossroad of Central Street - the famous scenic street in Harbin and Shangyou Street. The peculiarity of its geographic location and the exclusivity of its business items are applied as the main conceptive line for this project to carry out the creation of the exterior space forms and the endowment of the internal spirits.

In the long river of the years, everybody has some memory on time whoever they are. It can either physical or immaterial. In this case, what has maintained this memory is the “special space” woven by the wickers in convenience for witnessing the memory on the years together.

The inspiration of the design mainly comes from the word “special time” in same pronunciations but in different words. The word “special” implies the exclusiveness and unusualness of the space. However, the word “time” is interestingly divided into this time and that time. This space thus also has the so-called internal soul of “this is one time and that was another situation” and the “special” external form.

As we come from different situations and have different values and world views, we, just like the pieces of mirrors in different sizes and different locations as well as the space-time numbers filled with infinity on the wall which reflected and crossed with each other, most probably can find one piece of mirror and number here fit for ourselves, all in this and that time.

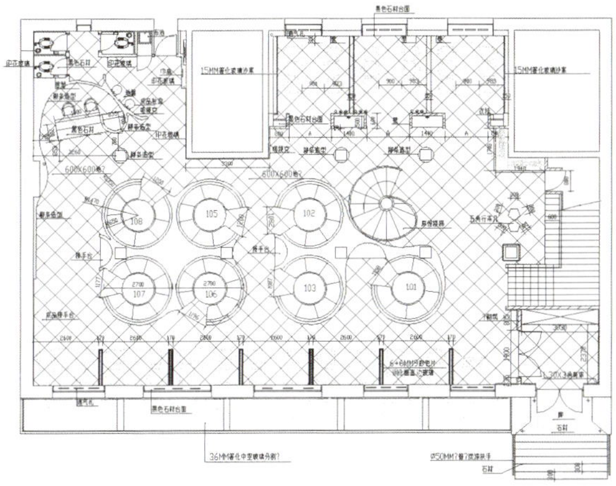

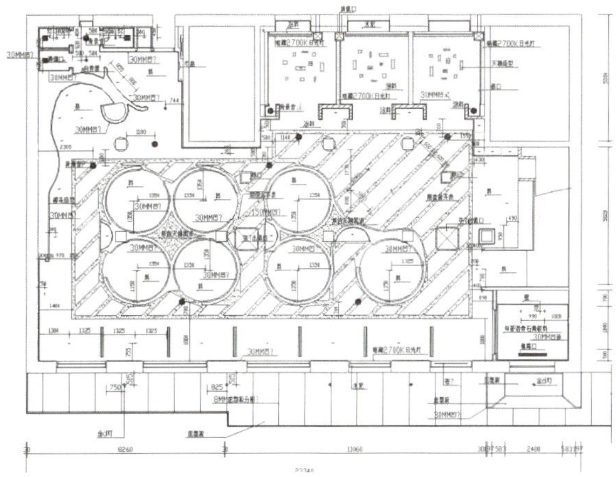

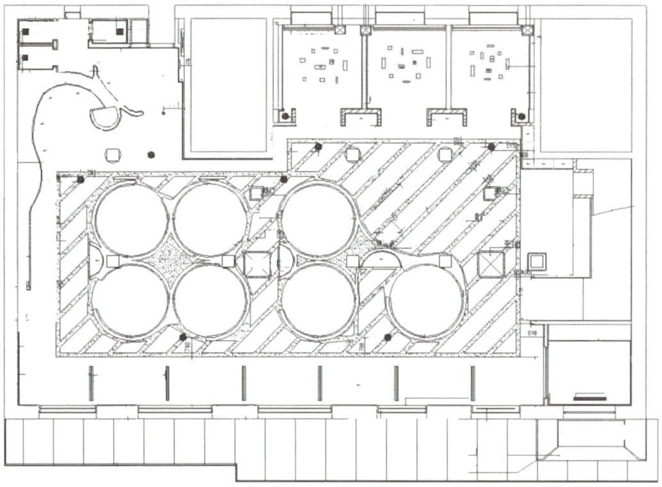

设计单位:龙域设计 / 设计:徐侯聪

蓝轩餐厅
Lanxuan Restaurant

面积:800㎡ / 坐落地点:上海 / 主要材料:毛石、青砖、镜面、地砖、木地板 / 摄影:胡文杰

建筑分两层。一层几乎都是落地窗，窗外的翠竹青砖、溪水潺潺尽收眼底。水天之间的白色建筑恰如其分地贴合江南淡妆素裹的优雅情怀。借由木制小桥步入一层大厅，门厅以东为操作区，以西为用餐区，一条圆弧形走道隔出两个下沉的就餐区。一面餐桌沿水面摆开，透过细密的竹帘，别有一番情致。二层主要是包间，需要通过室外楼梯单独进入。一、二层的共通空间，设计师大胆启用红色作为悬挂吊灯的主色，既中和了相对冷静的氛围，又为空间注入些许时尚气息。简单纯粹的设计内敛沉稳，尽显灵秀之气。

The building is divided into two floors with the first floor erected with almost complete French window. You can take in green bamboos, gray bricks as well as the gurgling water outside the window. The white building between the water and sky has properly matched the elegant feelings of the light make-up in white in the south of Yangtze River. You can enter the hall at the first floor by the wooden bridge. The west of the entrance hall is the dining area while the east is the operation area. Two sunken dining areas are formed by an arc corridor. The dining tables are arranged along the water with a revelation of a specal mood through the dense bamboo curtains. The VIP rooms are in the second floor with the separate entrance through the outdoor staircase. In the common space between the first and second floors, the designer has boldly applied red as the main color for the pendant lamps, which has not only neutralized the comparatively cold atmosphere, but also has injected certain sense of fashion. The simple and pure design is introverted and grand with full reflection of delicacy and beauty.

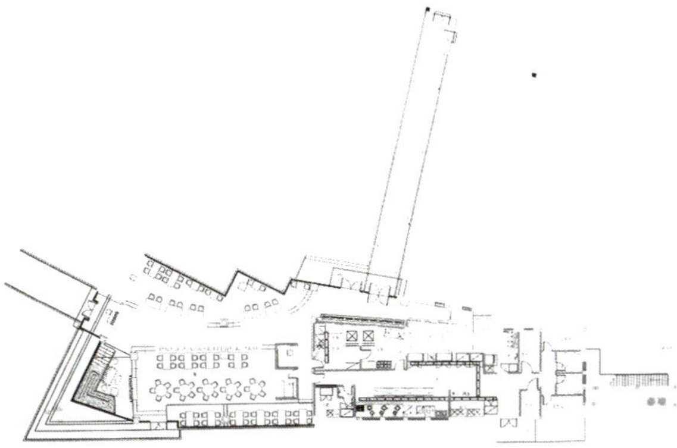

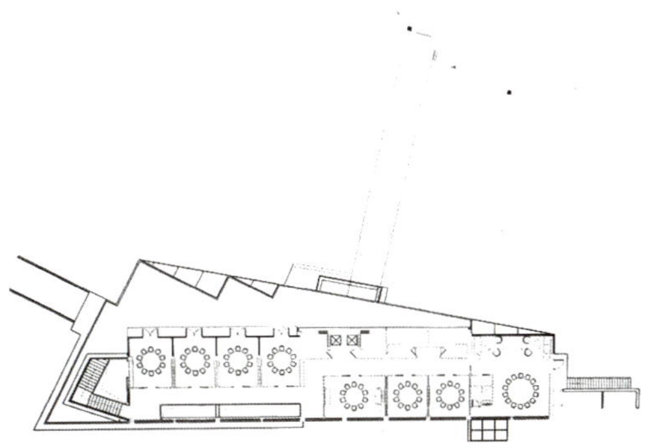

设计单位：哈尔滨唯美源装饰设计有限公司 / 设计：王仲文

涮库
Shuan Ku

面积：1000m² / 坐落地点：哈尔滨新阳路 / 主要材料：灰色砖、冷轧铁板、铁网饰面、不锈钢马赛克 / 摄影：张奇永

“士翔涮库”这个名字在设计初期就已确定下来，设计师从“库”字入手，运用有关的符号语言加以运用。

主入口处的入户门用冷轧铁板做方形凹格处理突出了工业感，二层外阳台用铁网封闭加装洗墙灯，形成一个虚空间，丰富了外立面的效果又突出了主题。

进入大厅后，右面的墙用灰色砖直接砌筑而成，墙面上一排排的方孔增加了各空间的通透感。在左面临窗一侧用角铁做骨架，表面用铁网饰面做隔断，隔断下部安装射灯来突出铁网效果。这样既可以使客人之间不直接接触，又丰富了空间的细节。大厅地面全部处理成水泥压光，在通道中间镶嵌不锈钢马赛克导引线，起到空间导引作用。在主墙面悬挂具有现代风格的壁画，随墙面的凹凸做了起伏变化，增加了分割后的时代感。

The name of “Tu Xiang Shua Ku” has been decided at the preliminary design stage. The designer has started with the word “Ku” and applied with the relevant symbolic languages.

The entry door at the main entrance is made of cold-rolled iron sheets in a form of square concave grid to address the industrial sense. The exterior balcony at the second floor is closed by the iron mesh with additional wall washer lamps to form a visional space that not only enriches the effects of the exterior facade but also emphasizes the theme.

After entering the hall, the wall in the right side is built directly by the gray bricks. The square holes in rows on the wall have increased the penetrating senses. In the left beside the window, the angle steel is used to make a skeleton with the iron mesh facing as the partition on its surface. The spot lights are installed at the lower part of the partition to highlight the effects of the iron mesh, which not only can avoid the direct contact between the guests, but also enriches the details in the space. The ground in the hall is arranged as cement grinding and stainless steel guiding lines for mosaics are inlaid in the middle of the passage to guide the spaces. The murals in modern styles are hung on the main wall with the undulating changes along with the roughness of the wall to increase the times senses after the division.

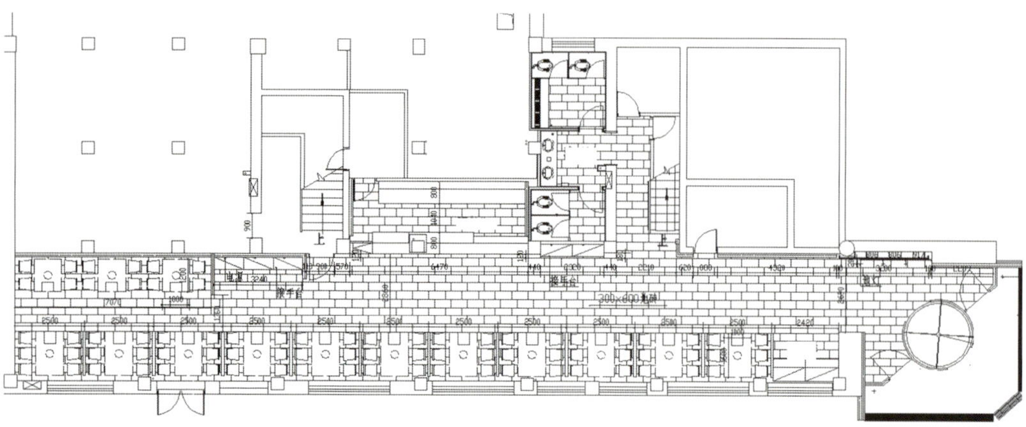

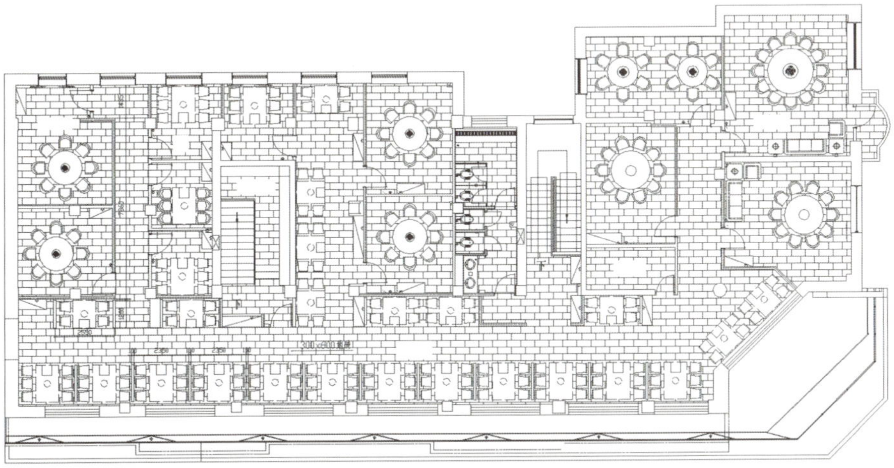

设计单位:宁波市高得装饰设计有限公司 / 设计:范江 / 参与设计:卢忆

天一国际豪食汇美食中心
Sky International Grand Cuisine

面积:1200m² / 坐落地点:宁波 / 摄影:范江

天一国际豪食汇美食中心位于宁波天一广场天主教堂的底层,梁柱多、管线多、采光差,建筑极其不规则,与要经营的自助餐厅的空间条件严重不符。设计师根据建筑的不规则,选取以圆形柱网为中点,贴切地划分了不同种类的备餐区,有主食区、饮品区、点心区、水果区等等,用餐区有能看到整个烹调过程的卡座、能欣赏表演的座席、家庭座、可上网的区域等等。有此设计,空间的缺点变成了特点。

设计师想营造有别于带商务性质或进驻酒店的那种自助餐厅的程式化风格,它是浪漫的、怀旧的,有着淡淡的文化气息与较高的品位。其中一面墙挂着世界各地的教堂的黑白照片,与外部环境(教堂)形成一种默契的和谐。因原先的采光不好,现在索性处理得低调而有情致,定做的折皱布灯做为吊灯、壁灯,通过不同的组合方式显现统一、显现变化。

卫生间及其前面的通道,虽处在比较偏的位置,但设计师却处理得格外用心。6个卫生间无性别差异,主背景是整幅黑白摄影作品,座便器在画中凸现,有个性但不突兀。靠边的墙面是大块银镜犹如试衣镜,在这个小空间给人以艺术的虚幻感。正对卫生间的一排墙面是大玻璃嵌着五幅20cm×20cm黑白硬板画,这些硬板画由一幅幅《乱世佳人》、《蝴蝶梦》、《小玩意》、《七年之痒》等经典剧照构成,如长长的电影胶卷,让人有一种熟悉的感觉,这是一个一去不复返的年代。

Sky International Grand Cuisine is located at the ground floor of Catholic Church of Sky Square in Ningbo. It has many beams and columns, pipelines, bad lighting as well as extremely irregular building, which are greatly in unconformity to the space conditions required by a buffet restaurant to be operated. The designer has selected the round column net as the central point according to the irregularity of the building to appropriately divide different types of food preparation areas including main course areas, drink area, desert area and fruit area etc. In the dining area, there are the seats where you can see the whole cooking procedure and enjoy the performances, as well as the family seats and the area where you can browse the internet. The shortages of the space become the characteristics though the design.

The designer intends to create a style different from the programmed type of buffet restaurant of business characteristics or those stationed in the hotels. It is romantic and nostalgic with the light cultural atmosphere and quite high taste. One of the walls is hung with the black-and-white photos of the churches around the world, which forms a tactic harmony with the external environment (the church). As the original lighting is not good, now it is directly arranged in a low-profiled and appealing way. The customized folded cloth lamps are installed as the pendant lamps, wall-mounted lamps, which has displayed a uniform and obvious change in the different ways of arrangements.

Although the toilets and their front passages are located in a quite remote area, the designer has arranged them with cares. Six toilets have not sexual differences and the main background is a view picture of a black-and-white photograph. The toilet seats become prominent in the pictures. The wall directly opposite to the toilets is a big glass embedded with five 20cm × 20cm black-and-white hardboard pictures, consisting of the classic stage photos such as "Gone with the Wind", "Rebecca", "Little Toys" and "Seven Year Itch " et. It is just like a long cine film and the people feel familiar with it. It is an era that never comes back again.

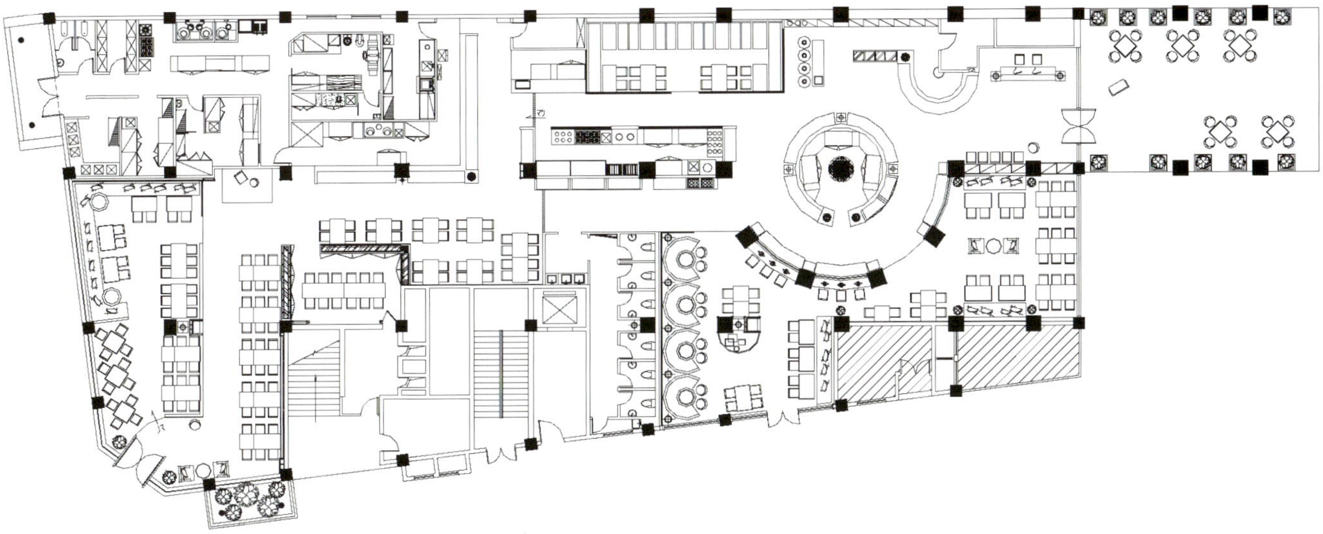

设计:Mauro Lipparini

上海天地一家外滩6号店
Shanghai Tiandi Restaurant, Bund No.6

面积:1300m² / 坐落地点:上海 / 摄影:胡文杰

业主希望，“天地一家”在每个城市都能体现那个城市最浓郁的地方文化。于是，在视觉元素上，体现了上世纪二三十年代上海最兴盛时期的装饰艺术派（ArtDeco）建筑风格成为本餐厅的视觉设计定位；而餐厅希望表达的气质，则是上海女人那种开放、优雅、精致、华贵的气质。外滩、ArtDeco、女人，成了这个餐厅的关键词。

气质——开放的气质体现在平面设计上，是平面布局的开敞和空间的序列感。精致，是空间所有的视觉元素都被很用心地控制住，相互呼应。优雅，是整体感觉大方简洁。华贵，是一些闪亮的东西吸引人，就像女人佩戴的珠宝。颜色——那个时代女人们用的妆粉的色调，调子之间过渡非常细腻。视觉元素——Art Deco作为平面的视觉元素，由特别为外滩店设计的Logo演绎出的平面造型出现在壁画、天花、地面、隔断、门把手、盘子、筷子架、菜单甚至是牙签盒上。图案的主题和变形出现在各处，相互呼应。玻璃和水晶装饰的方式借鉴了珠宝的切割，纯手工制作，将光线迷离地折射开。

电梯出来，黄浦江景扑面而来，窗前是舒适的沙发休息座，收罗的古董Art Deco家具带来了一丝怀旧的感觉。右手通向VIP包间，通直纹路的黑檀护墙板与房门融为一体。走道尽头是酒窖，艺术玻璃让酒窖显得年代久远。左手是进入餐厅的过厅，空旷将整面石雕Logo的主题墙拉出了纵深感，而精美的手工刺绣服饰带来了精致与华丽的视觉感受。餐厅的入口是设计师最满意的地方，暗色的柱和柱间的光带形成了透视的效果，正中的艺术玻璃将餐厅内部的景象折成马赛克效果，让人联想到上世纪初大西洋航线上那些豪华邮轮的内部设计。进入餐厅，最引人注意的当然是那八根水晶光柱，轻盈透亮而又有坠感，在素雅的空间中熠熠生辉；白色的圈椅是赖特的设计，简约但又极富形式感，不愧为20世纪初的经典作品。为了体现Art Deco的岁月沧桑感，餐厅中的灯具和一部分的家具是设计师从法国和意大利的古董店中淘回来的1920年代的真正的古董。VIP包间的地面，用的是鸡翅木地板，人工打磨烫蜡，很有点做旧的感觉。在窗帘、台布甚至餐巾的选择上，都遵循了咖啡色的色调，和墙面的颜色呼应，也是对女人肤色的衬托。

在餐厅里，弥漫的是一种淡淡的怀旧，精致而舒缓的气息。时间仿佛停住了，停在黄金的年代。在这个什么都快的时代，慢就是一种奢侈。静下心来，慢慢地品。就像老上海的女人，没有什么惊天动地的业绩，却用女人的精致和细腻衬托出上海滩那些波澜壮阔的故事。

The client hopes that “Tiandi Restaurant” can demonstrate the densest local culture of the city where it is located. Thus, in aspect of visual elements, to demonstrate the architectural style of Art Deco in the most prosperous period of Shanghai in 1920s~1930s has become the visual design orientation of this restaurant; however, what the restaurant wants to deliver is the open, elegant, delicate and luxurious bearings of Shanghai women. Bund, Art Deco and women become the key words of this restaurant.

The bearing —— open bearing is reflected on the plane design. It is the opening of the plane layout and the sequence of space. Delicacy is that all visual elements in the space are deliberately controlled properly and echo with each other. Elegance is that the whole looks grand and simple. Luxury is some bright things to attract people, just like the jewels worn by women. Colors —— the makeup hues used by women at that time. The transition between the hues is very exquisite. Visual element —— Art Deco is taken as a visual element. The plan modeling deduced from Logo designed for this restaurant has appeared on the frescos, ceiling, floors, partitions, door handles, dishes, chopstick racks, menus and even toothpick boxes. The main subjects and transfigured forms appear everywhere and echo with each other. The decorative method of glass and crystals has drawn on the experience of the cutting of the jewelry, made by hand completely, to refract the lights outwards mistily.

The scenery of Huangpu River will greet you immediately once you step out of the lift. In front of the windows are the comfortable sofas for rest. The collected antique Art Deco furniture has brought a sense of reminiscent feelings. It is directly connected to VIP rooms on the right. The ebony clapboards with straight grains have integrated with the doors. Walking to the end is the wine cellar, which looks to a long history due to the artistic glass. On the left is the hallway to the restaurant. The spaciousness has extended the depth of the thematic wall with the whole stone-carved logo while the delicate hand-embroidered dresses have brought with the delicate and luxurious visual feelings. The entrance of the restaurant is most satisfactory to the designer. The dark columns and the light strips among the columns have formed into a perspective effect. The artistic glass in the center has refracted the scenery inside the restaurant into a mosaic effect so that the people associate with the interior design of those luxurious cruisers along the sea routes of Atlantic Ocean in the early of last century. When you are walking into the restaurant, the most attractive are of course the eight crystal light columns, lightsome, bright and with a sense of decoration. It looks brilliant in the simple but elegant space; the white round-backed armchairs are the design by Wright, simple but rich with sense of forms. It is undoubtfully a classic work in the early of the twentieth century. To reflect the sense of vicissitudes of years, the lighting fixtures and part of the furniture in the restaurant are the real antiques of 1920s sought from the antique shops in French and Italy. The floors in VIP rooms are Panga Panga floor boards which are polished and waxed by hand to provide a sense of old treatment. The selection of curtains, table cloths and even table napkins has always followed the coffee color to echo with the wall color, which is also a setoff to the women' s complexion.

The restaurant is permeated by a gentle reminiscent, delicate and relaxed sense. The time seems to be stopped in the gold ages. In this era that everything is fast, the slowness is a kind of luxury. Calm down and enjoy slowly. Just like the women in old Shanghai, they do not have any shocking achievement, but they have set off those magnificent stories in Shanghai Bund with the women' s exquisiteness and delicacy.

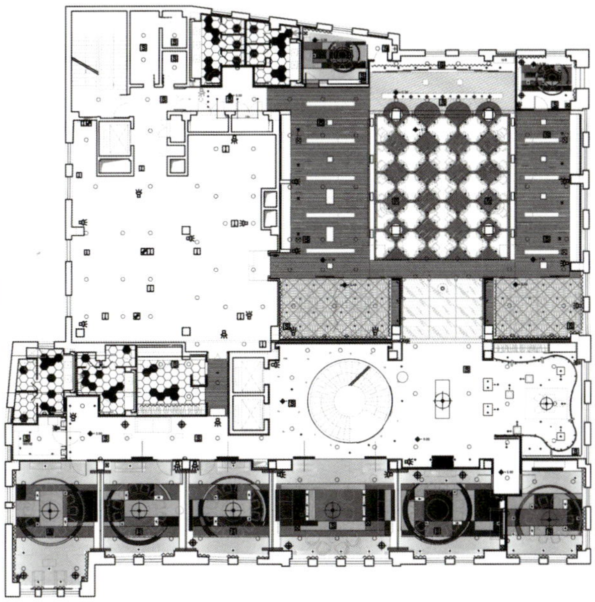

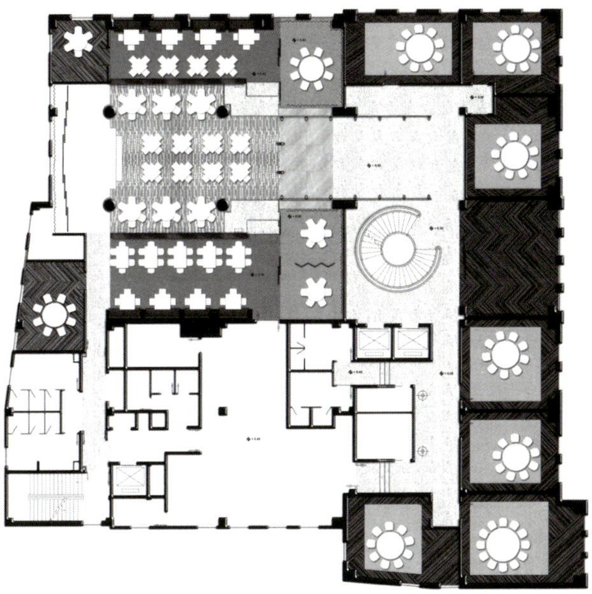

设计单位：无锡市上瑞元筑设计制作有限公司苏州分公司 / 设计：林孝江

福记名仕会高级餐饮会所 Fuji Celebrity Club

面积：2500㎡ / 坐落地点：苏州 / 主要材料：镜面不锈钢、木化石、水晶砖 / 摄影：文宗博

该项目业主投资定位为顶级餐饮会所，以包厢消费为主，辅以小型宴会厅。因此，我们界定了以江南地域文化、国际化的时尚表达为整个构思的出发点，整个空间的布局以江南园林平面构架为参照系，多灭点的布局手法让有限的空间出现相对丰富的视觉变化，表皮的处理手法力求营造出高档奢侈品专卖店的感觉。发光水晶墙所表达的珠宝盒的概念与抽象化的中式月洞门相得益彰，营造出梦幻般的体验感。

The owner has invested this project as a top catering club, mainly engaged in VIP rooms with small banqueting halls as a supplement. Therefore, we have oriented the fashion expression on the internationalization of the cultures in the south of Yangtze River as the starting point of the whole conception. The arrangement of the whole space has referred to the plan structure of the gardens in the South of Yangtze River and the arranging technique of multiple vanishing points has demonstrated the limited space with a comparatively abundant visual changes. The treatment of the surface strives to create a sense of a franchised shop for superior luxurious products. The idea of jewel case delivered by the luminescent crystal wall is properly echoing with the abstract Chinese-style moon door to create a dreamlike experience.

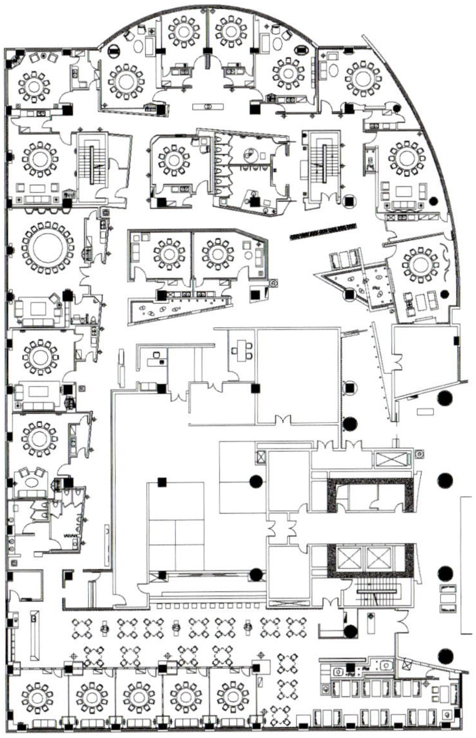

横看成岭侧成峰
Different Views at Different Positions

王琼 Wang qiong
教授 硕士生导师
金螳螂设计研究院 院长
苏州大学建筑与城市环境学院 副院长
中国建筑学会室内设计分会 理事
中国建筑装饰协会设计委员会 副主任
中国饭店协会设计装饰专业委员会 常务理事

何潇宁	宏安瑞士大酒店
叶铮	白玉兰宾馆
琚宾	唐悦酒店
郭珂蔚	上虞雷迪森万锦大酒店
上瑞元筑	云鼎花园酒店
陈石林	西湖四季酒店艺联店
殷艳明、胡威	杭州德嘉精致酒店
姜湘岳、吴海燕	金太隆国际酒店
高超一、张驰、康瑞德	豪生酒店
富元	香泉温泉度假酒店
叶铮	锦江之星——沈阳南湖店
	凯宾斯基酒店
王剑	皇廷国际大酒店
琚宾	富邦国际酒店
杨颜江、孙威、肖莹	云轩山庄

今年春节，我和家人与好友相约去马尔代夫度假，远离了工作的繁杂，远离了城市的喧嚣，远离了手机的骚扰，选择享受大海的温情。让我们感到惊讶的是，我们入住的悦椿庄竟然是一家没有电视机的酒店，Ravin的《Huvafen Fushi Maldives》从CD机里静静地流淌出来，瑰丽奇幻、充满海洋气息，让我们短暂的度假生活变得更加安宁与浪漫。

躺在松软洁净的床上，望着大海，放松而惬意。忽然间，我忘了自己是一个设计师，平时入住酒店时常常关注的功能性、规范、数据一概从脑中消失，它们变成了生活中的一种自然而然——触手可及的便捷、床的干净舒适、视线的开阔、风景的优美。这些，让我产生了一个想法：如果我不是设计师，我将如何评判这家酒店?

多年的室内设计工作让我养成了一个习惯，每次走进一家酒店，满脑子都是它的功能、流线、空间、造型、用材，经常会考虑如果让我来设计这家酒店，我将通过什么样的途径来吸引客人，如何满足不同客人的需求，更多是寻求一种酒店空间的表现力。如今每每念及此事，不禁莞尔，不知仅是职业习惯，还是已经发展到神经质的地步。

然而，在设计的过程中，一直有许多问题在困扰着我们，面对这些问题时，我们总是以设计师的身份来寻求解决方案，但是，当我们只关注设计层面的要素时，问题往往变得越来越复杂，反而是业主和管理公司的建议给了我们很多启示和帮助。这不得不让我们汗颜，让我们反思。

中国已经发展为全球酒店设计的最前沿、竞技场，青年旅社、国际性连锁酒店、高档会所、为特殊人士服务的精品酒店，五花八门，门类齐全。大家更多地考虑到自己作品的风格、个性表达、将其作为一个T型台，表现出自己强烈的个性，不断做秀，正在走着或即将走向形式主义的道路，却忽视了设计对市场、对人的服务要素。对于室内设计师来说，换个角度去观察、去思考酒店设计的问题，恐怕也是一种必然趋势。

设计本身是为“人”服务的，如果暂且抛开作为设计师的职业习惯，从“人”的角度去体验、感受一家酒店，去了解它如何为顾客提供更贴近“人”的服务，反而会对酒店设计得到更全面、更接近真实的认知。

衡量一个酒店的设计，不一样的人可能会有不一样的理解，但是他们对这个酒店的认识有一个共性，这便是酒店的功用性与商业性：一方面是顾客在酒店中能否得到安全的、卫生的、便捷的、舒适的服务；另一方面是业主的投资能否迅速得到回报。

设计来自体验，艺术源于生活。如果将来再次踏入一家酒店，作为一个设计师，我们是否可以尝试着以普通顾客、业主、管理公司、企业高管、资本家、艺术家等其他身份来进行一次消费与享受，来进行一次职业习惯以外的评价与思考呢?

My family and I have made an appointment with my friends to take a vacation in Maldives during this Spring Festival, keeping far away from miscellaneousness of the work, noisiness of the city and disturbance of mobile phones to enjoy the warmth of the sea. To our surprise, Angsana, what we have been staying is a hotel without television. “Huvafen Fushi Maldives” by Ravin is running out of CD player quietly, magnificent, fanciful and full of ocean smells. In this way, our short vacation becomes more peaceful and romantic.

Lying in the soft and clean bed and watching the sea, we feel relaxed and comfortable. All in a sudden, I forget myself as a designer and all the functionalities, standards and data that are usually paid with attention when staying in a hotel have completely disappeared from mind. They become a natural and touchable convenience, clean and comfortable bed, wide vision, and beautiful scenery. All these have developed an idea in my mind: if I were not a designer, how would I assess this hotel?

Many years of interior design work has made a habit to me. Every time when I enter a hotel, what are filled in my mind are its functions, streamlines, spaces, modelling and materials. I often consider which kind of methods I will apply to attract the guests and how to meet the demands of different guests if I was appointed to design this hotel. The expressivity of the hotel space is what I seek more. Whenever I talk about these, I can’t help smiling: it is not only a professional habit, but also has been developed to certain neurotic extent.

However, during the design, there are many problems that have perplexed us. When confronted with theses, we always try to find the solutions on behalf of a designer. But, when we are only concerned about the elements at the design level, the problems are getting more and more complicated. Instead, the suggestions from the clients and management companies have provided us with many inspirations and helps. We can not but feel deeply ashamed about it and think it over and over.

China has developed into a furthest front line and an amphitheater for the global hotel design. The back packers Inns, international chain hotels, top clubs and exquisite hotels for special personages etc., are kaleidoscopic with complete categories. What we think more about is the style and the demonstration of individualities of our work. We have deemed it as a T stage to deliver our strong individualities and make shows continuously to be following or going to follow a formalistic road. But we forget the service elements of the design on the market and people. To an interior designer, it seems to be a definite trend to observe and consider the problems about hotel design in another point of view.

The design itself is a service to “human”. If you go to experience and feel a hotel from “human” point of view temporarily without consideration of the professional habits as a designer, you will have a more complete and realistic acknowledge on the hotel design.

In assessment of a hotel design, different people may have different understandings. But, they have a common understanding on this hotel. That is the functionality and commerciality of the hotel: on the one hand, whether the guests can receive a safe, hygienic, convenient service; on the other hand, whether the clients’ investments can achieve the quick returns.

The design comes from experience while the art derives from life. When we enter a hotel again, as a designer, can we attempt to take a consumption and enjoyment as a normal guest, client, management company, senior manager of a company, capitalist or artist etc. to have a assessment and consideration out of a professional habit?

设计单位:顶贺环境设计有限公司 / 设计:何潇宁

宏安瑞士大酒店
Swissotel Grand

面积:50000㎡ / 坐落地点:上海 / 主要材料:肌理米黄石、洞石、金箔、黑檀木、木地板、地毯

上海宏安瑞士大酒店，位于南京西路一侧，航站楼旁，总建筑面积是50000㎡，管理公司为北美菲尔蒙特酒店管理公司旗下的“瑞士”品牌。瑞士酒店的全球其他酒店以现代风格著称，在上海的首家瑞士酒店也同样要求走现代风格路线，但经过同业主的反复沟通，认为在现代风格的大前提下，仍需要体现当地特色，即“上海”味。19世纪中后期，上海开始成为东方文化与西方文化交流的重要窗口之一，形成具有特色的“海派文化”。因此，“海派文化”的提炼及表现形成了本案的突破点。通过提炼石库门的一些经典设计元素，搭配西式的彩绘玻璃以及提炼后的西式柱廊和中式月亮门等手段，浓缩了“海纳百川、中西合璧”的海派文化精髓，使酒店的空间设计独具特色，不同凡响。

Swissotel Grand Shanghai is located at one side of Nanjing West Road beside the terminal with a total floor area of 50000m2. Its managing company is a “Swiss” brand under North American Fairmont Hotel Management Company, famous for its modern style in the other global hotels of Swissotels. The first Swiss hotel in Shanghai is also required for its modern style. But after the repeating communications with the owner, it is agreed that the hotel also needs to reflect the local features in the big premises of modern style: “Shanghai” style. In the mid-and late nineteenth century, Shanghai has become one of the important windows between Chinese and western cultures. Since then, the eastern and western cultures are intercommunicated here and formed into a characterized “Shanghai-style culture”. Therefore, in the design of this hotel, the abstraction and presentation of “Shanghai-style culture” become the breakthrough of this design. With reference to large quantities of documents and repeating communications with the Owner, the design has extracted some classic design elements of Shikumen in arrangement with the western-style stained glass as well as the combination of the abstracted western-style column corridor and Chinese-style moon door etc, the essences of Shanghai-style culture which have extracted “the sea can hold hundreds of rivers and the Chinese style can match the western style properly”. All these have made the design of the hotel space unique and outstanding.

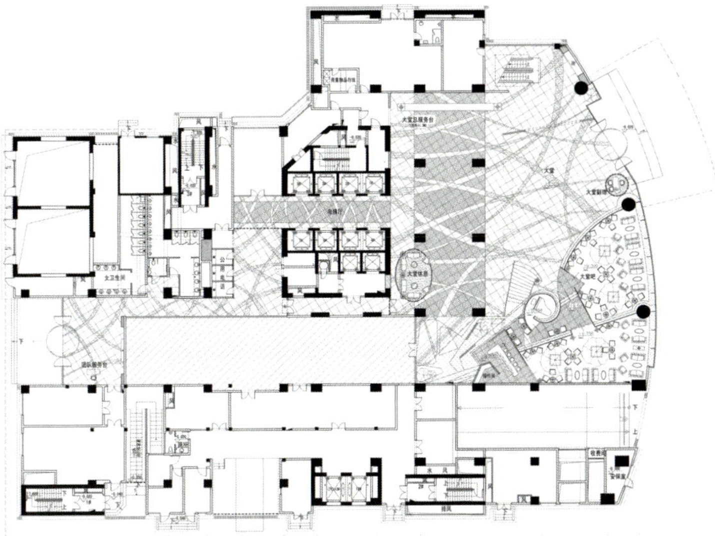

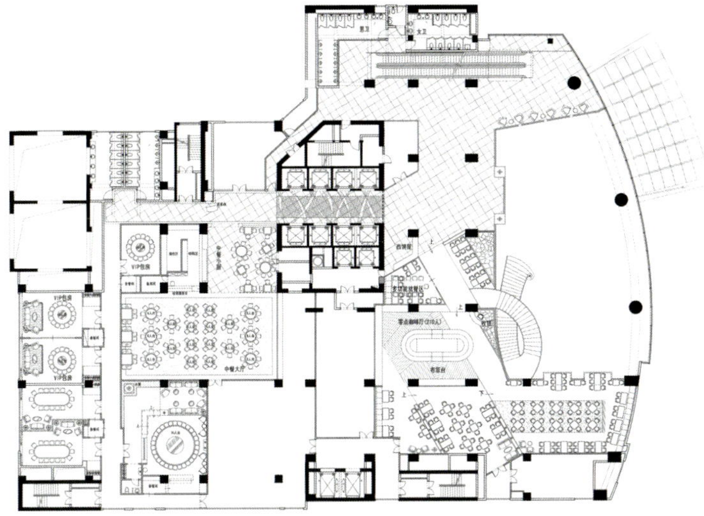

设计单位:上海泓叶室内装饰有限公司 / 设计:叶铮 / 参与设计:郑思南

白玉兰宾馆
Magnolia Hotel

坐落地点:上海

白玉兰宾馆始建于上世纪80年代初，历经岁月流逝，已到非改建不可的境地，锦江集团决定从白玉兰宾馆入手，开创一个新系列品牌——锦江白玉兰。改建后的宾馆主要包括首层大堂、咖啡厅、酒吧、商务家政中心、大中餐厅、包间、多功能厅、VIP接待、各式会议厅、客房、套房、高级套房等。

整体设计采取较为折中的风格，融现代与古典、东方与西方之间，并使圆形平面更有序列感和层次感。

Magnolia Hotel was built in the early 1980s. After years passed by, it has to be modified. Jinjiang Group decides to start from Magnolia Hotel to create a new series brand –Jinjiang Magnolia. The main functions after the modification include the hall at the first floor, coffee house, bar, business and housekeeping center, big and medium–size restaurants, VIP rooms, multifunctional hall, VIP reception, various types of conference halls, guest rooms, suites and superior suites etc.

The whole design has been applied with a quite compromised style, integrated modernity with classics and eastern style with western style. At the same time, the round plane looks more orderly and hierarchic.

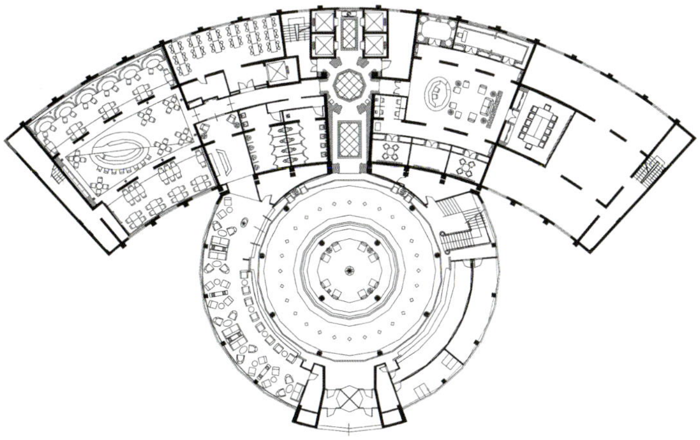

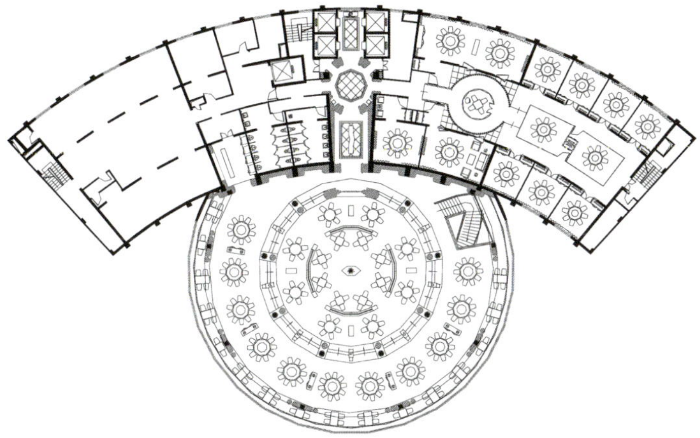

设计:琚宾

唐悦酒店
Tangyue Hotel

面积:6000m^2 / 坐落地点:北京 / 主要材料:玻璃、GRG、涂料 / 摄影:琚宾

一次偶然中的启发，常常会让我们不知所措，也可以使你起死回生。北京唐悦酒店的设计就是一次偶然中的意外收获。

在前期方案构思阶段，试图寻找一种既有传统文化特点又可以进行现代演绎的设计方式，这种创意的狂热几乎使设计师处在江郎才尽的状态，于是就去超市疯狂购物。水果是他的最爱，在把挑好的水果放到购物车的当儿，一只石榴在仓惶逃跑中摔到地上裂开，结果使设计师茅塞顿开。

石榴本身对设计并没什么重要的启迪，但石榴内部的那些因拥挤而产生的形态，恰恰是优美的触点。由此展开，果实、叶、茎、花，包括色彩和生长状态，一时间都成了原始的设计元素。用传统的图案手法演绎出新的感觉，可以说是一种尝试。以石榴为主题蔓延出来的唐悦酒店，正以信息的方式被保存被传播。

最大的慰藉莫过于看到接踵而至的宾客，以及为他们精心打造的家以外的家的概念，在那些惊讶的表情中得到肯定。

酒店的暂时性居住特点应允许个性的张扬，让人在短时间内感受好奇与激情。立足于传统文化时就如面对一个挖掘不尽的资源，把多种立体的艺术手法运用于设计之中，在特定的表现方式下创造出可以引起强烈震撼的空间设计。

An enlightening chance by accident often either makes us at a loss or makes you back to life. The design of Beijing Tangyue Hotel is an unexpected reward in an accident.

At the early stage of conceptual design, the designer tries to find a design method both of traditional cultural features and modern demonstration. The enthusiasm on this creation almost ended the designer' s wits. So he went shopping in the supermarket crazily. The fruits are his favorite. When he put the selected fruits into the shopping cart, a guava hastily fell to the ground and got split, from which the designer suddenly saw the light.

Guava itself does not have any important inspiration on the design. But the form generated due to the crowdedness inside the guava is just the touching point of the beautifulness. As developed from here, the seeds, leaves, stem and flowers including the colors and growing status all in a sudden become the original design elements. a new feeling demonstrated by the traditional patterns can be called as an attempt. Tangyue Hotel generated from the guava as the main subject are being preserved and spread in a form of information.

To our greatest comfort it can be nothing more than the customers one after another, the idea of home outside the home created for them carefully, as well as the affirmation received from those expressions of surprise.

Just as the designer has said, the temporary staying feature of the hotel shall allow the aggression and make people curious and enthusiastic in a short time. Basing on the traditional cultures is just like facing a source that can never be dug thoroughly. Multiple kinds of tridimensional artistic techniques are applied to the design so that a space design which can arouse strong shocks has been created in a specific demonstrative way.

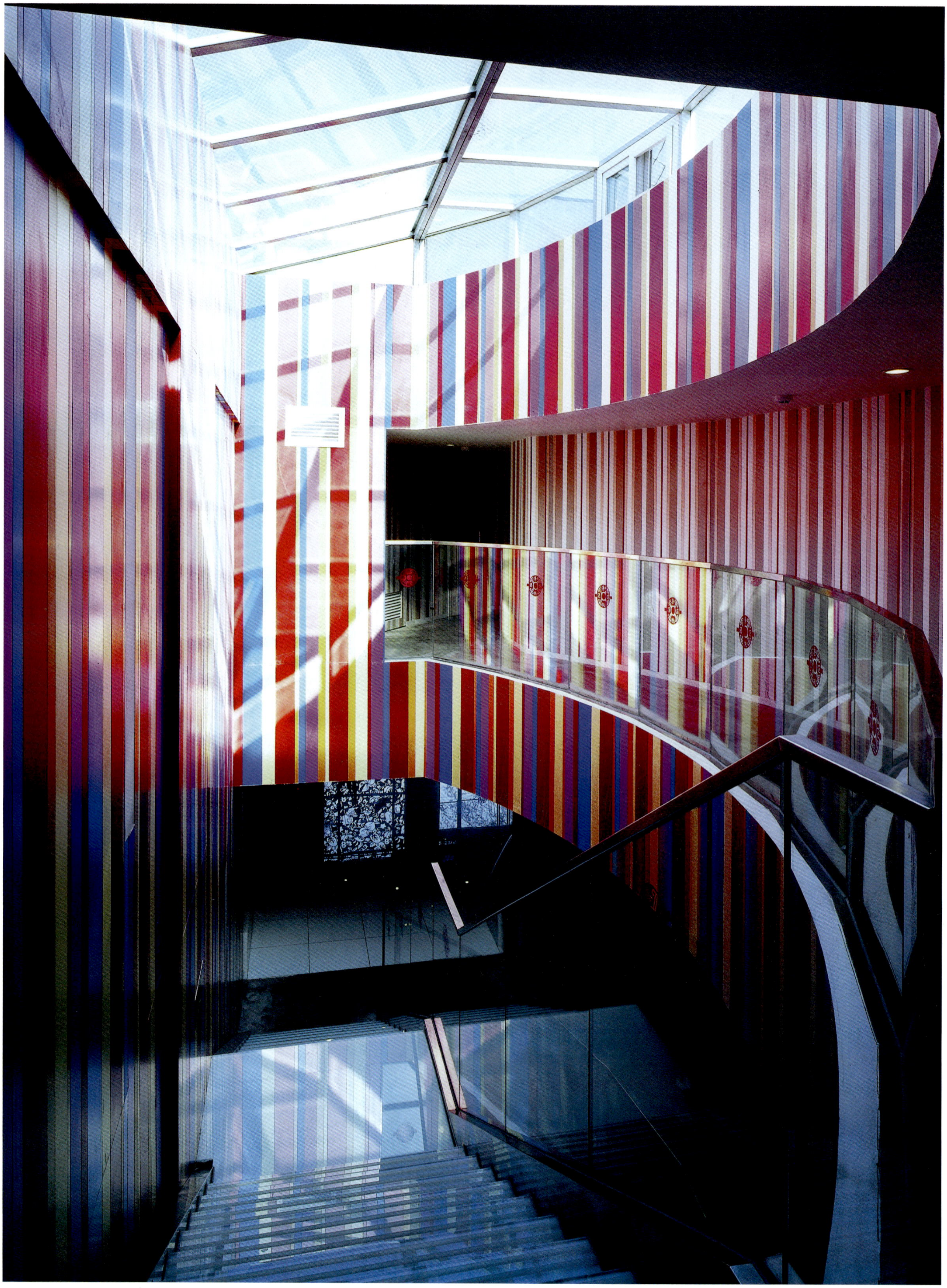

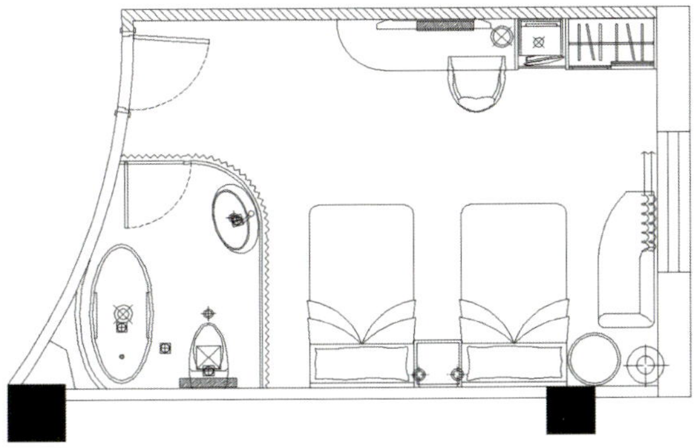

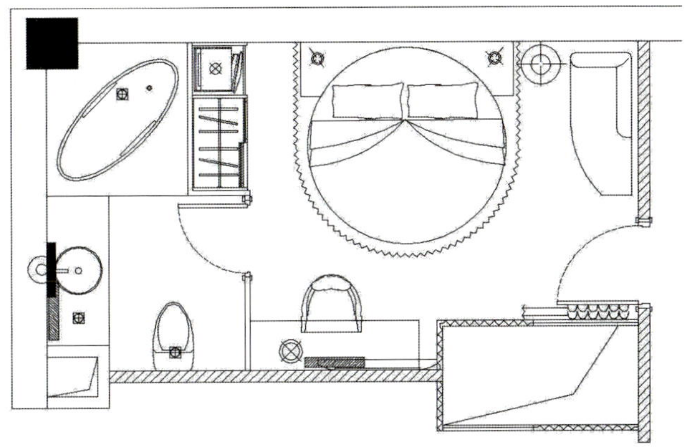

设计单位：金螳螂浙江设计研究院 / 设计：郭珂蔚 / 参与设计：孙洪涛、司马东兰

上虞雷迪森万锦大酒店
Leidisen Winning Hotel Shangyu

面积：48000㎡ / 坐落地点：上虞

上虞雷迪森万锦大酒店位于上虞城北新区市民大道与凤鸣路交叉口，是上虞CBD商务区域核心地带的标志性建筑。整体设计趋于成熟，把酒店设计的品位与细节控制得比较好，相对来说，“创新、文化、艺术”等等现在国内酒店设计师热衷的东西却有所控制，而是充分考虑了它们和商业的平衡性。

大堂设计的第一步就是创新能力和功能之间的平衡的关系，雷迪森酒店管理公司的“系统标准”在所有的行为之下作为依托，使这个酒店设计以及现在的酒店运营是在有序的、有效的过程中进行。大堂设计必须结合酒店室内外整体功能及流程设计，所谓的“成熟度”首先就是要看这个。而合适、细腻的“细部设计”则是高端酒店设计最见功底之处。

通过将中西方古典元素进行提炼重组，同时运用在行政酒廊空间中。在酒店中反复运用的“中式云纹图案”作为统一元素，通过不同材料不同形式运用到不同功能空间中，演绎国际品位的豪华与精致。

为了提升中餐厅的国际品位，设计融合了西方的古典高贵和东方的审美情趣，把家具和配饰的设计精致到每一个小细节，处处显示出卓而不凡的尊贵气质和高雅品位。不拘泥某个具体风格，但又表现出其最具价值的文化原生形态。

Leidisen Winning Hotel Shanyu is located at the crossroad of Shimin Avenue and Fengming Road in New District of North Shangyu city and also a symbolic building in the core zone of CBD in Shanghai. The entire design tends to be mature and the tastes and details of the hotel design are controlled properly. Comparatively speaking, the subjects such as "innovation, cultures and arts" etc. that the domestic hotel designers are crazy for are indeed controlled. The balance between them and the business have fully considered instead.

The first step of the lobby design is to the balanced relationship between our designers' creativities and the functions. "System Standards" of Leidisen Winning Hotel Management Company is the base for all activities to ensure that the design of this hotel and the current hotel operation can be carried out in an orderly and effective procedure. The design of the lobby shall integrate the integral interior and exterior functions of the hotel and the procedural design, which the so-called "maturity" first put emphasis on. The proper and exquisite "detailed design" is also what we pay attention to. Whether a superior hotel is designed well mostly lies on the conditions of its "detailed design".

We have abstracted and reorganized the classic eastern and western elements and also applied them to the space of administrative wine shop. "Chinese-style cloud patterns" that are repeatedly applied in the hotel are taken as the uniform element to demonstrate the luxury and delicacy of the international tastes by applying different materials in different forms to the different functional spaces.

To upgrade the international taste of the Chinese restaurant, our design has integrated the western classics and nobility with eastern aesthetic tastes tighter to make the design of the furniture and ornaments delicate to every small detail and demonstrate the excellent but unusual nobilities and elegant tastes. We do not rigidly adhere to certain specific style but demonstrate its most valuable original cultural forms as well.

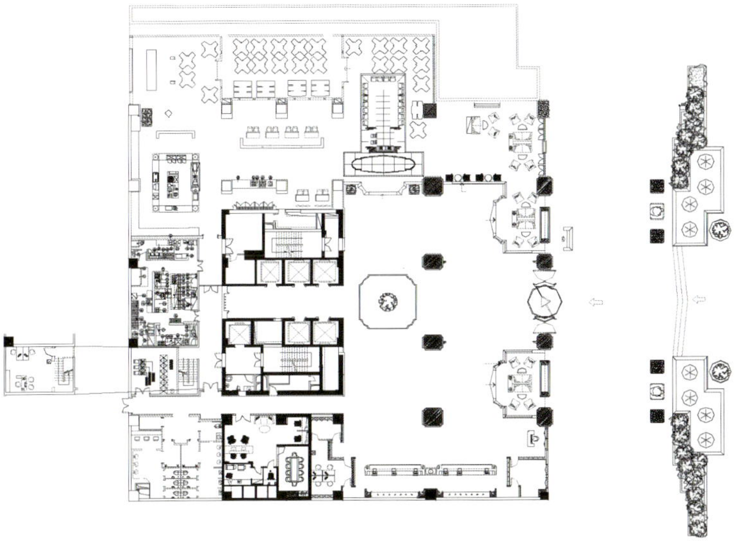

問余何事棲碧山笑而不
答心自閑桃花流水窅然
去別有天地非人間

设计单位：上瑞元筑

云鼎花园酒店
Yunding Garden Hotel

面积：1700m² / 坐落地点：苏州 / 主要材料：黑白根石材、中花白石材、肌理涂料、马赛克、皮革、玻璃、镜面不锈钢 / 摄影：文宗博

从商业的角度来看，它不是一个属于现有酒店概念的商业酒店，它只能说是一个具有某些酒店功能的小型商业会所，这完全是根据业主多年经营与其相关行业的市场需求来设计的商业杂烩，既有精品酒店的一些概念，也综合了一些夜场设计所特有的形式和特征。酒店位于苏州新区市中心，经营位置极佳。整个酒店的平面布局具有极强的针对性，完全由业主提出经营要求，然后由设计师来融会贯通。整个平面以弧形走向展开，为的是曲径通幽的神秘感，由此而达到客户所要求的娱乐及住宿私密性，由于公共空间极其简单（几乎没有公共空间），所以在公共通道中设置了连门厅在内的三个转折空间，以便可以破沉闷之感，同时又使空间有层次感。

房间的设计采用了全开放的卫生间，睡床区及休息沙发区，都以虚空间来划分，相互穿透、重叠，满足了业主要求，大房间中小空间，既有精品酒店的体会，又有KTV酒吧的感官享受。

云鼎酒店的设计是将业主的具体要求由设计师做了很多的融会贯通，很有分寸地表达了业主想要的一种特殊的商业空间。

From the business point of view, it is not a business hotel with the conception of the existing hotels. It only can be called a small business club with the certain hotel functions. It is a mixed business stew designed completely according to the market requirements of the related industry engaged in by the client for many years. It not only has some conceptions of the exquisite hotel, but also combined some forms and characteristics unique to the design of some night clubs. The hotel is located in the center of Suzhou New District with an excellent business location. The plan layout of the whole hotel is carried out in a targeted way. It is upgraded completely according to the business requirements by the client and achieved through a comprehensive study by the designer. Since the entertainments and accommodation have privacy, the whole plan is expanded in an arc way to obtain a mystic sense of a winding path leading to a secluded place and achieve the privacy required by the customers. As the public space is very simple (there is almost no public space), three turning spaces including the hallway are set up in the public passage to break the boringness and endow the space with a hierarchic sense.

The complete open toilets are designed for the rooms. The bedroom and lounge are divided in virtual spaces. They are intercrossed and overlapped to meet the requirements of the client. Some spaces are in the big room, where you can feel a sense of the exquisite hotels and also have the pleasures of senses of KTV bars.

The designer has made many comprehensive studies and adjustments according to the specific requirements of the client for the design of Yunding Hotel to decently deliver a special business space demanded by the client.

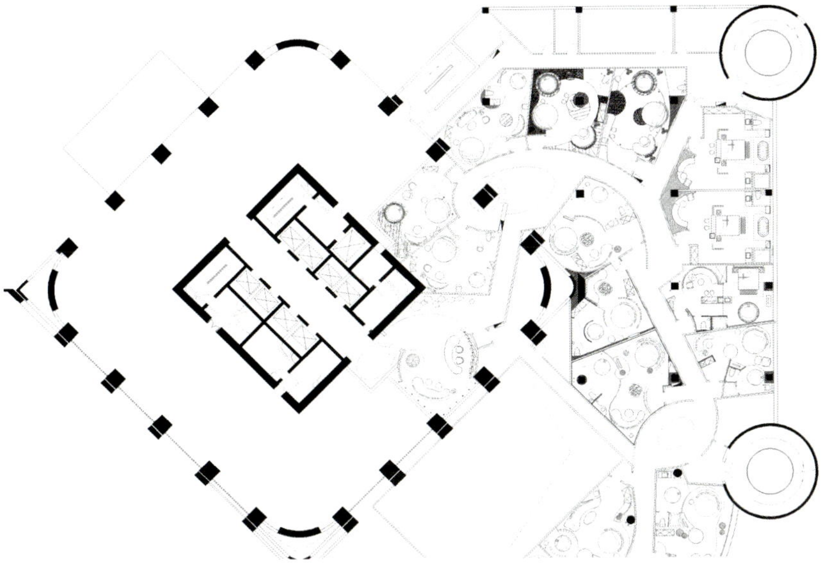

设计单位:杭州正午装饰设计有限公司 / 设计:陈石林

西湖四季酒店艺联店
SSAW Hotel,Yilian

面积:6200m² / 坐落地点:杭州学院路29号 / 主要材料:灰色板岩、红橡木夹板饰面套色、光面灰麻拉丝板、竹木地板

酒店位于杭州一条不起眼的小巷内，闹中取静。设计师利用原有建筑的特殊环境，以东南亚风格来定位酒店的整体设计，将室内设计和外环境融为一体，将酒店打造成喧嚣都市中的一方绿洲。

在整个酒店的设计中，以沉稳的棕、灰两种色调相互搭配为基调，再以一些宝蓝或者紫色系等色彩鲜艳的装饰为点缀，来体现极富东南亚风格的视觉感受。考虑色彩及质地的选择，所搭配的东南亚家具的色泽以原木色调为主，多为褐色等深色系。在布艺色调的选用上，以东南亚风情标志性的深色系为主调，沉稳中透着一点贵气。精致的细节贯穿了整个酒店，温馨小巧的门厅，浓情的咖啡厅，灯光柔和的走廊，简洁舒适的客房，木饰面的运用和东南亚风格艺术品的点缀，使酒店的每一个角落都透露出浓浓的异国风情和舒适温馨的自然气息。

The hotel is located in an undistinguished alleyway, quiet in a busy place. The designer has utilized the special environment of the original building and the whole design of the hotel is orientated in the Southeast Asian style to integrate the interior design and exterior environment together so that the hotel is developed into an oasis in the busy city.

In the whole design of the hotel, the mutual arrangement of the steady brown and gray is the basic tone, decorated with the vivid colors such as sapphire blue and purple series to reflect the visual feelings rich with the Southeast Asian style. In consideration of the colors and textures, the color of the matched Southeast Asian furniture is main in the natural wood color and most in dark colors such as brown. In the selection of the colors of the cloth arts, the symbolic dark colors of Southeast Asian style are the main tone, revealed with some nobility in the steadiness. The delicate details have run through the whole hotel: warm and delicate hall, passionate coffee house, the corridor with soft lights, simple but comfortable guest rooms, as well as the application of wooden facing and decoration of artworks in Southeast Asian style, which have revealed every corner of the hotel with a strong foreign style and comfortable and warm natural atmosphere.

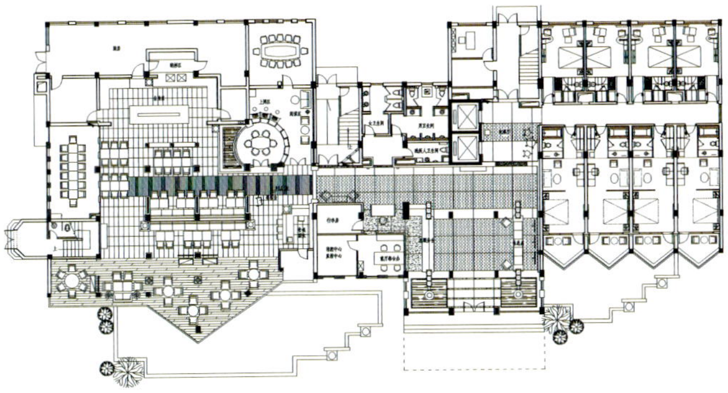

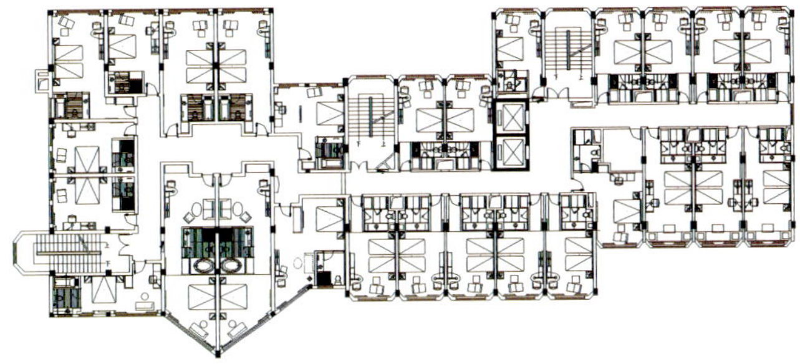

设计：殷艳明、胡威

杭州德嘉精致酒店
Hangzhou Dejia Hotel

面积：8000㎡ / 坐落地点：杭州 / 主要材料：石材、不锈钢、马赛克、壁纸、云石灯片、丝印玻璃、木饰面、彩色玻璃 / 摄影：臧兵

在设计中，我们意图通过大量自然元素与空间造型、材质、图案的结合，营造出一个独特、时尚的精品酒店新形象。

建筑外立面是设计思想的起点，渐变的橙黄成为建筑与自然光线对话的中介，赋予了建筑整体明快、自然、多层次的丰富情感。

设计思想在公共空间中进一步展开，改变了大堂原有格局，从进深和高度两个纬度围合出一个曲线自然、流畅、通透的小型中庭；S状伸展的水晶珠帘，映衬S状伸展的弧形木制隔断，形成了相对独立、富有动态美感的中餐空间；在系列包房中，木格栅的天花、墙面的延伸、竹与木的对话，在红色玻璃的映衬下彰显出不同的气氛和时尚魅力；多功能厅则巧妙地运用金色镜面的反射及延续使结构柱和梁得到了很好的消隐，又通过柱、梁的分割使空间更加灵活。

人性化的客房体现了设计思想的延伸与细化，“拟人化”的设计手段赋予九种客房不同的“人物性格”：理性硬朗的商务客房、浪漫体贴的情侣房、也有休闲雅致的咖啡语茶套间……显示出酒店体贴入微的情怀。

In our work, we purport to build a new image of this hotel through the combination of massive natural elements, the spatial conformation, and the various materials in the design.

On the external of the building, the change gradually orange yellow was chose to be the dialogic medium between the construction and the natural light, it also made the construction in the sprightly, the natural, and the multi-level rich emotion.

The design of the public space has many highlight. We first change the lobby original arrangement; extend the area from the height and depth to reconstruct a curvy nature and clear space. The crystal bead droplight in the S shape which extends complements the arc wooden podium which extends in the S shape too, forms a relatively independent, rich dynamic and esthetic sense Chinese meal space. There are also many serious individual rooms for people to have dinner, we use many natural elements in design such as the bamboo, the wooden grid on the ceiling, the mirror reflection which extended the wall in the vision in order to appear the different atmosphere and the fashion charm in the rooms. In the multi-purpose hall, we ingeniously utilize the gold mirror' s reflection to reduce the feeling of the structure column and girder, and flexibly divide the space also through the column and girder.

The guest rooms seem quit human by our design. The different color and pattern which we use manifest the different characters in the nine kinds of guest rooms: some are rational and calm which suitable for the businessman, and some are romantic and thoughtful for the lovers, there are also some apartments which full of leisure and elegance... ...All of these demonstrate the hotel' s consideration for the people who will live here.

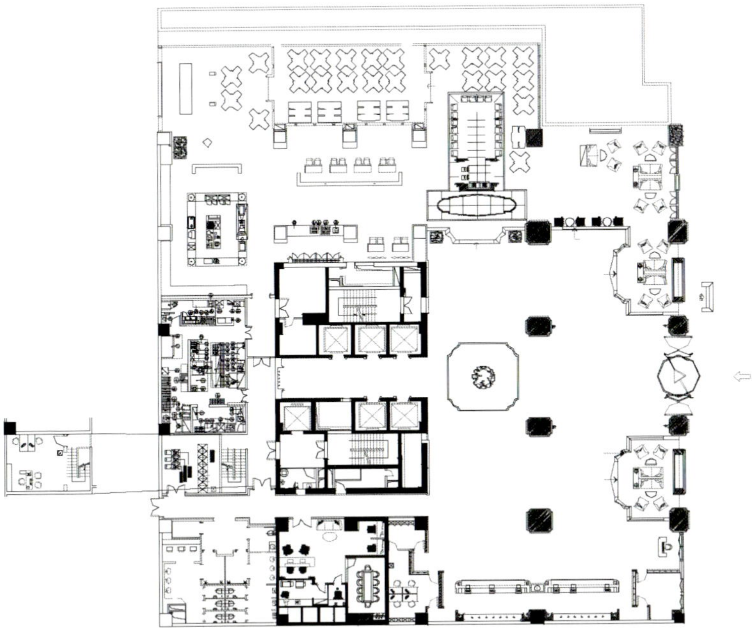

设计单位:海岳建筑装饰工程设计有限公司 / 设计:姜湘岳、吴海燕 / 参与设计:袁娟、纪红、刘玲、潘婷

金太隆国际酒店
Jintailong International Hotel

面积:18000m² / 坐落地点:南京 / 主要材料:地毯、木雕家具、大理石、墙纸 / 摄影:陈乙

设计师确定以新古典欧式主义为设计风格，且以黑色为主色调。

接待大堂的功能设计，将西餐厅和大堂吧合为一体，功能偏向于大堂吧。减弱西餐厅的功能，通过高低的空间错位，将两个功能区微弱地区分开来。通过半透明并且镂空的水晶天花板，把大堂和大堂吧微妙地区分开，让两个空间在同一水平面但同时又是相互联系的。

大堂两边的墙面采用发光玻璃，让墙面从黑色中跳跃出来，也让路过的行人一下子被这种舞台剧似的气氛所吸引。

总台对欧式的古典家具进行了夸张，让家具的尺度放大夸张到极致，现代感顿生，总台上方的雕刻镜柜也是在通常的比例上放大。通过这些夸张的比例，将古典往现代转变。

二楼的中餐厅，由于缺少自然的光线，使得设计师对室内光线的控制反而更为有利，通过对自下而上的发光玻璃墙面将空间表现得更为神秘和具有戏剧性。整体的色调是厚重的，通过红绿来相互对比，通过黑色进行中和协调。

进入电梯厅，空间仍旧是幽暗的。到了客房打开房门，眼前会顿觉一亮，白色的空间会让客人感觉仿佛到了属于他自己的空间。

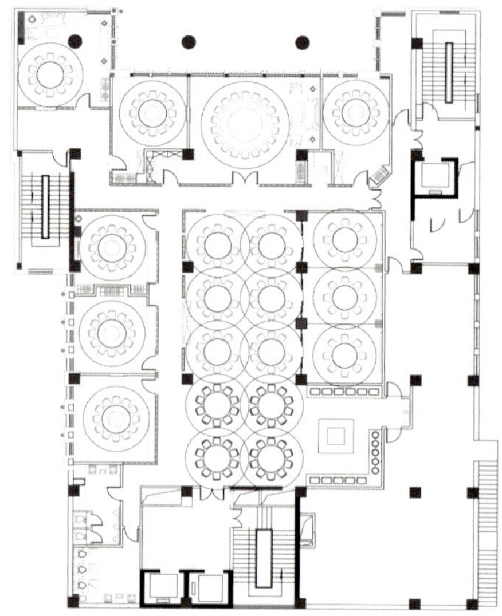

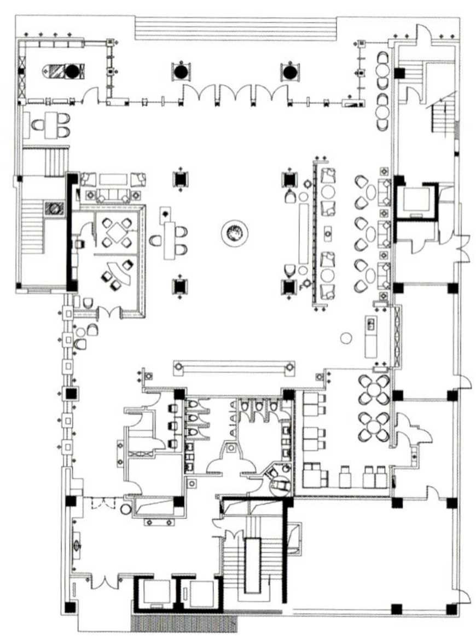

The designer has decided that the new classic continental synthetism and black is the main color.

The functional design of the reception hall has integrated the western restaurant and lobby bar together while its function emphasizes more on the lobby bar. The function of the western restaurant is weakened and the two functional areas are separated slightly by the height differences of the spaces. The lobby and lobby bar are divided delicately by the semi-transparent and hollowed-out crystal boards so that the two spaces are in the same plane and also be interconnected.

The walls on both sides of the lobby are applied with the luminescent glasses to pop out the wall from black and abruptly attract the pedestrians in East Zhongshan Road with such a dramatic atmosphere.

The receptionist desk has exaggerated the classic continental furniture and the size of the furniture has been exaggerated to the utmost with a sudden appearance of modern senses. The carved mirror frame above the receptionist desk is also enlarged in the normal scale. Through these exaggerated scales, the classic is changed to the modernity.

Although the Chinese restaurant in the second floor lacks the natural lights, it instead becomes an advantage to the designer to control the interior lights. The space is shown more mysterious and dramatic through the luminescent glass wall from bottom to top. The entire color is heavy, but counteracted in harmony through the black with the inter-contrast between red and green.

The space is still dim once entering the elevator hall. But once you arrive at the guest room and open the door, it will brighten your eyes at a glance and the white space will make you feel you are in your own space.

BOSS

设计单位：上海金螳螂环境设计研究有限公司 / 设计：高超一、张驰、康瑞德 / 深化设计：陈尤新 / 软装设计：那格九号 朱岚

豪生酒店 Howard Johnson Hotel

面积：3000m² / 坐落地点：上海张江 / 主要材料：艺术地砖、地毯、红影木、花梨木饰面、艺术壁纸、乳胶漆 / 摄影：臧兵

本案为展讯中心酒店配套的西餐厅及酒吧设计，风格定位为中西合璧，既有中国文化传统精髓，又有西方高贵简约的元素，二者要充分融合。整体要求低调、内敛、不事张扬而自然高贵。一层为西餐厅，顶部运用了大面积的红色是整个餐厅的一大特点，通过材质的不同和灯光的变化来划分不同的用餐环境，自然过度，和谐统一。二层为包厢、雪茄吧及酒吧，适合商务会议、会餐。酒吧以黑色及红色为主题，既休闲、浪漫，又略带神秘。

It is a design for the western restaurant and bar of Howard Johnson Hotel Zhangjiang Shanghai. The style is oriented in a combination of Chinese and Western, both with the traditional Chinese cultural essences and the western noble and concise elements. They shall be fully integrated. The whole requires low-profiled, introverted, understated, but natural and noble. The first floor is used as the western restaurant, where the large area of red is applied to the ceiling is a big feature of the whole restaurant. The different dining environments are divided through the material differences and light changes, which are naturally transitioned, harmonious and unified. The second floor is for VIP rooms, cigar bar and bar, which is suitable for business conferences and dinner parties. Black and red are applied as the main color of the bar, leisure, romantic and slightly mysterious.

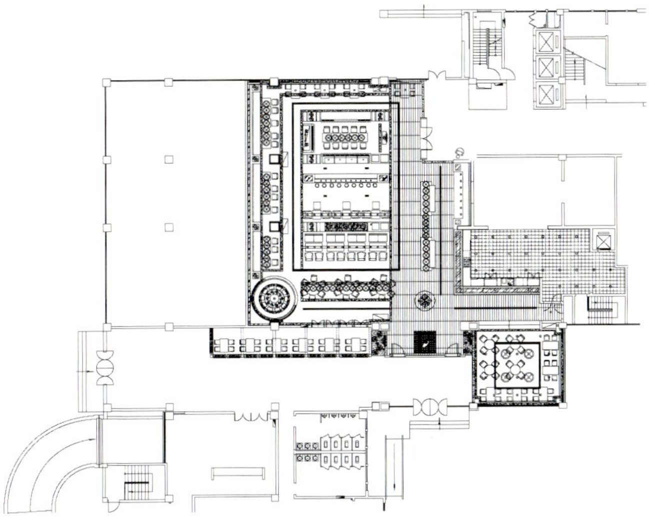

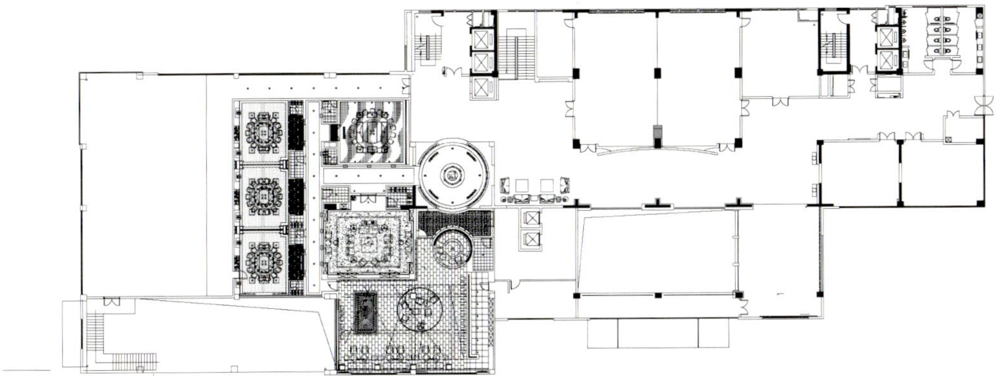

设计单位:海南富元设计公司 / 设计:富元

香泉温泉度假酒店
Xianquan Spring Holiday Inn

面积:24000㎡ / 坐落地点:安徽 / 主要材料:砂岩石、自然毛石板、柚木实木、青瓦片、茅草、竹子、土粗布

主楼大堂入口门厅的顶部全部采用实木网架结构，配以皖南地区民居的青瓦做装饰，加上火山岩与鹅卵石结合的柱子，给人一种复古自然的印象。门厅顶部瓦面的三组流水跌入地面上三个巨大的石砵中溢流直下，汇入毛石砌筑的水渠池塘之中。水渠两侧的木栈桥，既方便客人流通行走，也增添了水、石、木、人之间的灵动。

大堂内部的正面全部是实木玻璃的折叠门，平时可以全部敞开，便于通风采光，使室内外融为一体，让人有一种置身大自然的感觉。大堂顶部采用斜屋面的做法，全部用实木板条，中间实木包梁的澡井上配以粗麻布与细毛竹切面制成的灯饰品。大堂中间吊以巨大木格栅和砂岩石雕壁画，同特有的梯形石刻柱子加上土建改造时保留下横纵向的四道横梁，显得更加自然与粗犷。大厅墙面以米色粗涂料为主，墙面上带有木雕的壁龛，吧台立面人工凿制的砂岩石材与实木，一切都与自然生态相结合。吧台背景用松木横切断面的装饰造型，可用来存放洗浴用钥匙牌的小洞，也可以起到极强的装饰肌理效果。

大堂在整体设计理念上充分体现人与自然，温泉文化的完美结合，创造自然而随意的休闲度假环境。

客房和套房为了追求巨大的视觉空间效果，采用空间互借的手法，全部采用开敞式卫生间。卫生间、洗浴间与卧室的开敞与互通，使居者完全的放松。造型与色彩处理具有独特的异域风情并与皖南建筑完美结合，使之更突出地域文化和温泉文化的独特意境。

The top of the hallway at the lobby entrance of the main building is completely applied with solid wood grid structure complete with the grey tiles in the residences of the south Anhui province as the decoration. It has left an archaic natural impression on people together with the columns built by volcanic rocks and cobbles. Three groups of flowing water from the tiles on the top of the hallway cascade into three huge stone bowls on the ground and overflow downwards to the canal and pond built by rubbles. The timber trestle bridges on both sides of the canal are not only convenient for the guests to walk by and also increase the interaction among water, stone, timber and human beings.

The facades of the interior lobby are all the folding doors made of solid wood and glass. Usually they are all completely open for easy ventilation and lighting as well as the integration between the interior and exterior so that the people will have a feeling of staying in the nature. The top of the lobby is built in a way of sloping roof. It is completely made of timber battens. The caisson ceiling with the beam enclosed by solid wood in the middle is arranged with the light ornaments made of bamboo sections. The huge wood grid and sandstone carving frescos are suspended in the middle of the lobby. It looks more natural and wilder together with the peculiar stair-shaped stone columns and the four horizontal beams kept during the civil modification. The wall surface in the lobby is mainly applied with the cream-colored parget. The alcoves with wood carvings on the wall, the sandstone materials and solid woods on the facade of the bar counter done by hand are all integrated with the natural conditions. The background of the bar counter is decorated with the pine sections cut horizontally. The small holes that can be used to put the key card used for bathing also have an extremely strong decorative effect.

The integral design conception of the lobby has fully reflected the perfect combination between human beings and nature, as well as the spring culture so that a natural and casual environment for recreation and holiday is created.

In pursuit of the huge visual space effects, the guestrooms and suites have all been applied with the opening toilets by means of borrowing the spaces from each other. The toilet, bathroom and bedroom are open and interlinked so that the residents feel completely relaxed. The modeling and colors are arranged in a unique foreign style and also is perfectly combine with the buildings in the south of Anhui Province so that the peculiar atmosphere of the local and spring cultures becomes more prominent.

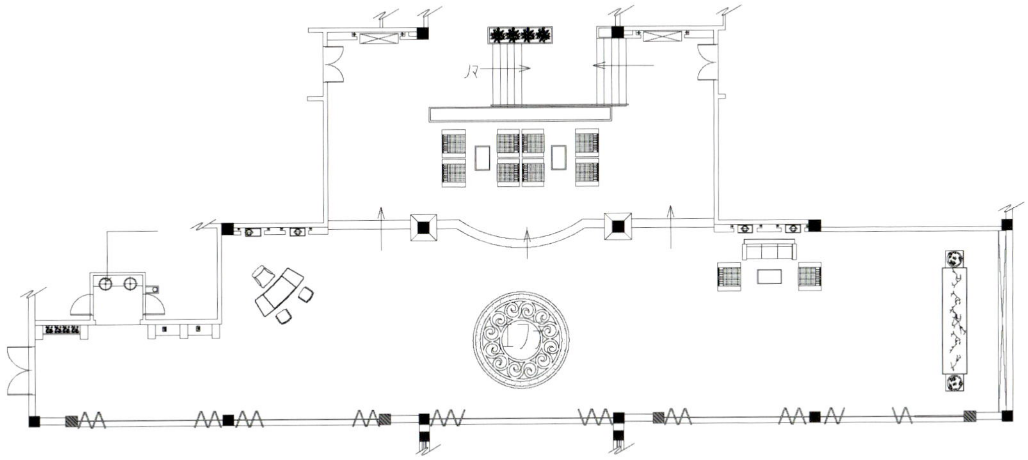

设计单位：上海泓叶室内装饰有限公司 / 设计：叶铮

锦江之星–沈阳南湖店
Jinjiang Inn – Shenyang Nanhu Hotel

坐落地点：沈阳 / 摄影：文宗博

本案是锦江之星连锁酒店的又一个新成员，也是沈阳地区第五个连锁店，室内设计部分主要为大堂、餐厅、会议室、电梯厅等公共区域。整个设计概念在空间上主要强调了一个连续延展的层面，并打破传统意义上天、地、墙的界面分割，使之在室内空间表达上更为浑然一体，并从原空间框架的意义上分离而出，表现其层面的独立性特征。黑、白、灰的纯净，构成了空间在视觉上的色彩基调。理性而又时尚。

It is another new member of Jinjiang Inn Chain Hotels and also the fifth chain hotel in Shenyang Area. The interior design mainly includes the public areas such as the lobby, canteen, conference rooms and elevator hall etc. The whole design conception has mainly emphasized on a continuously expanding surface. The traditional division of the interface between the sky, ground and wall has been broken to demonstrate the independent features of its skin. The purity of black, white and gray has visually constituted the basic color of the space, rational and fashionable.

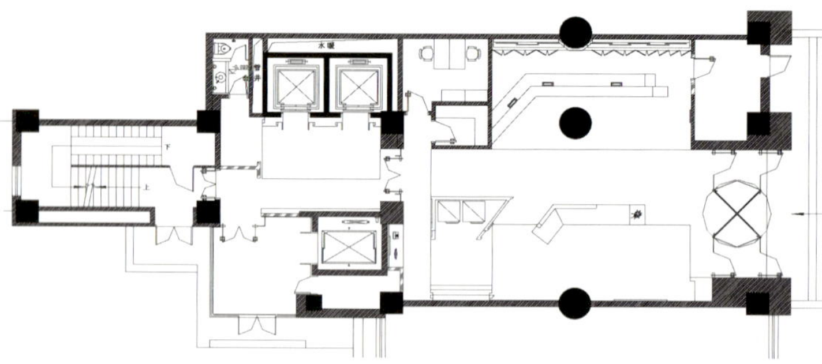

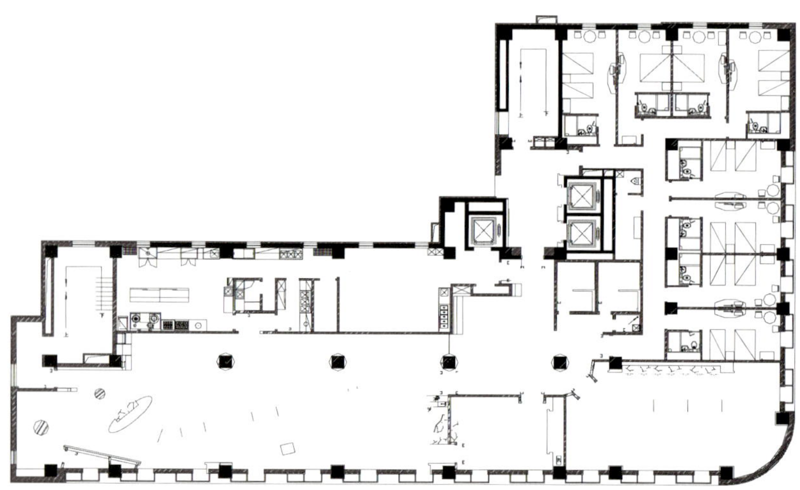

凯宾斯基酒店
Kempinski Hotel

坐落地点：苏州市工业园区国宾路168号 / 摄影：文宗博

本案位于苏州园区美丽的金鸡湖和独墅湖间，毗邻金鸡湖高尔夫球场和金鸡湖大酒店。

大堂吧汇聚融合了不同的设计元素，可在轻松惬意的氛围中尽享各式美酒佳酿、饱览美丽的独墅湖景，度过怡然自得的时光。整个酒店的设计风格完全符合追求时尚与品质生活的高端人士的品位。各式套房布置得宽敞优雅，搭配窗外绚丽的景色，令每一位客人在苏州的迷人浪漫氛围中享受到奢华的高品质服务。德国特色的普拉那啤酒屋和全景雪茄坊更让人充分感受到来自西方的古典和奢华。另外，酒店还设有各类休闲娱乐场所，方便忙碌而紧张的工作后的人们释放压力、荡涤身心。

This case is located between the beautiful Jinji Lake and Dushu Lake in Suzhou Industrial Park, close to Jingji Lake Golf course and Jinji Lake Hotel.

The lobby bar has gathered and combined different design elements. You can enjoy the various types of good wines, drink in the beautiful scenery of Dushu Lake and spend a happy and leisure time in an easy and cozy atmosphere. The design style of the whole hotel is in full compliance with the tastes of the elites who pursue the fashion and quality life. The various types of suites are arranged wide and elegant. Together with the splendid scenery outside the window, every guest can enjoy the luxurious high-quality services in the attractive and romantic atmosphere in Suzhou. Paulana beer house in German characteristics and full-view cigar store make you better feel the classics and luxury from the western. Furthermore, the hotel is provided with various types of recreation and amusement places so that people can relax themselves and calm down themselves conveniently after the busy and intense work.

设计单位：金螳螂设计研究院南京设计分院 / 设计：王剑

皇廷国际大酒店
Palace International Hotel

面积：18000㎡ / 坐落地点：溧阳 / 主要材料：玫瑰木、白橡木、米黄石材、艺术玻璃 / 摄影：潘宇峰

酒店设计是将特定生活质量和现代交际环境融合的创造过程，设计师把对形、色、光的追求融入到设计细节中，打造出了一个舒适、精致、含蓄的酒店空间，并多方捕捉现代时尚元素，借浓厚而大气的笔墨尽情渲染公共空间，借精细而婉约的笔调刻画私密空间，步移景移，相得益彰。同时运用唯美并具人性化的设计手法将表情丰富的材质进行多维组合，使之呈现出强烈的视觉效果。不同材质的色彩关系，相同材质的不同质感交相辉映、美轮美奂。精心挑选的精致艺术品提升了空间品质，而当地文化精髓也被不留痕迹的糅合在整体环境之中。

The design of the hotel is a creative process of syncretizing specific life quality and modern social environment. By merging the pursuit of shape, color and light into each detail, the designer has created a comfort, exquisite and implicit hotel space. The public space is characterized by various modern fashion elements, whereas the private space is depicted by elaborate as well as graceful-and-restrained style. The scene changes with the move of steps, which brings out the best in each other. A strong visual effect is presented by the multi-dimensional combination of materials rich in expression by the aesthetical and humanistic designing techniques. The color relations of different materials and the different quality of same materials reflect and shine mutually. Delicate artworks by aborative selection have exalted the quality of the space, meanwhile the essence of the local culture has been naturally and harmoniously integrated in the whole environment.

设计:据宾

富邦国际酒店
Fortune-Land International Hotel

面积:22000m² / 坐落地点:北京 / 主要材料:艺术壁纸、GRG、丙烯颜料 / 摄影:孙祥宇、据宾

当你走进北京首家艺术设计酒店——富邦国际酒店，紧紧缠绕着你的是一种独特的氛围。也许是来自莲花的诱惑，抑或是设计师在构思时的那种亢奋的延续？不管怎样，审美高潮的不断蔓延是整个空间给人留下的第一感觉，然后慢慢品味方知个中味道。一种化生的理念正是莲花想对人们述说的情节。那些向人们展开的诱惑也许在传递一种繁衍个性的概念，审美个性与个性审美在此共生，来自一种宽容和谅解。不要谴责设计师给你制造了那么多的被迷惑的机会，也不要让自己在激动中遗漏每个被感染的环节。取莲花作为设计的原点，来自对佛性虔诚与浪漫的感悟，以及对超度和宁静的体现。莲花有四德，一香、二净、三柔软、四可爱。世间花卉先开花后结实，莲花则在开花同时，结实的莲蓬已具备，“华实齐生”的特质，莲花因此被佛家视为能同时体现过去、现在、未来。而皎洁、信用、佛、君子的说法是对莲花的另一种评价。以传统文化表现性感、婉约、独特、宁静、时尚，这就是富邦国际酒店在打造时的初衷。

设计师在以简约的美学手段构筑一个具有多重含义充满激情与爆发力的空间同时，也致力于引导顾客尝试一种新的生活方式。在摆脱千篇一律的酒店模式的审美退化以后，能让你用心去品味那些隐藏于不经意间的妩媚。

设计师在夸张色彩表现的当儿，没有忘记用婉约的手法向人们娓娓道来性感的时尚。那些于叶、于茎、于藕、于大珠小珠落玉盘之间的迂回，抑或写实抑或幻化，那些夸张与严谨的对话使得空间具有了欢乐和唯美的表情。传统文化的魅力在这里被演绎得恰到好处，漫步其间可以重温精神的浪漫。这也正是本设计完成后迅速被审查苛刻的Boutique Hotels International全球酒店认证集团吸纳为会员，并推荐给如麦当娜等各界精英名流居住的因素，而这一切都与根植于本土文化的设计密不可分。

Once you enter the first artistic design hotel in Beijing - Fortune-Land International Hotel, what closely surrounds you is a kind of unique atmosphere. May it come from the lure of lotus or the continuity of the excitement from the designer during his conception? Whether or no, the continuous spread of the esthetic climaxes is the first impression the whole space has left on people. And then they can enjoy slowly and be aware of the feelings. A kind of incarnate conception is just the plot that the lotus wants to tell people. The lures exposed to the people maybe are delivering a conception of individuality continuity. The aesthetic individuality and individual aesthetics can get accreted because of a kind of tolerance and understanding. Neither condemn the designer to create such a lot of opportunities of luring you nor let yourself missing any affected part in the excitement. Taking the lotus as the source of the design comes from the devotion to Buddhism and the awareness of romance as well as the reflection of releasing the souls from suffering and quietness. The lotus has four virtues: fragrance, purity, gentleness and loveliness. In the world, the flowers bloom first and seed later. However, the seeded lotus pod already has had the feature of "Flower and seed develop together" while the lotus is blooming. Therefore, the lotus is deemed to reflect the past, presence and future. The view of "brightness, credit, Buddha and gentleman" is another appraisal on the lotus. The presentation of sexiness, gracefulness, uniqueness and fashionableness by traditional cultures is the just original intention of building up Fortune-Land International Hotel.

While the designer is applying a simple and concise aesthetic technique to create a space with multiple meanings and filled with enthusiasms and explosive forces for you, he is also devoted to introducing the customers to tray a new living style. You can enjoy those unconsciously hidden charms attentively after breaking away from the stereotyped hotel-type aesthetic degradation.

While the designer is exaggerating the demonstration of the colors, he also remembers tirelessly telling the sexy fashion to the people in a graceful way. The space is endowed with the pleasant and aesthetic expression through the roundabouts of the leaves, stems, roots as well as big and small pearls on the green leaves, real or illusory, as well as the dialogues between the exaggeration and precision. The charm of the traditional cultures is demonstrated perfectly and the spiritual romance can be reviewed while walking among them. These are also the reasons why this hotel is accepted as a member by Boutique Hotels International which has strict evaluation after the design is completed and why it is recommended to the elites and celebrities such as Madonna. All these are closely connected with the design deeply rooted on the local cultures.

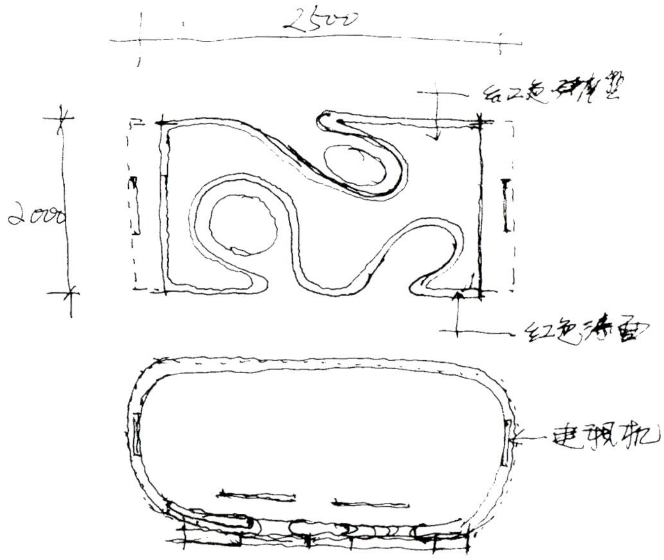
2500
2000

设计单位：苏州金螳螂建筑装饰股份有限公司设计七分院三所 / 设计：杨颜江、孙威、肖莹

云轩山庄
Yunxuan Shanzhuang

面积：7000㎡ / 坐落地点：南京 / 主要材料：大理石地面、织物硬包、墙纸、木饰面 / 摄影：文宗博

作为一家高端的商务会所，大堂入口没有像一般设计的金碧辉煌。深灰及黑色主打的背景墙、服务台，低调内敛，银白色的蝴蝶吊饰，深厚中透露出冉冉古秀之气。包间内毛泽东的《人民解放军占领南京》等诗词歌赋嵌于屏风之内，为背景添饰一份人文雅韵。公共就餐区以点光源分布于各个需要加强照明的地方，以镜面围合的柱子丰富了餐厅的空间内容。客房布置得典雅舒适，折射出淡淡的中西合璧的影子。信步所至，都能感受设计师将独具匠心的设计理念淋漓于人景合一的尊荣和显贵。

As a high-grade business club, the entrance of the lobby is not designed resplendent and magnificent as usual. The background wall mainly in dark gray and black and information desk are low-profiled and introverted. The silvery white butterfly suspending ornaments are revealed with the slight classic graces in the profundity. The poem "The People's Liberation Army has occupied Nanjing" written by Mao Zedong in VIP room are embedded in the screens to add certain human cultural elegance to the background. The point light sources are distributed in the public dining area where the lighting shall be strengthened. The columns enclosed by the mirror finishes have enriched the space contents in the restaurant while the guest rooms are arranged elegant and comfortable with the refraction on the slight shadows of a combination of Chinese and Western. As long as you walk, you always can feel that the designer has incisively and vividly applied the unique design conception to the glory and nobility of the combination of human and scenery.

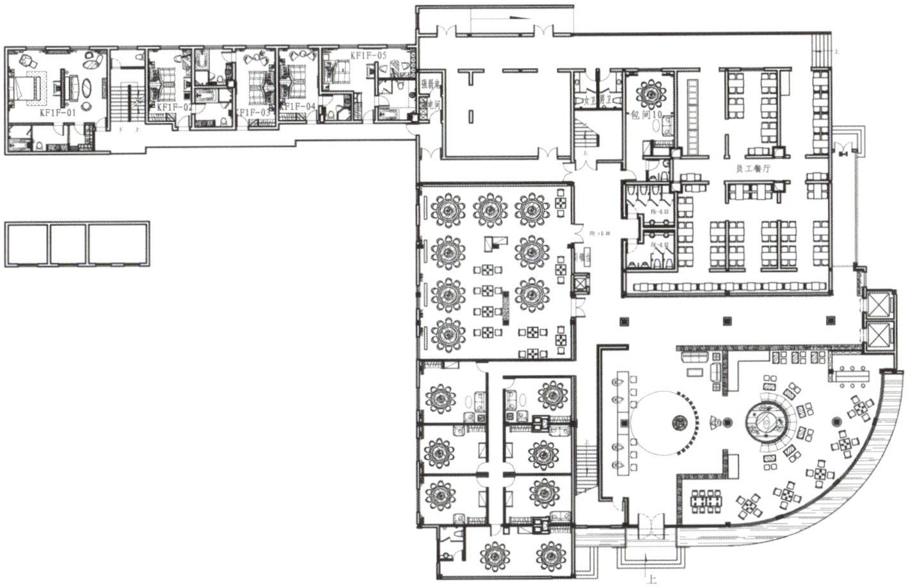
KF1F-01
KF1F-05
包间10
员工餐厅

在体验中认知
recognition through experience

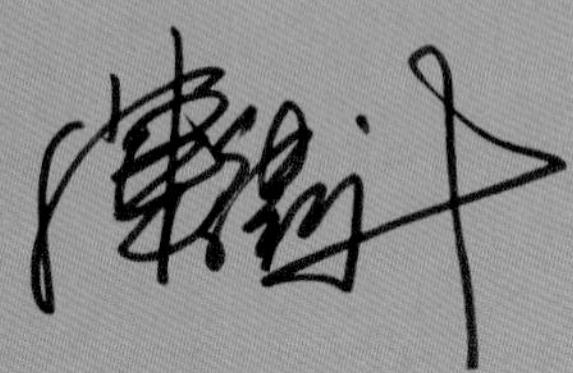

陈卫新 Chen weixin

C+S筑内空间设计顾问有限公司总设计师

高级室内建筑师

潘开富
仲松
琚宾
余平、董静
耿治国 、 Gustaf Kan
范江
小川训央
陈旭东
卢文伟

丰硕茶楼
“颂”音乐酒吧
“唇”酒吧
瓦库2号
CJW酒吧
月波楼
DE LA COAST
一品茗茶楼
常州颐和会馆

BAR TEA HOUSE

随着国内业主与消费者在设计风格的接受程度上的进一步"跃进"，酒吧·茶楼项目的设计风格更具有多样性的倾向，从中国古典风格，到具有当代性的现代风格，大家做设计的切入角度是多种多样的，也反映出目前中国市场的现状，反映出决策者、思想者，以及所有的空间参与者的包容性。在我看来，无论是国际化的视角还是基于中国传统文化的东方式设计，要达到两者完美结合，才是真正的挑战。视觉上的惊喜与享受，是短暂的？还是耐人寻味的？各种设计观念的自如表达，正以"海选"的姿态迎合时代潮流的审美评判。

茶楼，从古到今，它的存在意义都是相似的，但功能却在不断的时代更替中丰富多彩。今天的茶楼，更多的是包涵餐饮娱乐休闲的一个大的概念，特别是20世纪90年代后，港式茶餐厅以及随后的浙江式茶楼，对国内茶楼、茶餐厅的影响巨大，使原先茶楼的概念更多的具有了"时尚、复合"的新意义。相对茶楼来说，酒吧设计更多地陷入一个困境。国内近几年酒吧像雨后春笋般纷纷涌现，酒吧空间常利用天花的升降，地坪的高差，以及围栏、列柱、隔断等进行多元次的空间分割。构成特殊的环境空间，似隔非隔，隔而不断。在这样的动态空间引导大众从动的角度看周围事物是十分重要的。所以此类项目的平面规划及动线设定是生意成败的一个重要因素。决策者对设计者的信心完全取决于过去"业绩"的成功。许多设计师过多地追求新材料、眩灯光、酷造型的所谓"场所感"，而导致设计被形式所缚，作品的人性化与舒适感较差。可以这样说，在形式美的设计已不再很困难的当下，人们更关心的是空间，以及体验过程中的种种感受。好的设计不是一晃而过，更要让人不知不觉中深思，自然而然地呼应参与，以至成为空间设计最终效果的参与者之一。在此过程中产生的身份认同或归属感完全来自人对空间的认知与交流。只有好的设计才会让人在不自觉的过程中，听到自己心灵的声音，并且懂得，时光在这样精致小型商业空间里一点点逝去的意义。

茶与酒的两大归宿，表面看来，一静一动，因人而异，似乎体现人们在文化上的分野，实质却是异曲同工。一流的室内设计是以个性的出类拔萃与功能划分的完美统一赢得普遍喝彩的。一个城市的建筑的美感大处在于"天际线"，小处就在酒吧、茶座等个性空间的细致关怀。无论对这个城市的居民还是过客，这两处同样重要，这是整个城市文化的品位倾向在建筑上的具体反映。

With the further "leap" of the acceptance extent on the design styles by the domestic clients and customers, the design styles of such project works have a tendency of more diversity. From Chinese classic style to the contemporary modern style, the starting points the designers have made during the design are various and diversified, which also have reflected the current status of Chinese market, and compatibility of the decision-makers, thinkers as well as all space participants. In my point view, no matter how you assess them from the internationalized viewpoint or the eastern-type design based on Chinese traditional culture, only the perfect integration of both sides is a real challenge. Are the visual surprise and enjoyments transient or thought-provoking? The free expression of various design ideas are faced with the aesthetic evaluation of the times trends in posture of "mass selection".

Tea house from ancient times to the present has similar meanings in its existence, but its functions have been wonderful in the constant replaced times. Today's tea house is more like a big conception including catering, amusement and recreation. Especially in the late 1990s, the Hong Kong style cafes and subsequent Zhejiang style tea houses have had great influence on the domestic tea houses and cafes to endow the conception of the existing tea houses with more meanings of "fashion and composition". As compared with the tea houses, the design of bars has more got into a dilemma. The domestic bars have emerged suddenly like the bamboo shoots in spring in recent years. The bar spaces are normally divided through the rise and drop of the ceilings, the height difference between the floors as well as the rails, columns and partitions etc. to constitute special environmental spaces, which look separated but actually not. It is very important to guide the people to view the surrounding objects from a dynamic angle in such a dynamic space. Therefore, the plane planning of such projects and the setting of dynamic lines is a key element to the success of a business. The confidence of the decision-makers on the designer completely depends on the success of the previous "achievements". Many designers have excessively pursued the so-call "sense of place" with new materials, brilliant lights and cool styles so that the design is confined by the forms and the humanization and comfortableness of the works are quite bad. It can be said that the beauty of forms is not the most difficult any more. What people are concerned more about are the various feelings of the spaces and experience procedures. A good design shall not flash away. More importantly, it can enable people to think deeply unconsciously, respond and participate naturally and finally become one of the participants in the final effects of the space design. The recognition of the identification or sense of belonging generated during this procedure completely comes from the acknowledgement and communication with the space by people. Only a good design can enable people in an unconscious course to hear the sound of his/her soul and also understand the meaning of which the time passes away little by little in this small delicate business space.

As for the two end-results of tea and wine, superficially, one looks quiet and the other looks active and varies due to different people, which seems to have reflected the people's divergence on the cultures. But it is actually different in approach but equally satisfactory in result. The first-class interior design has won the universal applauses through the excellence of individualities and perfect integration of function divisions. The aesthetics of the buildings in a city mainly lies on the "horizontal line" while it is the detailed care of the individual spaces in the bars and tea houses etc. from small point of view. Irrespective of a resident or a passer-by in this city, both are equally important. It is the specific reflection on the buildings by the cultural taste of the whole city.

设计单位:南京智点设计顾问有限公司 / 设计:潘开富 / 参与设计:徐高林、戴婷

丰硕茶楼
Fengshuo Tea House

面积:630m² / 坐落地点:张家港 / 主要材料:五莲花、实木、亚克力

享受自然的美景，倾听鸟儿的歌声、观赏鱼儿的游动……整个空间使人感觉悠然自在。

对于肌理墙面的创意，设计师运用现代的艺术手法，通过不同材质，不同组合方式，不同的纹理走向，同时使其具有某种使用功能，最终传达某种文化、色彩与精神层面的内容，这也许就是设计师想要展示给顾客的。

朦胧的灯光让那一份静谧流淌在每个角落，更富有一丝神秘，置身其中，你会享受那一份放松与浪漫。

Enjoy the beautiful natural sceneries, listen to the songs of the birds and watch the swimming of the fishes…… the whole space makes people leisurely and comfortable.

As for the creation of the textured wall surface, the designer has applied the modern artistic techniques to finally deliver certain kind of culture and the spiritual contents through different materials, combination methods and direction of different textures while making it with the certain functions. This may be what the designer would like to display to the customers.

The dim lights flow the quietness over every corner, even with a mystery. Staying in it, you can enjoy that relaxation and romance.

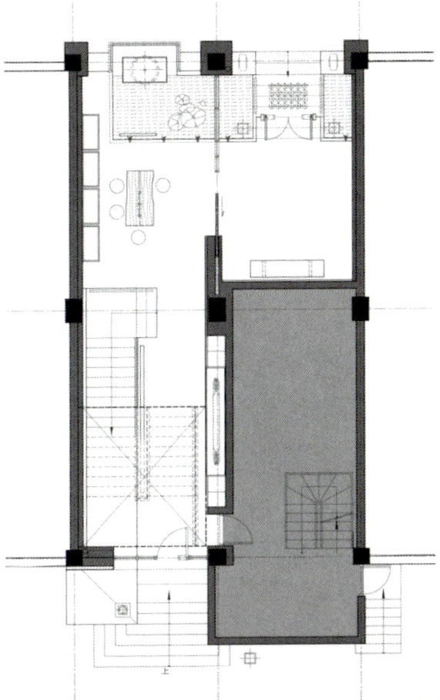

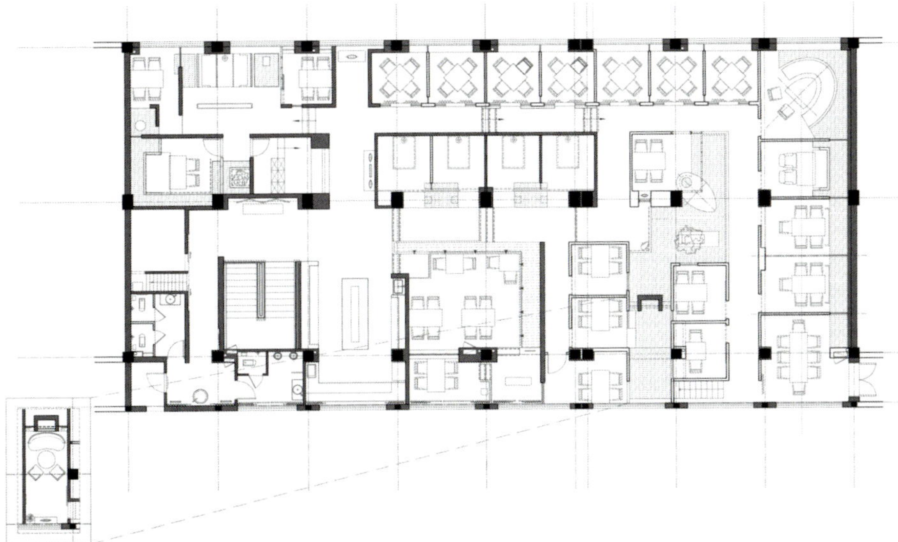

设计：仲松

“颂”音乐酒吧
“Song” Music Bar

面积：410m² / 坐落地点：北京 / 主要材料：玻璃、木板 / 摄影：莫尚勤

“颂”酒吧坐落于北京新兴的时尚购物中心“世贸天阶”地下一层，它与众不同的空间设计意念与三位来自不同领域的时尚人士密切相连。一位是著名的音乐人、一位来自餐饮业，还有一位即设计师仲松。

本案的设计理念和手法相当纯粹。设计师将对雕塑艺术的敏锐和认知，结合自身的判断力和实践在酒吧进行一次疯狂地“雕琢”。项目定位一个为音乐而生的空间，通过声乐变化原理与空间结合，设计成连绵起伏的“盆地景观”。层叠错落的大量木制曲面造型，不仅使空间变成“山间梯田”，同时从音乐的功能性衡量，它营造了完美的声场，乐曲如同木制曲面般缓缓展开，流畅的线条萦绕在整个空间。白色的座椅依地势起伏分布，餐台下发光的不规则弧形有如沼泽。不同的高差空间使人们的行为也随之充分地释放。

空间的另一特别之处是在“盆地景观”高层，分别设计了两个钢结构制成的空间作为多人进行交流和休憩的独立场所，它们依旧被包裹在流动的曲线间。这两个空间原本是整体空间的不规则边缘地带，设计师利用这里的地形，通过与整体线条的融合和材质及家具的选择，使其呈现出别样的海洋游艇风格——白色宽大的沙发，数控技术雕刻中木制密度板形成流畅的线条空间，仰望天花板，白色也包裹其间。此时，昏黄的灯光和众多现代的实物，在酒与音乐的熏陶下，仿若徜徉在大海之中……

“Song” Bar is located on the first underground floor of the new developed fashion shopping center in Beijing. Its unique space design idea has been closely connected with the three fashion personages from different fields: one is a famous musician, one from catering industry and another is the designer Zhong Song.

The design idea and technique of this case is quite pure. The designer has combined the acuity and recognition on the sculptural art with the individual judgment and practice to make a crazy “carving” in the bar. The project is orientated on a space born for music. Through the combination of the theory of the vocal musical changes with the space, it is designed into a stretched and undulated “basin landscaping”. The large quantity of the modeling in the cascaded curved wooden surfaces not only makes the space to be a “terrace in the mountain”, but also has created a perfect sound field in view of the functionality of the music. The music has extended slowly like the curved wooden surfaces and the smooth lines have lingered over the whole space. The white chairs are arranged along the topography while the irregular luminescent arcs under the dining tables look like a swamp. People’s behaviors are consequently released through the height differences of the spaces

The other special point of the space is the high position of the “basin landscaping”. Two spaces made of steel structures are designed separately as the independent places for people to communicate and take rest. They are still wrapped in the flowing curves. The two spaces used to the irregular edges of the whole space. The designer has applied the topography here to demonstrate a special style of ocean yacht through the integration with the entire lines and the selection of the materials and furniture - the wide white sofas. The wooden MDF boards carved by digital control technique have formed a space with smooth lines. Looking up to the ceiling, the white is also enclosed inside. Meanwhile, the dim lights and various modern material objects look like roaming in the sea under the edification of wine and music……

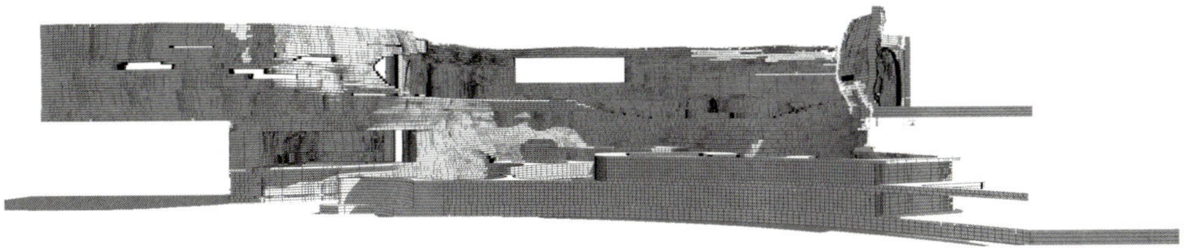

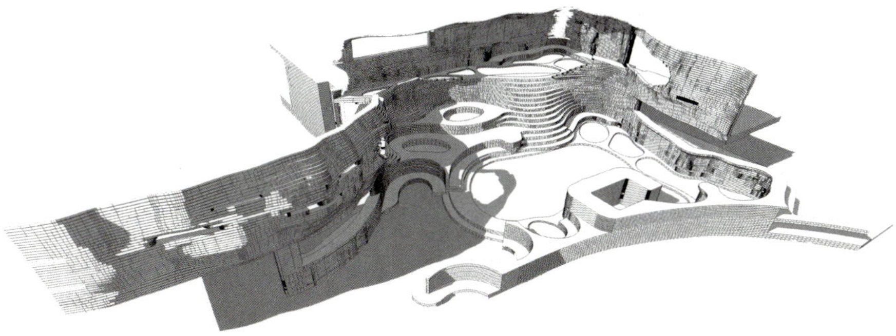

设计:琚宾

"唇"酒吧
lip Bar

面积:230m^2 / 坐落地点:深圳 / 主要材料:白色微晶石、木材、GRG、地砖 / 摄影:方震鹏 琚宾

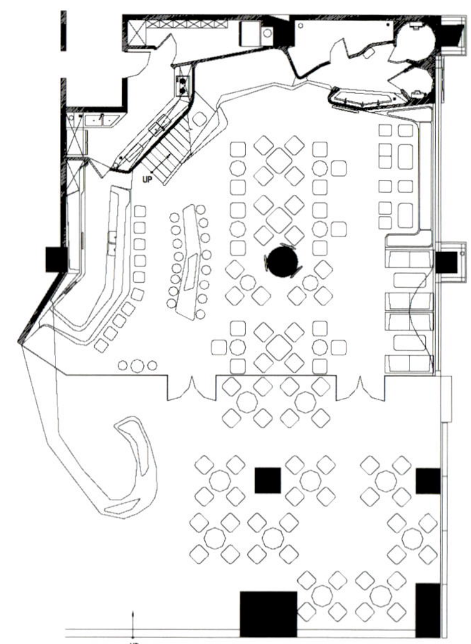

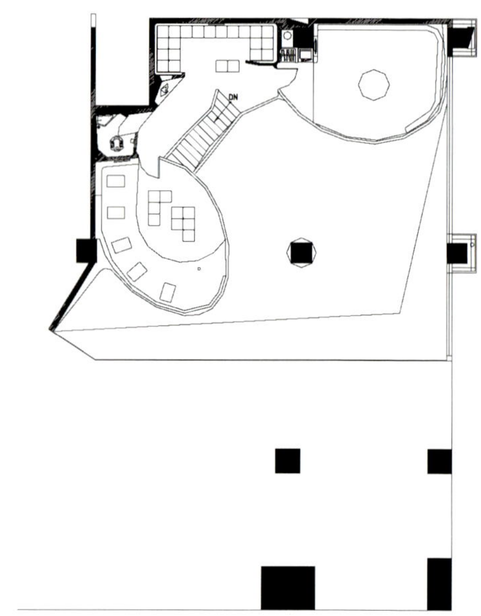

设计和施工这个酒吧的过程是一次很难忘的经历，虽然项目比较小，对于做习惯了几万平米酒店的我，从空间的角度上来说，像是被限制了发挥一样。但此项目的承接前提是我的完全主导，这样就比较有意思了。

刚开始设计的时候，自己也不知道要做成什么样子，完全是主观的概念。看过现场后，心中已经有了想法。色彩的统一性，形体的变化在空间中形成装饰的艺术。通过灯光的变化来体现酒吧的多变性。的确，它很干净，除却灯光甚至有点像太空舱。干净，有种很纯粹的安静的感觉。在所处的那个比较繁华、物质的环境里，也许应该用比较小资来形容。是梦幻？或者是空灵？

白色的确是最完美的底色。这和个人喜好有直接关系，色彩不断的变幻，整个空间也跟着不断的变。变出萤火虫，变出星空，变得华丽，变得暧昧，直至迷离，直至忘记所处的环境。这种变化实现了对于设计的理解——空间的多样性与体验性，极端的安静与夸张的妩媚。

当初给我灵感的是那幅《我去世的兄弟》。自夸地说，作者达利大概怎么也想不到他的画能被我的作品这样实现从而“引领”时尚。

在设计的最初并没有想去颠覆哪种东西。寻找的念头应该是来自对原有状态的厌倦。也谈不上是对时尚的引领或当下的使命。但也许当你伴着音乐和不断变化的灯光徜徉在这个能给你体验感的空间时，才能慢慢体会到在设计最深处的东西和想表达的感受。光与声、视觉和听觉相互交融的同时也许会影响你的心情，也许主导的意味会更浓些。

或许“唇”这名字及名字给人带来的遐想就将这个酒吧带到一个比较“高端”时尚的定位。于是，新型材料对形态的体现再加上完全白色的空间，用一种纯洁中的宁静安抚那些疲惫或孤独的灵魂。或许只有在你的主观被这种空间打乱并迷惑后才明白，生命价值的很大一部分是来自对原始的怀念。

The process of the design and construction of this bar is a very memorable experience. Although the project is quite small, as for the one that has been used to designing the hotels of over ten thousands of square meters, with the viewpoint of the space, I feel as if I was confined with my development. But the premise that I undertake this project is that I can fully control the design, which will be quite interesting.

When the design starts, I have no idea of how to design it and it is a completely subjective conception. After visiting the site, I already have hade some ideas. The uniformity of the colors and the changes of the forms and structures will form into a decorative art. The variability of the bar is demonstrated through the changes of the lights. Certainly, it is very clean. But it looks like a space capsule without the lights. It is clean with a kind of very pure quiet atmosphere. In that very prosperous and physical environment where it is located, maybe it should and can be described as quite capitalism, dreamy or vacant?

White certainly is the most perfect background. It is directly related to the individual preferences. The whole space is also changed constantly as the colors change continuously. It either is changed into glowworms or the starry sky or becomes ambiguous or even misty till you forget where you are. This kind of change has realized the understanding on the design - diversity and experience of the space as well as the extreme quietness and exaggerated charm.

What has inspired at the beginning is the paint "My Dead Brother". Boastfully speaking, the painter Dali probably never supposes that his paint can be realized by my work in such a way to "lead" the fashion.

In the preliminary design, it is never thought of overthrowing anything. The seeking idea shall come from getting tired of the existing status. It neither can be regarded as leading the fashion or the current task.

However, probably only when you are wandering about in this space that can bring you the enjoyment along with the music and the constantly changing lights, you can gradually experience the things in the innermost of the design and the feelings to be expressed. The integration of light and sound as well as the senses of sight and hearing may affect your mood or more dominates you.

Maybe the name of "Lip" and the fancy brought by this name can orientate this bar in a quite "superior" fashion. Thus, the demonstration on the forms and structures by the advanced materials, together with the completely white space, has appeased those tired or lonely spirits with the peacefulness in a kind of purity. Maybe only when your subjectivity is disturbed and confused by this kind of space, you will come to realize that the most of life values lies in cherishing the memory of the original.

设计:余平、董静

瓦库2号
Wa Ku No. 2

面积:300㎡ / 坐落地点:西安 / 主要材料:瓦、草批墙、青红混色陶地砖 / 摄影:董克

“瓦库2号” 位于西安大雁塔脚下通易坊仿古一条街上，延续“瓦库1号”，同以瓦为元素的茶馆设计。“瓦库”不需分辨什么是古典、现代，什么是东方、西方，只需品茶论瓦，说故事。

“Wa Ku No. 2” is located in the themed street in Tong Yi Fang street modeled after ancient times beside Grand wild Goose Tower, Xi’an. It has continued “Wa Ku No. 1” and is also a tea house with the tiles as the design element. “Wa Ku” does not require distinguishing between the classics and modernity or between eastern and western. You only need to enjoy the tea, talk about the tiles and tell the stories.

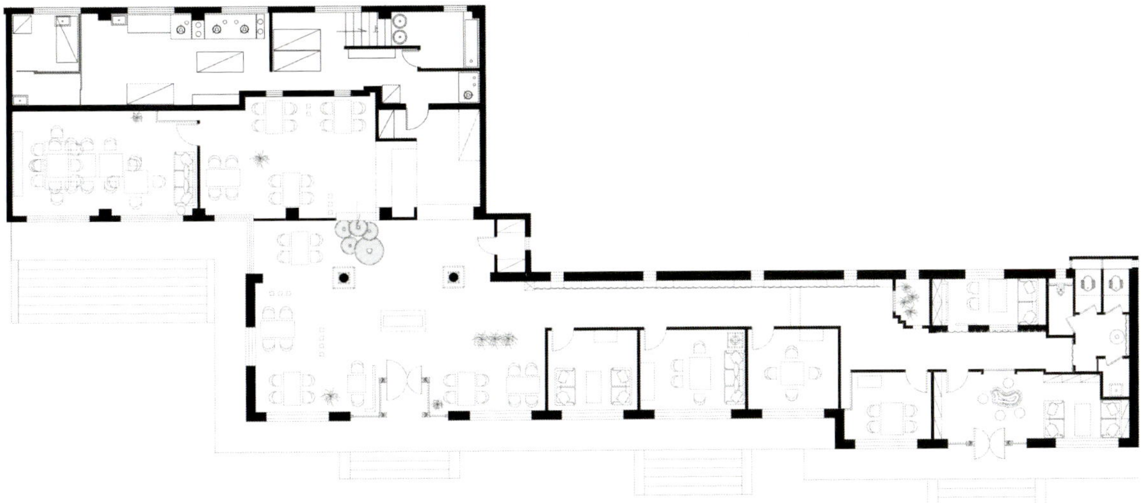

设计单位:飞形设计事业有限公司 / 设计:耿治国、Gustaf Kan / 参与设计:马立荣

CJW酒吧
CJW Bar

面积:1800m² / 坐落地点:北京CBD中心 / 主要材料:木花格、丝绒窗帘、金属、玻璃、壁纸、香柏木 / 摄影:Nacasa & Partners Inc

设计师为全案覆罩剧场式的华丽场景，让每座场域像似彼此扣叠的盒体，预示多层次的舞台效果。从户外露天区透过落地玻璃窗望去，室内像一座炫丽璀璨的珠宝盒。室内，舞台两侧置入对称的盒状量体作为lounge和餐厅位置，DJ台定身于舞台正对面，共同围塑中心低伏的舞池区。两端对峙的爵士live与DJ秀酝构着轮相竞演的双主轴型式，各据一方的lounge和餐厅又彼此在盒框内上演交谊动态。通往包厢的昏黄长廊，左右侧墙面各自铺陈深红布幕与黑色壁纸，延续大厅醇厚的剧场氛围。透过不断层叠的舞台意象，让宾客能恣意选择浅尝游逛或停逗，回应品牌与时俱进与复合娱乐等特色。

运用现代简练手法与素材糅合东方语汇，回应北京浓烈的文化幽思及刻画时代新意等多样面貌。舞台两侧的木盒从古典窗花再演绎，细木组织的密网成为天花板与立面的皮层，并为空间界定出墙屏、窗扇、洞口等造型，浅木色细网托衬着深色背景，在形影交错之间引发天棚深邃的遐想。门厅、lounge、餐厅、酒柜运用古典月门作为联系语汇。百宝格酒架与龙纹木雕吊灯，座椅穿着中国传统花布，借由不同时代的特征缩影与经典文化展开密切回应。

The designer has covered this design case with a theatric luxurious scene and made every area like the boxes overlapped with each other to indicate a multi-level stage effects. Viewed through French window from the outdoor open area, the interior is just like a brilliant and bright jewelry box. On both sides of the interior stage, the symmetrical boxes are set up as the lounge and the location of the canteen. DJ desk is just positioned directly opposite to the stage and is enclosed into a dancing area where the center is quite low. Jazz live and DJ show confronted with each other on both ends have been arranging the competitive performances in a dual major role type. The lounge and canteen occupying each individual side are carrying out an interaction with each other in the box frame. Along the dim long corridor direct to VIP rooms, the walls on both sides are covered with dark red cloth curtain and black wall papers, which have continued the strong theoretic atmosphere in the hall. Through the constant overlapped stage images, the visitors can freely choose to have a try and stroll around or be indulged in it to respond to the characteristics that the

brand catches up with the fashion as well as composite amusement.
The modern simple techniques and materials are applied in combination with the eastern languages to respond to various aspects such as the strong cultural mediations in Beijing and depiction of the times' new ideas. The wooden boxes on both sides of the stage are deducted from the classic paper-cut. The dense nets formed by the fine wood become the skins of the ceiling and facade and define the wall screens, window sashes and holes etc. for the space. The light fine nets set off the dark background and lead to the fancies on the profoundness of the sunshade. The classic moon gate is applied as the connecting vocabulary for the hallway, lounge, canteen and liquor cabinets. The carved dragon image lamps and treasure boxes in Chinese style as well as the chairs covered with Chinese traditional printed cloth are closely echoing with the classic culture by means of the characteristic epitomes in different times.

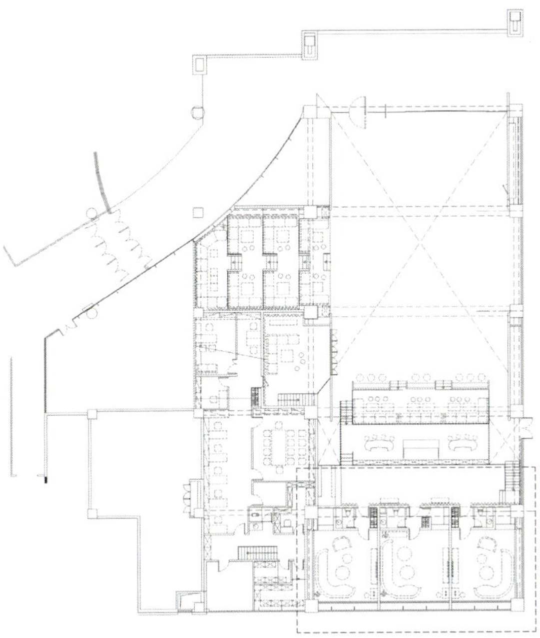

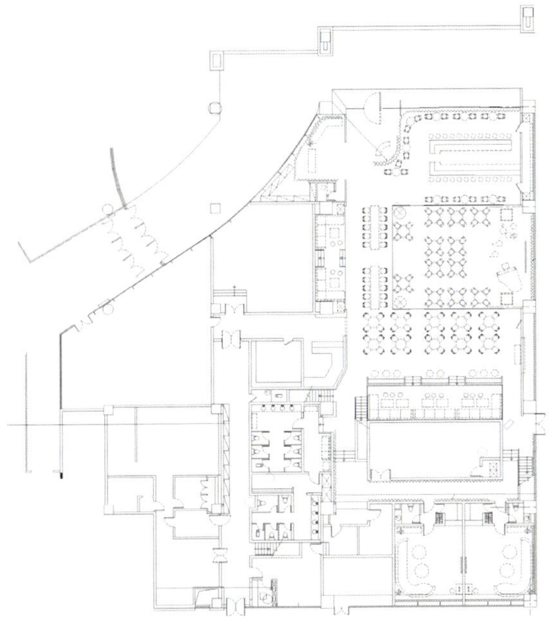

设计单位：宁波市高得装饰设计有限公司 / 设计：范江 / 参与设计：郑磊

月波楼
Moon Light Pavilion

面积：500m² / 坐落地点：宁波 / 摄影：文宗博、潘宇峰

位于宁波东钱湖旁的月波楼，原是南宋史浩的藏书楼遗址，现重新修整打造成精品茶楼。设计师从营造文化气息入手，宋朝始有活字印刷，设计师把以前历代歌咏东钱湖的诗句打散进行重新排版，做成竹木雕刻的字出现在主景中，但有几句却是完整的，并且以金色突出，可以朗朗上口咏诵，表现历史在失散中渐渐得以复原。展示品亦如书架、各色茶叶、茶点类在玻璃罐中被文雅地展示着。主材是黑胡桃木，墙布是丝绸质地，华贵中透出优雅。前尘往事，仿佛看到宋代的骚人墨客吟诗喝茶，在动荡的年代文事却依然兴盛，生活依然奢迷讲究。

湖光山色，波光涟漪，茶馆的楼上做为贵宾接待，可以开个小型的茶会，楼下观湖品茗，湖上阳光轻洒，微风吹过。可闻悬挂的金属片互相撞击发出的叮咚声，可见微微的金色灿灿。玲珑精致、茗香四溢，此刻的惬意是不需用言语来描绘……

Moon Light Pavilion, located beside Dongqian Lake in Ningbo, was formerly the ruins of Shi Hao's Pavilion Library in South Song Dynasty and is renovated to be an exquisite tea house. The designer has begun with the creation of a cultural atmosphere. Since the movable-type printing has appeared in Song Dynasty, the designer has broken up and recomposed the poems that appraise Dongqian Lake in the past dynasties and made them carved in the bamboos and woods to appear in the main scene. But some sentences are complete and made prominent in gold which you can read aloud to demonstrate that the history gets restored gradually in the loss. The exhibited articles such as book shelves, teas and snacks are displayed elegantly in the glass jars. The main material is black walnut and the wall cloth is silk, luxurious but elegant. The past and old stories can be imaged here, as if you see that the poets and writers in Song Dynasty were reciting the poems and drinking teas. The cultures were still prosperous and the life was still luxurious and dainty even in that turbulent time.

The lakes and mountains are beautiful while the waves are rippling. VIPs can be received upstairs in the tea house, where small tea gatherings can be held. People can view the lake and enjoy the teas downstairs. The sun lights are spraying on the lake with breeze passing by while the suspended metal pieces bump with each other to give off the tinkling sounds with view of the light glittering gold. It is exquisite and delicate, overflowed with tea fragrances. The pleasure at this moment does not need any language to describe……

月波楼

设计单位：上海英菲柯斯设计咨询有限公司 / 设计：小川训央 / 参与设计：佐佐木 力

DE LA COAST

面积：492m² / 坐落地点：上海 / 主要材料：马来漆、烤漆玻璃、石材、地毯、乳胶漆、异形孔板、通花板、木甲板、马赛克、铁片刻花 / 摄影：胡文杰

这次项目进行中的焦点是外滩夜景的展示方法。在同一区域内逐日激化的开店风潮中，作为少数的日资企业参与者，如何来理解这个得天独厚的景观优势，怎样向世界上各个不同的人种描述外滩的夜景，成了一个重大的课题。

驻足在上海的街头，外滩几乎是上海城中一个永恒的存在，当我们考虑为何大家要来到这里消磨时光时，发现选择了这里的每一个人，都是为了亲身体验某些“非日常”的感觉。而若是把这个着眼点换成我们店面设计的视点，就会看到，不管正式也好休闲也好，我们希望当天客人不管是怎样的心情怎样的目的，都能在本店度过一段轻松惬意的时光。那么，怎样达到这个目的，是至关重要的一点。至于在店内用何种方式消磨时间，则全由客人自己选择。因此，由两层构造而成的店面内，我们尝试着在各层各区域，在每一处都以某个仿佛度假胜地般的非日常性为主题，通过色彩与夜景观赏方式的变化，让客人在店内各处都享受到不同的乐趣。通过把外滩的夜景本身作为店内设计的一部分嵌入整个空间内，营造出仿佛与外滩的晚风以及夜空融为一体的美好气氛，让人感觉远离了尘世日常的喧嚣与烦恼。

The ongoing focus in this project is the display method for night scenery of the bund. In the gradually intensified shop opening current in the same area, as one of the few Japanese participants, how to understand this unique scenic advantage and how to describe the night scenery of the bund to the different people from the world have become an important subject.

Standing in Shanghai' s street, the bund is almost the everlasting existence in Shanghai City. When we are thinking why everybody comes here to kill time, we find that everybody here would like to experience certain "very Japanese" feelings in person. However, if we take this aspect as the visual point of our shop design, we can see no matter formal or leisure, we hope that every guest can enjoy a period of relaxing and comfortable time in our shop no matter what kind of feeling he/she has that day. Then, how to achieve such target is the extremely important point. The guests can choose by themselves how to kill time in the ship. Therefore, in this two-storey shop, we have attempted to take certain holiday resort -like non-daily type as the theme in every area of each storey so that the guests can enjoy different pleasures everywhere in the shop through the changes of the colors and ways of watching the night scenery. The night scenery of the bund itself has been taken as part of the interior design and embedded into the whole space to create a beautiful atmosphere in which it seems that the night wind at the bund has been integrated together with the night sky so that the people feel far away from the daily noisiness and annoyances in this world.

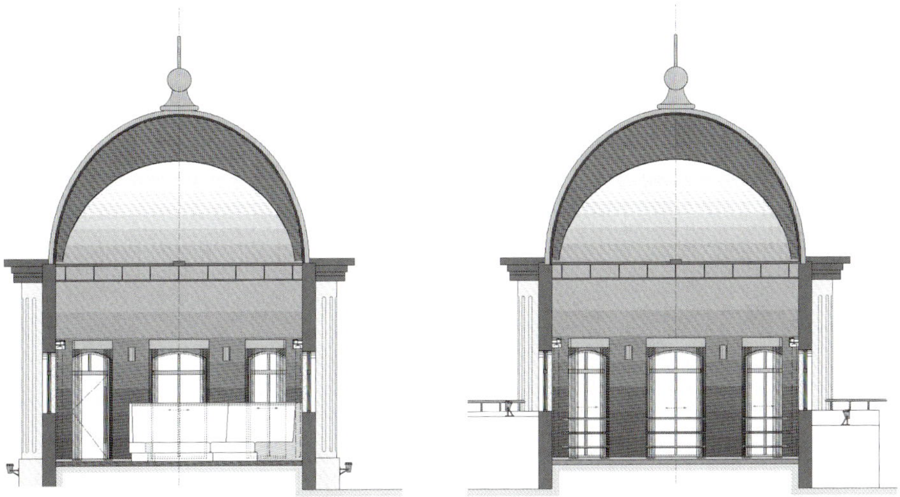

AURORA

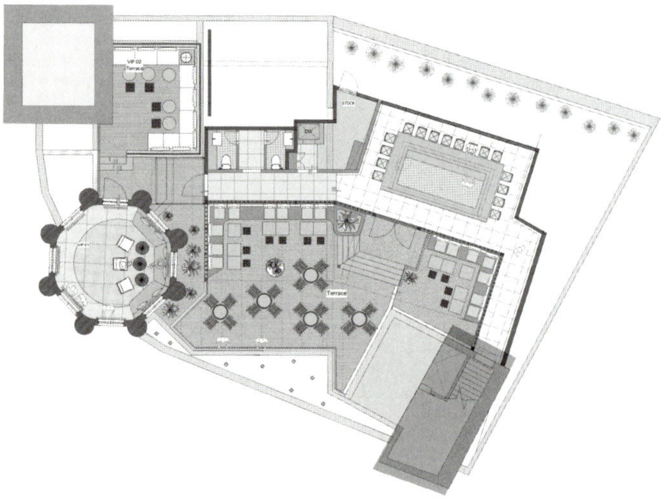

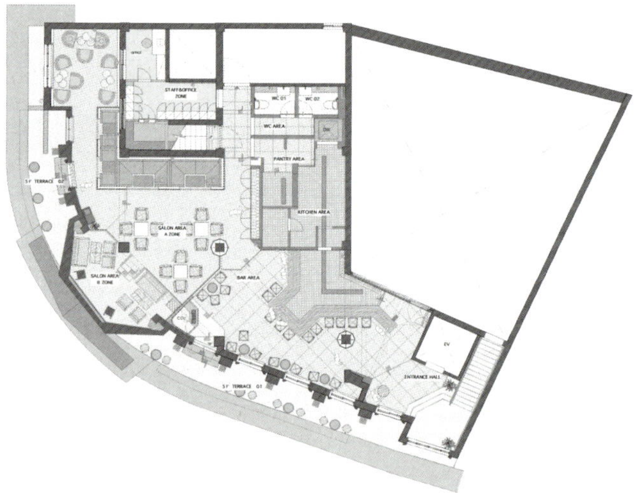

设计：陈旭东

一品茗茶楼
Yipinming Tea House

面积：2000m² / 坐落地点：长春 / 主要材料：地砖、花岗岩、柚木、玻璃、马赛克

本案中设计师只通过部分陈设和点缀强调中式意境，把设计的重点放在营造“意向自然”的境界上，材料的选用以自然为范本，突出人文回归的生态理念。色彩也以材料自身的固有色为主，只有少数地方用了一点中国红，使空间在大的调合中得到局部对比。中式的语境则通过那些垂挂的灯笼、刺绣茶巾、甚至窗前的一抹竹影来体现。

茶楼营业区设在三四楼，一下电梯就是个比较狭窄的空间，因此设计师在电梯对面的墙上安装了通体的镜片，使空间得到深度扩展，为防止大面镜片造成的直白、呆板效果，镜墙用无数方形木格分开，像中国古建筑中的花窗一样。走过镜墙就进入一个渐走渐宽的扇形空间，产生的透视感让视线向前扩展延伸。人造小瀑布是走廊的焦点，不仅增添了过厅的情趣，而且也给隐于其后的茶室带来了生气。茶室以瀑布为分隔，与过厅隔开。没有“瀑布优势”的包厢也巧妙利用玻璃、屏风等材料，既保障了各包厢的私密性，又可感受包厢以外的空间，产生心理上的连续感。茶室中重点的陈设、景观以及茶几都用装饰性灯光打亮，为突出这些强调的“点”，茶楼中没有设基础灯光，只有在室内墙体前后错开的地方，为强调形体的层次关系，在中间加了光源灯带，使得形体在光的映托下层次更加分明、清晰。

In this case, the designer emphasizes on the Chinese-style artistic conception only through partial display and ornament and stresses the design on the condition of "intentional nature". The materials are selected with the nature as the sample to lay stress on the ecologic idea of humanistic return". The colors are mainly the inherent colors of the materials and only few places are applied with Chinese red so that the space can be found with the partial contrast in the big reconciliation. The Chinese-style context is reflected through those suspended lanterns, embroidered tea cloths and even the bamboo shadows before the window.

The business area of the tea house is located in the third and fourth floors. There is a quite narrow space once stepping out the lift. Therefore, the designer has installed a complete mirror opposite to the lift so that the space gets expanded in depth. In case that the large-area mirror will create a direct and stiff effect, the mirror wall is divided into numerous square wood grids, just like the rose windows in the ancient Chinese architectures. Passing through the mirror wall, there is a sector space which becomes wider and wider as waling. The generated perspective sense extends and expands the sights forward. The artificial cascade is the focus of the corridor, which not only increases the interests to the hallway, but also brings the lives to the tea house hidden behind. The tea house is separated from the hallway by the cascade. VIP rooms without "advantage of the cascade" are also applied with the materials such glass and folding screens artfully, which ensures the privacy of each VIP room. At the same time, the space outside VIP rooms also can be felt and a sense of continuity is generated in the mind. The key furnishings, landscapes and tea tables are lighted up by the decorative lights. To lay stress on these "points", the tea house is not installed with the basic lights. Only in the places where the interior walls are staggered, to emphasize the hierarchical relationship of the structures, the light source strips are installed in the middle to reflect its hierarchies more distinguished and clearer by the setoff of the lights.

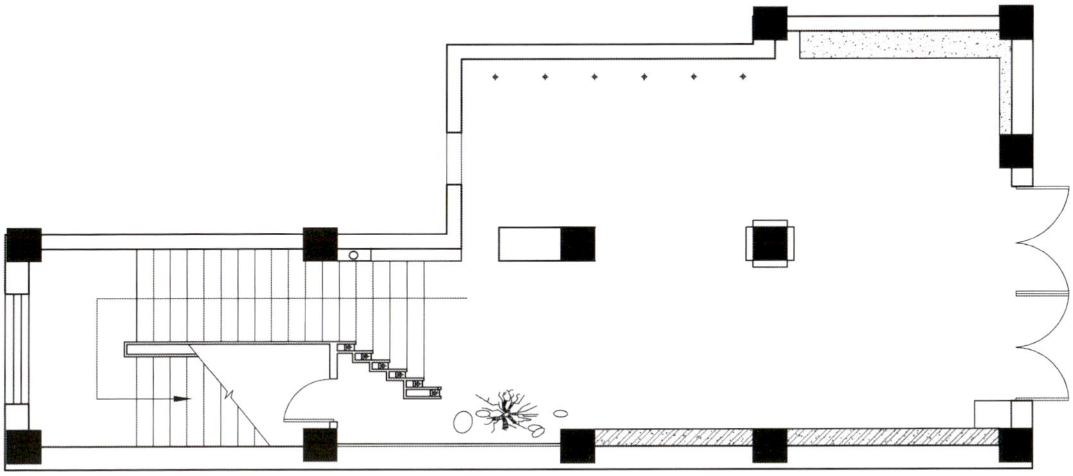

设计单位:杭州历程装饰设计有限公司 / 设计:卢文伟 / 参与设计:沈建方

常州颐和会馆
Changzhou Yihe Club

面积:1500m² / 坐落地点:江苏省常州市 / 主要材料:胡桃木 / 摄影:文宗博

本案具有颐和会馆徽派民居的特征元素，如“三雕”，同时综合了会馆的特定文化内涵及功能需求。

走进颐和会馆，可以充分地感受到江南民间的吉祥图案、宗教人物、戏曲故事、山水、花鸟、虫鱼等。在此有限的空间内，运用了徽派的山地特征，风水意向和美饰倾向。布局的中轴线对称，面阔三间，中为厅堂，两侧为室，厅堂前方天井的采光通风，亦有“四水归堂”的吉祥寓意。在装饰方面，大都采用砖、木、石雕工艺，如砖雕工艺的门罩，石雕塑的漏窗，木雕塑的窗棂、楹柱等，使其室内建筑精美如诗。进入门厅，古老金砖地花的上空悬挂着八角藻井，下方安坐着如来的佛像，实感寂静及空灵。砖雕门楼在取食空间内的表现，表达了业主的哲学理念和生活情趣。一座砖雕门楼，内涵如此丰富，构成一部中国民俗学的百科全书，真不愧有“东方凯旋门”之美誉，现为颐和会馆的镇馆之宝。

坐下来细细品味杭州虎跑泉水冲泡的西湖龙井和各地名茶，在古筝、各种戏曲的演奏表演中一饱耳福。在缕缕茶香和幽幽古曲中，慢慢地欣赏以往必须跋山涉水、远赴皖南才能欣赏到的古色古香，真正领略到徽派文化的艺术魅力。

This case has the characteristic elements of Hui-style residences in Yihe Club such as “three carvings” and also is integrated with the specific cultural meanings and functional requirements of the club.

Once entering Yihe Club, you can fully feel the lucky patterns, religious characters, stories in traditional operas, mountains and water, flowers and birds well as well insects and fishes in the folk of the South of Yangtze River. In this limited space, the features of hilly counties, geomantic meanings and aesthetic tendencies in Hui-style have been applied. The layout is symmetric along the central axis. It consists of three rooms with the hall in the middle and rooms on both sides. The lighting and ventilation of the courtyard in front of the hall also delivers a lucky message of “four types of water gather in the hall”. With aspects of decoration, brick, wood and stone works are mostly applied, such as the door shelter in brick carving workmanship, stone-sculpture ornamental perforated windows, and wood-carving window bars and columns etc. make its interior building delicate like a poem. In the hallway, octagonal caisson is overhung in the ancient golden brick floor ceiling while the real image of Buddha is seated below, which really looks quiet and unpredictable. The presentation of brick-carving gate tower in the food-fetching space has shown the philosophic ideas and living enjoyments. One brick-carving gate tower has such wonderful meanings and has become an encyclopedia of Chinese folklore. It is really worth its good fame as “Eastern Triumphal Arch” and now is the top treasure of Yihe Club.

Sit down to attentively taste the West Lake green tea and the famous teas from various places that made by the water from Tiger Spring in Hangzhou and enjoy the performances of a Chinese zither with 25 strings and various kinds of operas. In the traces of tea fragrances and faint ancient operas, you can slowly enjoy the antiqueness and actually appreciate the art charms of Hui-style culture that you can enjoy in the past only when you had to take a long way to the south of Anhui Province across the mountains and rivers.

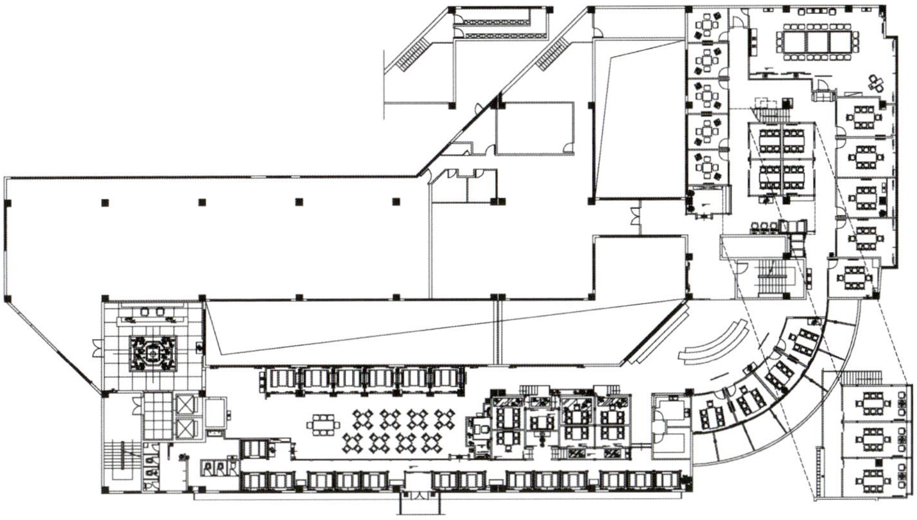

承恩數上南薰殿
退朝擎出大明宮

TRANSFORMATION OF THE OLD BUILDING

老建筑改造的观察与思考
Observation and Consideration on the Modification of Old Buildings

满登 Man deng
特约资深设计评论人
上海满登空间设计事务所创意总监，
上海市装饰行业协会环艺设计专委会副主任
中国建筑装饰设计专业委员会专家委员
IFDA国际室内装饰设计协会上海分会会长
中国室内设计研究院特聘专家

谷腾	灰域
朱岚	那格九号工作室
内建筑设计事务所	达利国际
毛桦	二十四城会所
登琨艳	比利时小馆
童明	苏泉苑
张成喆	复城国际销售中心
Morita	Visus
张成喆	IADC设计工作室
陈旭东	茶怡轩
冯志业	无限度广场改造

老建筑家族里不仅涵盖了名人故居、经典历史遗迹，还有旧厂房、旧仓库、旧民居等一些破旧的建筑，表面上老建筑的凋零破旧妨碍了观瞻，其实这里面却孕育了一个新的“设计生命”，她能使我们的城市文化“底蕴”更有品位。这些老建筑承载或见证了一段历史文化档案。假如因有碍城市的发展要拆掉它们，无疑就是“销毁”城市历史的“档案”。一个城市的发展和成长不能只看新建筑的规划，还要看老建筑历史和改造能否为我们的城市形象增添新的活力和文化底蕴？一个城市的崛起和认同是少不了老建筑历史的“铺垫”的。

因此，老建筑改造就成了设计界最为关注的地方，当然，也是最具挑战设计师的思想境界的地方。那么，如何看待老建筑改造？老建筑改造的设计语言又是什么？老建筑要成为城市记忆的一张新名片，我们又应如何去面对……

如果从城市建设多元化方向看，老建筑改造既可以是“修旧如旧”的，也可以是“修旧创新”的。前者是对老建筑的怀念和尊重，而后者则是老建筑要焕发“青春”的一种时代呼唤。也就是说，老建筑既有“时间”的记忆，又有历史文化沉淀的设计，也有现代气息和文化时尚的需求。两种老建筑改造的思维模式我认为是不存在谁“是”谁“非”或者谁比谁好的问题。如果说“修旧如旧”是还原老建筑的“原汁原味”，是对老建筑保护和怀念的最佳诠释，那么“修旧创新”则是对老建筑“老宅新生”的一种期待，一种现代人对老建筑改造的现代“情结”。它们传播的是“殊途同归”——城市建筑文化与设计需求文明的多样性。

当然，老建筑改造传播的文化信息、历史信息和时代信息是最为丰富多彩的。所谓丰富：老建筑改造与生俱来就有很多对比元素的运用，比如旧与新、传统与传承、传统和现代、保守与开拓、东方与西方、古典文明与现代文明之间的对比和对接会激发老建筑的“青春”回归。所谓多彩：老建筑改造注定是多元的，多角度多方位的发展趋势，顺应了我们这个时代和城市精神文明发展对老建筑改造的自然需求，也见证了我们这个时代与城市对老建筑历史的怀念与创新。衡量一个城市的发展，不仅要看GTP增长数据，更要看这个城市的气质、精神和品位，而老建筑改造则是衡量城市历史文化“底蕴”最有品位的“精神盛宴”。它满足了城市精神文明历史传承，也延伸了城市精神文明的文化内涵，同时也能“检阅”设计师在驾驭老建筑改造的思想境界和操作“底气”，我们可从年鉴的老建筑改造诸多项目中发现它的魅力所在。

The family of old buildings not only includes the former residences of the celebrities and classic historic relics, but also some old broken buildings such as old factory buildings, old warehouses and old residences etc. It seems that the declining and wornness of the old buildings have disturbed the view, but in fact it has gestated a new “design life”. It can make our city’s cultural “details” more tasteful. These old buildings have borne or witnessed a period of historical and cultural archives. If they shall be dismantled due to their obstruction on the development of a city, it doubtlessly is “destroying” the “archives” of a city’s history. The development and growth of a city can not be viewed from the planning of new buildings, but also shall be seen if the history and modification of the old buildings can increase the new energies and cultural details of our city’s image. The grow-up and acknowledgement of a city also can not lack the “underlying” of the histories of the old buildings.

Therefore, the modification of the old buildings has become the most concerned places in the design circle, and of course an area that most challenges the ideas of the designers. Then, how to treat the modification of old buildings? What is the design language for the modification of old buildings? To make the old buildings become a new name card for the memory of a city how can we be confronted with it?

In view of the diversification of the city construction, the modification of the old buildings can be “repairing old building as old one” or “repairing old building and upgrading”. The former is a memory and respect for the old buildings while the latter is a call of an era that the old buildings shall take on an entirely “young” look. That means that the old buildings not only have a memory on “time” but also is a design of historical and cultural accumulations. It also has the demands on the modern senses and cultural fashion. As for the thinking modes on the modification of two kinds of old buildings, I think that there is no problem on who “is” right and who is “wrong” or who is better than whom. If it is said that “repairing old building as old one” is to restore an “honest-to-goodness account” of the old buildings, and also a best interpretation to the protection and memory on the old buildings, then “repairing old building and upgrading” is an expectation on “new life of the old buildings”, a kind of modern “complex” on the modification of the old buildings by the modernists. What they distribute is to “reach the same goal by different routes” – the culture and design of the city buildings demands the diversification of the civilization, which is my basic opinion during my study on the modification of the old buildings.

Of course, the cultural, historical and times information distributed by the modification of the old buildings are most wonderful. The so-called wonderfulness is: the modification of the old buildings has inherently had the application of many contrastive elements: for example, the contrast and connection between the oldness and newness, tradition and inheritance, tradition and modernity, conservation and exploration, east and west, classic civilization and modern civilization will inspire the recovery of the “youth” of the old buildings. The so-called colorfulness is: the modification of the old buildings is destined to be a diversified, multi-angle and multi-orientational developing trend and has complied with the natural demands on the modification of the old buildings by the developments of our times and the city’s spiritual civilization and also has witnessed the memory and innovation on the history of the old buildings by our times and city. The development of a city shall be evaluated not only from the data of GTP growth, but more from the temperaments, spirits and tastes of this city. It has met the continuity of the history of a city’s spiritual civilization and also has extended the cultural meanings of the city’s spiritual civilization. At the same time, it also can “inspect” the ideas of controlling the modification of the old buildings and “basic strength and confidence” on the performance. We can find its charms from many projects about the modification of the old buildings in the yearbook.

设计：谷腾

灰域
Hui Yu

面积：975㎡ / 坐落地点：上海 / 主要材料：混凝土、钢结构、玻璃 / 摄影：谷腾

在更富有创造力的工作方式下，极其严密的等级空间概念将不复存在。水泥、钢和玻璃成为这个旧厂房改造的三大设计元素。电梯口外一个黑色的大球体，它代表地球，黑色有些沉静悲观的色彩，世间万物中美丽和残缺的东西并存，简单而不浮躁。重新拾掇起这个废旧的厂房，给予它新的生命。

More creative ways of working in this way, the concept of space and its strict hierarchy will no longer exist. Cement, steel and glass into the three major design elements of the old plant. Elevator area beyond great wall of a large black spheroid, it represents the Earth, some remained pessimistic black color, the world of beautiful things and found things coexist, and not simple rash. Re-repair of the old plant, giving it new life.

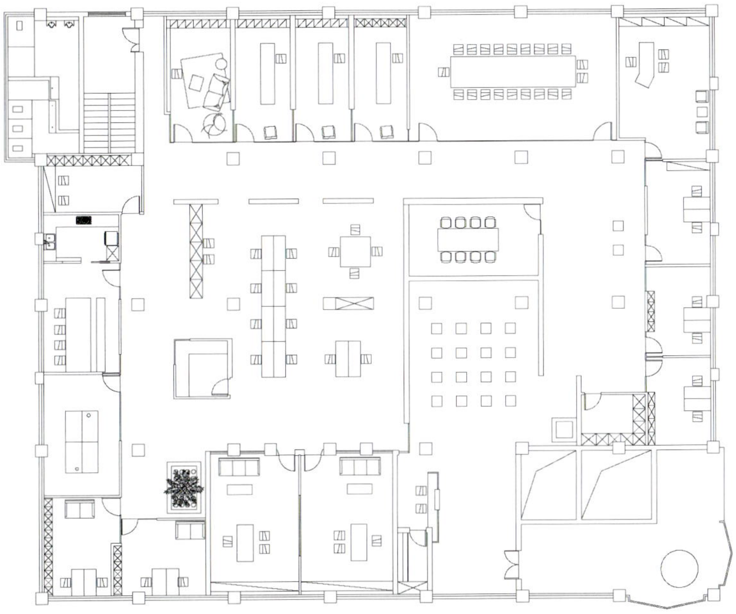

Wash
room

设计:朱岚

那格九号工作室
No. 9 Nage Studio

面积:170m^2 / 坐落地点:上海 / 主要材料:涂料、环氧地坪、防腐木地板、布、镜面、玻璃、陶瓷装饰品

本案原为羊毛衫厂的污水处理办公室，现改造为办公楼。在平面上大大调整了原来的布局，打掉了门口的临时搭建屋，一层入口从正中调至左侧，一层变得更为整体。

一层进门为中式元素的高背椅和定制的红色绢丝绡画，走道的红竹子与红色线帘在视觉层次上凸现出来，又有开门红的寓意。一层主基调定位为黑色，在幽幽红色灯光的引导下，进入了中厅。中厅设计为一个小型的视听区，定制的吧椅，侧墙的罗汉床，红色的丝绸帷幔，黑色的毛皮落地镜，显然成了party的中心位。内厅则侧重于产品的展示，内墙定制了整墙的黑色木架，木架下层用蓝色丝绒的布幔替代了移门，对墙后定制了玫红色钢琴漆的展示柜，展示柜在party时可调整为自助餐柜。洗手间在原来的过道上隔出，有整墙的镜面，一侧为在黑墙上手绘的白色新古典花纹，在party的夜晚释放感性的美。

沿着红竹子的过道，迈上黑白阶梯，进入二层，视野豁然平静。整片的白色基调，黑白绒面的窗帘，白色的百叶帘，间隔空间的白色纱帘，白色的羽毛灯，顿时让人在瞬间转换于心神宁静。外厅定制的白色藤椅是那格九号工作室最初的产品，或坐着，或用餐，或冥想，或喝茶，或发呆，是公司内人气指数最高的小空间。内厅的灰色丝帘与白色纱帘，间隔出一个小型的会议空间，整体的白，有着艺术品的点缀，时尚，简约。内厅是公司的办公区域，沿墙做了一排白色的台面，台面上方定制了对应的不锈钢灯槽管，沿用了简约的造型。

改动较大的是天台，原来是一个水处理系统用的，占了天台三分之一的面积。花了很大的工夫把它打掉后，顶面做了两次防水处理，铺了地板，维修了栏杆，砌了投影墙，才初现现在的风貌。

在如今这个多元化的时代，公司的办公环境也朝这个方向衍化，在地理位置、功能上不能说是一个最佳办公场所，但一定是一个能激发灵感的很悠闲的办公空间。

It used to be the sewage water treatment office of Woolen Sweater Mill and was taken over in June 2007 to be modified into an office building. The original layout was greatly adjusted form its plane: the temporary built house at the entrance was removed and the entrance at the first floor was shifted to the left side so that the first floor looks more integrated.

Just at the entrance of the first floor are the high-back chairs and customized red spun silk drawings. The red bamboos and red yarn curtains at the corridor are distinguished from the visual hierarchy and also have a message of being prosperous once the door is opened. The main color at the first floor is orientated at black. Under the lead of the faint red lights, you will enter the middle hall of the first floor. The middle hall is designed with a small audiovisual area. The customized bar chairs, the arhat bed against the wall, red silk valances as well as the black fur console mirror have obviously become the central location of the party. The internal hall at the first floor lays more stress to the exhibition of the products. The interior wall has been customized with a wall-size black wooden rack with the blue velour curtain instead of the sliding door at the lower layer of the rack. After being against the wall, the exhibition cabinet in the rosy piano lacquer is customized. This can be adjusted to a self-served kitchen cabinet. The toilets at the first floor are divided from the original corridor. The whole mirror surface with the white handmade neoclassic decorative patterns on the black wall has burst a perceptual beauty in the night of the party.

Along the corridor with red bamboos, you will step on the black-and-white staircases and enter the second floor. The visual field suddenly becomes quiet. The whole white main key, black-and-white velour curtains, white shuttle blind, white yarn curtain in the partitioned space as well as white feather lamps can make people peaceful in mind instantly. The customized white cane chairs are the initial productions of No.9 Nage Studio, where people either sit, eat, mediate, drink tea, or daze. It is the most popular small place in the company. The gray silk curtains and white yarn curtains in the interior hall at the second floor have enclosed a small conference space. The whole white is decorated by the artworks, both fashionable and concise. The interior hall at the second floor is the office area. A row of white table boards is made along the wall with the appropriate customized stainless steel lamp slot pipes above following the concise modeling.

The quite big modification is made on the rooftop. It used to be used by a water treatment system with an occupation of one–third of the rooftop area. It has taken great efforts to remove it. The roof surface has been made with two waterproof treatments and covered with floor boards. The guardrails are repaired and the projection wall is built up so that the present appearance just comes into being.

The company is currently beginning to research and develop the self–owned brands. The exhibition hall at the first floor is added with numerous exhibition cabinets and the layout is also adjusted slightly. In such multiple times, the office environment of the company is also developing towards this direction. It can not say the best office area from its geographic location and functions, but it must be a very leisure office space that can inspire you.

设计单位：内建筑设计事务所

达利国际
High Fashion International)

坐落地点：萧山 / 主要材料：铝条、不锈钢、U型玻璃、木板 / 摄影：潘杰

办公楼由两幢旧有的厂房改造而来。为改变厂房建筑刻板沉重的外立面，设计除了将原建筑外墙变为浅灰色外，还以铝条建构新的表皮。通过透光度计算，铝条被打乱重组，形成网状外挑结构，如同给整个建筑外罩了一只镂空盒子，除了起到调节光线的作用外，丰富的光影变化也使室内空间变得生动有趣起来。建筑正面入口周围的一圈水道，上铺不锈钢网板通行，赋予办公楼主体轻盈和浮动的视觉感受。在表皮与建筑之间的过渡地带，设计通过树木的植入增加空间活力与生气，同时也将自然的气息引入室内，融合了内外空间那道原本生硬的转换界限。

办公楼内，设计利用厂房层高的优势，将空间分为上下两层，一层以陈列和办公为主，二层则主要作为会议和洽谈区域。在此，设计师引入集装箱的理念，实现空间的功能吞吐。在开敞的大空间中，一、二两层各有一些由U型玻璃和铝条围合出的独立小室串联穿插其中，作为办公室、洽谈室或会议室使用。虽然大空间里套小空间，但经典的工业装置样式具有的流动感，却让空间仿佛一个运转中的港口一般，并无碍流畅性和整体感，较之方正的隔间反而更有趣味。

针对企业流行发布的需要，设计处理得很巧妙，利用楼梯与走道，将交通干线轻易转换为展示舞台。46m的木楼梯连接着一、二两层，台阶宽达6m，演出时就成了长长的缓坡T台，天花顶部马道安装舞台灯光，提供照明。台阶从中间高出台面再架一道稍窄的白色木阶梯，每隔一段，立起木架，可以摆放衣物，展示产品。楼梯的尽头连接着二层贯穿整个大厅的走道，依着梁柱架起，100m的距离，两边是透明的玻璃围栏，又是一处简炼纯净的秀场地。

The office building is modified from two old existing factory buildings. To change the stiff and heavy exterior facade of the factory buildings, the design not only has changed the exterior building walls in light gray, but also has applied the aluminum strips to construct the new surface. Through the calculations of light transmittance, the aluminum strips are disarranged for rearrangement to form into a net cantilever structure, just as if a hollowed-out box is covered over the whole building. In addition that it can adjust the lights, the change of the rich lights and shadows also makes the interior space vivid and interesting. A circle of water path around the entrance of the frontal building is paved with stainless steel net boards as a passage, which has endowed the main body of the office building with a visual lightsome and floating feeling. At the transitional area between the surface and the building, it is designed to be planted with the trees to increase the vigor and vitalities in the space and also bring the natural smells into the interior, which has integrated that originally stiff transitional interface between the interior and exterior spaces.

In the office building, the design has utilized the advantages on the storey height of the factory building to divide the space into the upper and lower stories: one story is mainly for display and office while the second story is mainly for the conference and discussion areas. Hereby, the designer has applied the idea of the container to realize the functions of the space. In the big open space, some small independent rooms enclosed by U-shaped glasses and aluminum strips separately in the first and second stories are alternated in series among them as the offices, discussion rooms or conference rooms. Although the small spaces are included in the big space, the sense of flowing possessed by the types of classic industrial equipments has made the space lie running port with any influence on fluidity and integrity. Instead, it is more interesting than the square partitions.

To meet the requirements by the release of the enterprise' s trends, the design is done very smartly. The staircase and corridor are applied to easily change the main transportation line into an exhibition stage. The 46m long wooden staircase has connected the first story and the second story with the steps wide up to 6m. It becomes a long sloping T stage during the performance. The stage lights are installed in the riding tracks of the ceiling top to provide the lighting. The table protruded from the middle in the step has installed with a layer of narrower white wooden step. The wooden shelves are established every two sections for putting the clothes and displaying the products. The end of the staircase is connected to the corridor throughout the whole lobby in the second floor, where the transparent glass enclosure on both sides are erected in a distance of 100m along the beams and columns. It is another simple and pure show stage.

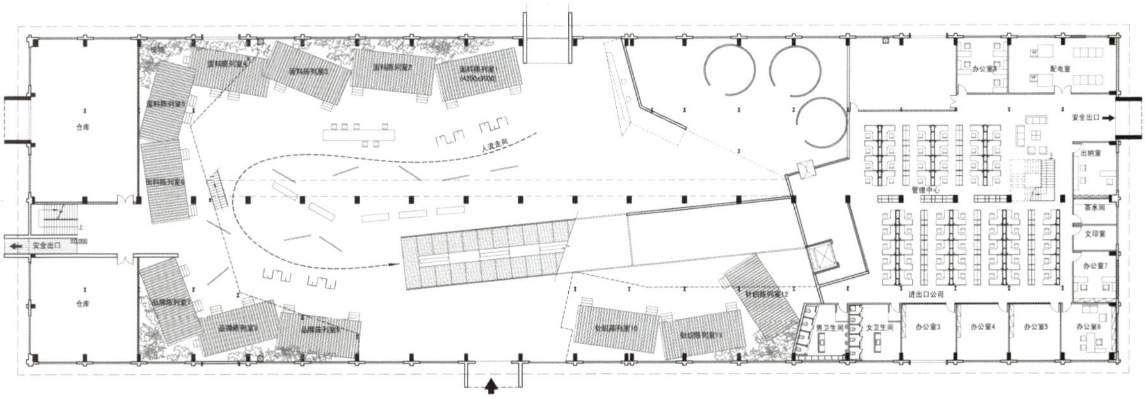

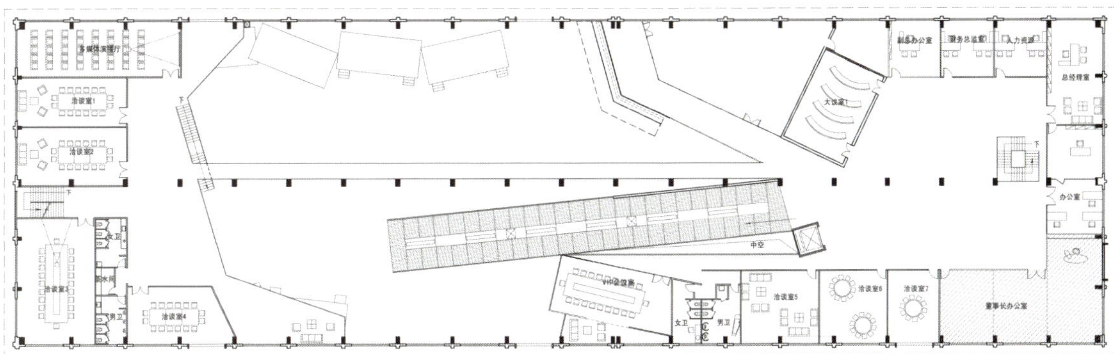

设计单位：于强室内设计师事务所 / 设计：毛桦

二十四城会所
Twenty-City Club

面积：3900㎡ / 坐落地点：成都

区别于普通意义的“简欧”，由“简约质朴的空间”配搭“欧洲时尚家私”来演绎突显艺术与人文的“新简欧”，并运用高科技手段提升观感，综合营造出亲和力强的具有家庭气氛的舒适环境。

一层以接待、展示及初级洽谈为主，人流量较大，故称为“动”层。空间的组成主要分为四部分：休闲阅览区、初级洽谈区、展示区、配套服务区。

接待台设在入口斜侧方，并与主入口之间形成一定距离，避免客人进入时产生心理压迫感。接待区围绕主服务台摆放于两侧，采用尺度宽大的家私营造尊贵、奢华的格调。休闲阅览区分为两种尺度的座位组合，即“大家庭位”与“小家庭位”，充分演绎家居感强的亲切气氛。水吧区设在整个休闲洽谈区的中心位置，呈圆形摆放，方便照顾到每一组客人。展示区包括模型组合展示、博物馆以及影音室，仿佛置身艺术精品店一般。配套服务区的洗手间为无障碍设计，有独立的“前室”空间，另设化妆室，大尺度的空间体现尊贵与奢华。

二层主要由办公空间及深度洽谈区组成，相对一层人流量较少，大部分为封闭空间模式，故称为“静”层。办公区分为大型多功能会议室、财务室、经理办公室及其他功能办公室。

深度洽谈区处于景观最佳的位置，大尺度的水吧营造高级咖啡厅的氛围，休闲舒适并不失尊贵奢华。地下层服务于物业及保安、保洁，空间组成较简单。分为各单位办公室、配套服务区、员工更衣室及休息室。

整个空间划分采取较现代自由的形式，更加满足售楼功能使用。“动”与“静”的划分使各个区域相对独立，互不干扰。公共卫生间空间宽敞，具有前室功能，体现奢华与舒适。

Different from the normal meaning of “simple continental style”, the “concise and simple space” and “continental fashionable furniture” are arranged to demonstrate and outstand the “new simple continental style” of arts and humanitarian; the hi-tech techniques are also applied to upgrade the impressions. All these have jointly created a comfortable environment with strong affinities and family atmosphere.

The first floor is mainly focused on the reception, exhibition and preliminary discussion with quite big streams of people. Therefore, it is called as a “dynamic” floor. The space mainly consists of four parts: recreational reading area, preliminary discussion area, exhibition area and matched service area.

The reception desk is located in the diagonal side of the entrance and has a certain distance with the main entrance to avoid that the customers have the sense of mental pressure. The reception area is located on both sides of the main service table and the furniture in big sizes is applied to create a noble and luxurious style. The recreational reading area consists of the seats of two different sizes, that is, “big family seat” and “small family seat”, which has fully demonstrated a friendly atmosphere with strong homely sense. The water bar area is located in the center of the whole recreational and discussion area and arranged in a round way so that it is convenient to serve every customer. The exhibition hall consists of model combination and exhibition, museum and studio, as if you are in a boutique store of arts. The toilets in the matched service area are a barrier-free design with the independent “front room” space and additional powder room. The large-size space has reflected the nobility and luxury.

The second floor mainly consists of office space and further discussion area with fewer people, as compared with the second floor. Most of them are closed spaces. Therefore, it is called as a “quite” floor.

The office space is divided into the large multi-functional conference room, finance room, manager office and other functional offices.

The further discussion area is located in a place with best sights. The large water bar has created an atmosphere of superior coffee houses, recreational and comfortable, but also noble and luxurious. The underground floor serves for the management service, security guards as cleaners with a quite simple space combination. It consists of the offices for individual companies, matched service areas, change room and rest room for the staff.

The division of the whole space is in a quite modern and free form to better meet the functions of selling the apartments. The division of "dynamics" and "quietness" has made each area comparatively independent and without interference with each other. The public toilets are spacious with the function of front room and demonstrate the luxury and comfort.

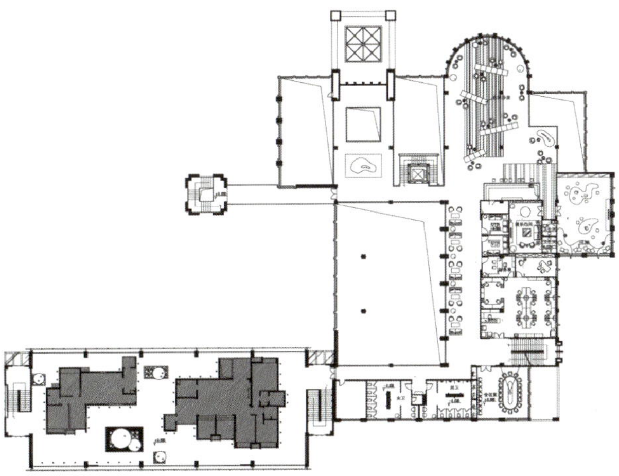

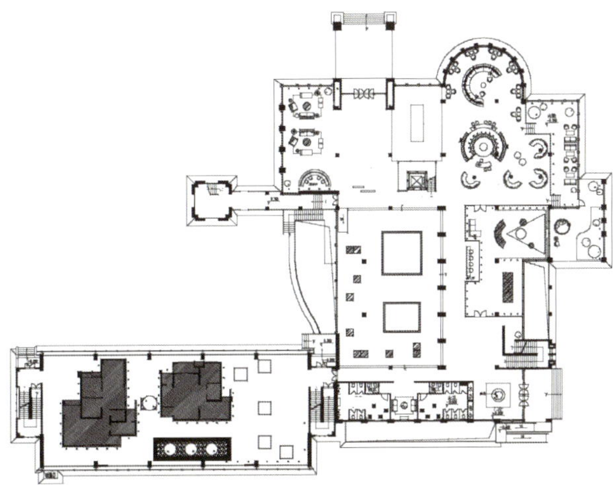

设计:登琨艳

比利时小馆
Belgium Little Restaurant

坐落地点:上海

自己为了圆却职业观点的理念，以一己之力投入上海旧城的更新改造的工作，在上海黄浦江岸的老工业区，规划投资更新再利用一座20年代的美国通用电气的老工厂。除了自己使用，也对外引进志趣相投者进来一起参与使用。一家建筑师的空间由他们自己设计改造，这家名为马瑞狮（Mcraftgroup）的公司，是做工业产品设计的比利时公司，有80位工作人员，选了五座已经不堪使用的老建筑，有男女澡堂、热水水塔、蒸汽锅炉间，加上一栋仓库，一栋小车间，周围环境植栽生长得非常美好。我个人本着对现代资源再利用，与对佛经所说的时间无边，没有过去，没有现代，没有未来的理意的体悟，将这几栋几乎任何使用者都不会再去使用，任何开发商都会拆掉重建的旧建筑，去芜存菁，拆掉实在已经不堪使用的部分，保留能够再使用的，并且将能够再利用的建筑构建材料保留返还再利用。

当然，使用者的现代功能是规划设计必须面对的，并且也在风水先生的强力建议下，让我在规划设计的时候，有点为难懊恼，结果当然都已妥善解决。而我自己现在的心灵状态对所谓老建筑再利用的信仰与观念，也是必然会在设计技法、工法与材料的使用上显露的，因为画图的笔在我手上，加上我自己又扮演一个服务的经营者。最后建筑改造的结果，呈现我的个人当下的心灵情境，也是必然的，就算我想隐藏，也没本事。

To fulfil the ideas from my professional viewpoints, I have been, by chance, dedicated to the renovation and modification to Shanghai’ s old city with my own efforts. In the old industrial zone beside Huangpu River in Shanghai, it has planned to invest, renovate and reutilize a 1920s old American GE factory. In addition to my own use, it also invites those with the same favorites to participate in the use. The space of an architects’ space is designed and modified by themselves. It is a Belgium company named M Craft Group and engaged in the design of industrial products. They have 80 staff and have selected five old buildings that can not be used, including male and female’ bathhouse, hot water tower, steam boiler room and together with a warehouse and a small workshop. The plants in the surroundings have grown prosperously. With aim to reutilization of the modern resources as well as my comprehension on the philosophy of the immense time without the past, present and future interpreted by the sutras, I have removed the dross and retained the elite of these several old buildings that almost any user would not use any more and any developer will dismantle by dismantling the parts that can not be used any more and retaining those that can be reused as well as keeping those building materials that can be reused for recycling.

Of course, the modern functions for the users are what the planning and design must be confronted with. Furthermore, with the strong recommendation from the geomancer, I feel a little bewildered and upset during my planning and design. The result is surely to make the settlement properly. While my present mental status about the belief and ideal of reutilizing the old buildings certainly will be revealed by the design techniques, construction methods and application of the materials, because the pen drawing the pictures is at my hand. Furthermore, I also play a role as the service provider. The results of the building modification definitely will demonstrate my current personal spiritual status. I have no such ability even if I want to hide.

设计:童明

苏泉苑
Suquan Yuan

面积:5613㎡ / 坐落地点:苏州十全街711号

苏泉苑是一个再开发项目，基地原先是苏州外事旅游车船公司的一片大型客车停车场地，期望改造成为一个购物和旅游地点。

该项目具有两个挑战：第一个是内部巨大的停车空间仅由一个不到6m宽的出入口与十全街相连接。一旦该地形被转变为商业使用，并且只能采用狭小的沿街面作为基地的出入口，这就产生了一种瓶颈性的不利因素。第二个是使用功能的不确定性。

如果从项目的识别性和意图上来考量，那么就没有比一个引人注目的建筑或者一个生机盎然的场景更加好的了。基于对传统建筑的印象和记忆，这座入口处的可能用作茶室的建筑被设计成了一个镶有长条木窗的青砖构成的匣子。建筑的立面可以根据使用情况发生改变。

它是一个纯净的匣子，充满光和影的游戏，展现材料和构造的质感。在结构上，它保持了多米诺体系，内部的空间分成服务区和辅助空间。这样的一个匣子能够适应将来更多未知的使用者，它将会是一个为未来上演的事件所准备的舞台。

Suquan Yuan is a redevelopment project which was planned to turn a huge parking space previously served for a tourism company into shopping and resort areas.

There were two challenging points for this project: one is the bottlenecked entrance which is only 6 meters in contrast to the big compound inside. Although it was quite appropriate for safeguarding and managing reasons, it would be very disadvantageous for a commercial space which was segregated from the street front.

The other is the uncertainty of the uses of the project. Like most speculative developments, it is almost impossible to have a final user or definite functional purposes for the building during design process.
If considering the problem of identity and indication for the project, nothing could be better than a building as an impressive sign or a lively scene for the whole site. Based on the impressions and memories on local traditional architecture, the entrance club, which is supposed to be a coffeeshop, was designed as a dark-brick box inserted with a wooden box, whose long windows could be opened and closed according to the interior using conditions.

It will be a pure box of light and shadows, materials and patterns. Also, the structure is simple to follow DOM-INO diagram, the spaces inside are distinguished as served area and servant rooms. It will be a box which could accommodate uncertain further users. It will be a stage anticipating occupation of future events.

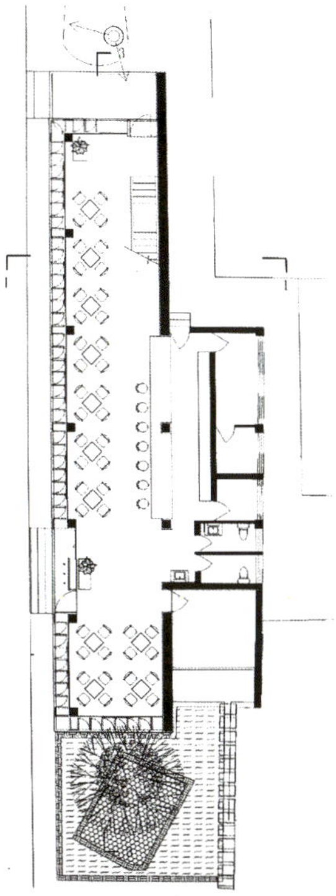

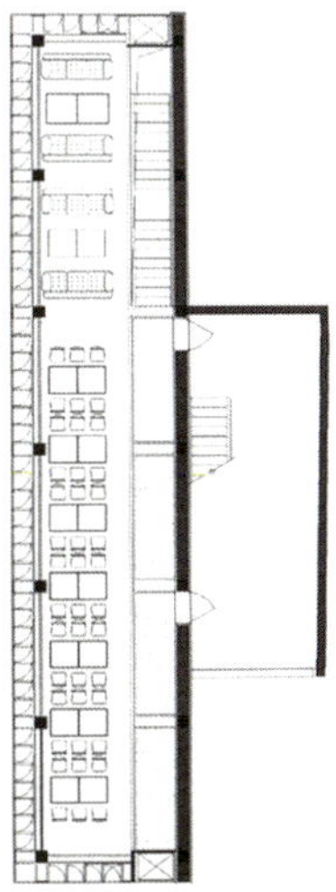

设计单位：IADC涞澳设计公司 / 设计：张成喆

复城国际销售中心
Sales Center of Fucheng International

面积：1500m² / 坐落地点：上海 / 主要材料：钢板做旧、拉丝不锈钢、玻璃、木材、石材 / 摄影：蔡峰

开发商希望本案例的设计应当体现旧工业厂房改造的特征：保护并延续基地历史，并表达后工业时代精神。仔细考察基地后，设计师并没有从破旧不堪的厂房中汲取更多的设计灵感，倒是被厂房外两行生机盎然的大树而深深感动了。他希望人们能继续享受这样的场景：宽广的树冠保护着建筑的宁静。由此，设计师通过环境以及环境对建筑空间的映射交融来延续场所的记忆，而"树"则成为本案设计的基本元素：树冠与建筑之间恰到好处的尺度比例关系，树木景观与建筑空间的内外交融，树木富有永恒意味的生长动姿与厂房建筑形态面临被时代淘汰的命运之间富有戏剧性的对比等等，都离不开这一基本元素的承载。室内北立面保留了原来的墙体和柱体，只作表面的清理整修，大面积的窗户亦保持原来的状态甚至省略了窗帘，厂房的历史似乎被定格，只有窗外的晨昏变化和生长不息的树木提示时间的缓慢进程。设计师更将浓淡虚实变换的动姿引入了空间。与之相对应的另一个立面设计中，设计师则汲取了皮影戏的手法，在墙体前约一米的距离设置了白色帆布背景，白色帆布具有一定的透光性，在阳光的投射下，产生了朦胧的逆光效果，与对面现实的自然形象产生了强烈而有趣的戏剧性的对话。销售中心的室内氛围延续工业厂房的钢铁味道，在光线的处理上追求在暗中凸显丰富的光影变化。立面采用钢架表面铁板做出铁锈的肌理效果，丰富材料的表现力和光影效果。桥的立面向上延伸在顶部进行有规律的交错连接，完成了向结构严谨而简单的原工业厂房天花的过渡，同时完整了桥的形象，使之成为室内空间的视觉焦点。

The developer hopes that the design of this case can reflect the characteristics of the modification of the old industrial factory building: to protect and continue the history of the base as well as to deliver the spirits of the post-industrial era. After careful observation on the base, the designer did not get more design inspirations from the wretched factory building. Instead, he was moved by two rows of living trees outside the factory building. He hopes that the people can continue enjoying this scenery: the wide crowns of the trees are protecting the quietness of the building. Therefore, the designer continues the memory on the location through the environment as well as the reflection and interaction on the building space by the environment. The "trees" become the basic element in this design case: the appreciate scale between the crowns of the trees and buildings, the internal and external integration between the tree landscaping and building architectural spaces, the dramatic contrast between the everlasting growing poses of the trees and the fate of being eliminated by the times confronted with the architectural forms of the factory building etc. All these can not live without this basic element. The interior north facade is reserved with the original wall and columns and only the cleaning and repair are made on the surfaces. The large-area windows are also kept with the original status and even the curtains are omitted. The history of the factory building seems to be stopped here. Only the changes of the mornings and nights as well as the everlasting growing trees outside the windows are pointing out the slow progress of the time. The designer even has introduced the poses of thick and thin as well as false and real changes into the space. In the other facade design corresponding to it, the designer has applied the technique of the shadowgraph and set up a white canvas background about 1 meter away from the wall. The white canvas has certain light transmittance and will create a dim backlighting effect under the projection of the sunlight, which generates a strong and interesting dramatic contrast with the opposite realistic natural image. The interior atmosphere of the sales center has continued the steel style of the industrial factory building. For the treatment of the sunlight, it pursues the abundant changes of lights and shadows in the darkness. The iron sheets on the surface of the steel frame are made as the facade to have a textural effect of iron rusts so that the expressive forces and light-shadow effects of the materials are enriched. The facade of the bridge is extended upward to have a regular cross-connection on the top so that the transition to the original factory ceiling with the precise and simple structure is completed and the image of the bridge is fulfilled to become the visual focus in the interior space.

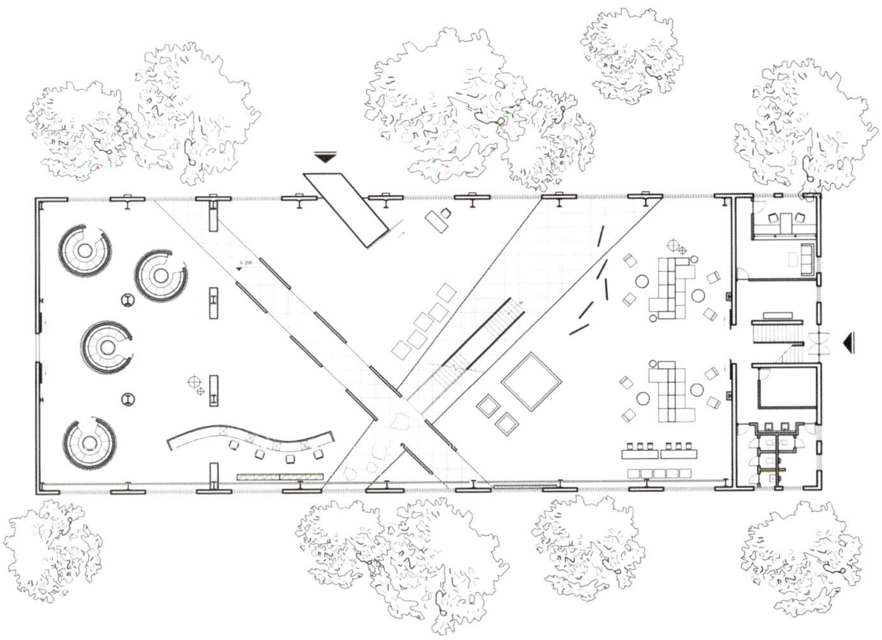

设计:Morita

Visus

坐落地点:上海外滩6号4楼 / 摄影:莫尚勤

整个Visus的内部设计保留了外滩6号大部分原来的风貌，沿用了玻璃窗边的倾斜式木结构，以及童话故事里才有的窗格子。昏暗的灯光下尽显浓郁浪漫、神秘、童话气息。墙上装点了许多镜子，视觉上放大了空间，坐在酒吧内，可以从不同角度领略到外滩风光，按设计师的意愿是希望身在Visus的人，可以从不同角度领略到外滩风光的同时，也能随着镜子中不停变换的人群而织就一道自然、多变的美丽风景线。

Indigo酒吧和Giromondo餐厅只是一墙之隔，餐厅有全透明的天窗，运用天然光线的效果，让就餐氛围也显得更轻松自如。往里面走又有另一番味道，尤其是贵宾房的装修设计里，马赛克拼贴而成的两个大型吊灯跃然墙上，在微弱灯光的照射下栩栩如生，貌似即将破墙而出。恬静，舒缓中呈现出一丝魅惑。身陷沙发，目光远眺，心灵不知不觉就沉静下来了。

Visus在拉丁语中解释为理想，这个名字或许很贴切，能在这样独特的地理位置去体验纯正的意大利式美味和奢华，这种感受就像追求理想的过程一样美好和忘我。

The interior design of the whole Visus has kept the most original appearance of No. 6 Bund. It has continued to use the sloping type wooden structure along the glass windows as well as the wind grids that only can be found in the fairy tales. It has fully revealed with a strong romantic, mysterious and fairy-tale atmosphere under the dim lights. Many mirrors are installed on the wall so that the space is amplified visually. Sitting in the bar, you can enjoy the bund sceneries from different angles. The designer hopes that the people in Visus can also create a natural and variable beautiful scene while enjoying the bund sceneries from the different angles of view.

Indigo Bar and Giromondo Restaurant are just one wall between. The restaurant has the completely transparent skylight. The dining atmosphere also looks more relaxing and easy by the application of the effects of natural lights. Walking towards the inside further, you will have another feeling. Especially, in the decorative design of VIP rooms, two large pendant lamps pieced together by mosaic appear vividly on the wall and look true to life under the projection by the dim lights as if they will break out the wall, peaceful and relaxing with certain fascination. Stuck in the sofa and overlooking, your spirits will calm down unconsciously.

Visus means ideal in Latin. This name may be very appropriate. You can experience the pure Italian type delicious food and luxury in such a unique geographic location. This feeling is as beautiful and self-forgetting as during the pursuit of the ideals.

设计:张成喆

IADC设计工作室
IADC Design Studio

面积:450m² / 坐落地点：上海 / 主要材料:环氧树脂漆、木地板、玻璃、钢 / 摄影:蔡峰

办公室由上世纪30年代的老房子改造而成。由三个楼面组成，入口的采光天井贯穿了一层和工作区，玻璃天顶解决了采光的不足，同时使入口空间具有了美术馆般的宁静气氛。围绕天井，二层采用整面玻璃墙体，使空间变得开放，通透，设计中，最大限度保留了老房子本身的结构元素。整个空间以白色为基调，化解了旧式建筑的阴暗和沉重，让空间变得轻盈、透明，并焕发出新的生机。

The office is modified from an old house in 1930's and consists of three floors. The lighting courtyard at the entrance has penetrated through the first floor and the working area while the glass ceiling has solved the problem of insufficient lights and endowed the entrance with a quiet atmosphere like an art gallery. Around the courtyard, the second floor is applied with the whole-piece glass to make the space open and clear. The design has maximally kept the structural elements of the old house. The whole space is mainly in white, which has lessened the darkness and heaviness of the old buildings and made the space lightsome and transparent with a new life.

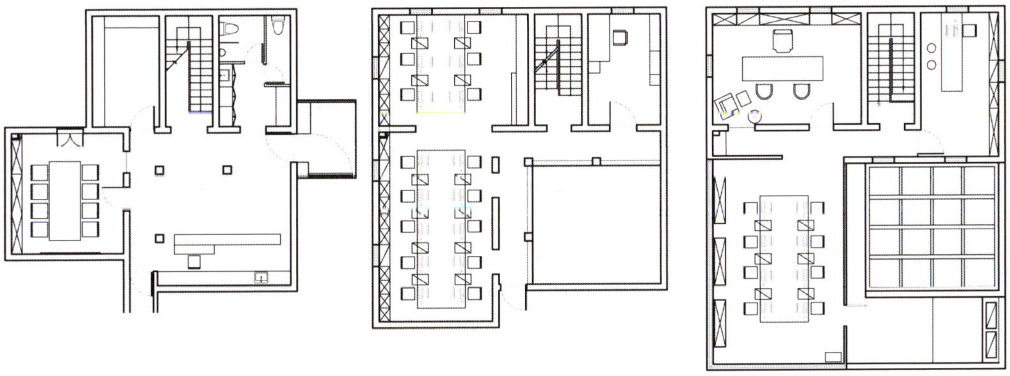

Italian Interior Design
London Apartments
Los Angeles Houses
OFFICES
FRENCH WORKPLACE DESIGN
SPADE
LUXURY HOTELS
STAIRS
室内经典案例
CLASSICS
Interiors: Designed to be Different
New German Interior Design
最新德国室内设计

设计:陈旭东

茶怡轩
Cha Yi Xuan

面积:200m² / 坐落地点：长春 / 主要材料:板岩、灰砖、玻璃、泰柚板、海砂石

为解决如何提升空间的趣味性和层次丰富性这个问题，设计师借用苏州园林中步移景异、对景借景的造园手法，在一楼大堂设了一个玻璃房，并把它架在空中，做成悬空的透明盒子，屋内与屋外的人互成观望景致，产生连绵不绝的观感。海砂石、灰砖、板岩等天然材料的组合体现出北方粗犷自然的生态面貌。

室内的灯具选择也与这种追求自然的理念相呼应。厅中吊灯用树藤编成灯罩，全无人工痕迹；茶桌上方的小灯选用藤球灯罩，点亮后在茶桌上投下深浅纵横的影子，活化了饮茶的气氛。

不止是材料细节，交通流线的布置也很巧妙。宾客从入口进入玻璃屋子，感受到的空间层层递进。楼梯平台的壁饰由工笔国画和镜片穿插组成，是垂直流线上的点睛之笔。向楼上行走的人影映入镜中，也同时迈进画里，镜与画，一实一虚进行着沟通与对话。

经过楼梯间进入二楼，交通流线相对简单。为增加空间的趣味性，包房与走廊之间的地上埋设有荧光灯带，使整个走廊产生飘浮感。与楼梯相隔的栏杆不是常见的从地上竖起，而是反其道而行之，从天棚悬下木栅栏，每个栏杆下都吊着一把小茶壶，与走廊的动势一起，浮在空中，即点明了茶楼的主题，又化解了小空间的沉闷气氛，不同质感的材料在环境中实现了调和与统一。

To solve the problem how to upgrade the interests of the space and richness of the levels, the designer has borrowed the gardening techniques of scenic changes while walking, scenic focal points and scenic borrowing in Suzhou Gardens. A glass room is set up in the sky of the lobby in the first floor, which becomes a suspended transparent box. The interior and exterior people become the watching scenery of each other so that a continuous impression is created. The combination of the natural materials such as sea sandstones, lime-sand bricks and slates has taken on a natural and wild north ecologic look.

The selection of the interior lighting fixtures also corresponds to this conception of pursing nature. The covers for the pendant lamps in the lobby are woven by the tree vines without any artificial traces; the small lamps above the tea tables are selected with the vine ball covers and the deep and shallow crossed shadows are projected on the tea tables after being lightened up to activate the atmosphere of drinking teas.

Beside the details of the materials, the traffic lines are also arranged very smartly. The visitors will feel progressive layer by layer once entering the glass room at the entrance. The wall decoration on the staircase platforms is assembled by the alternation of traditional Chinese realistic paintings and mirrors, which is a finishing touch on the vertical streamlines. The shadows of who walk to the upper floor are reflected in the mirror and also walk into the paintings. Mirror and painting, falsehood versus reality are communicating and talking with each other.

The traffic streamlines are comparatively simple after passing through the staircase and entering the second floor. To increase the interests of the space, the floor between VIP rooms and the corridor is buried with the fluorescent light strips so that the whole corridor creates a sense of floating. The handrails separated from the staircase are not up righted from the ground as usually, instead they are the wooden fences suspended from the ceiling. A small teapot is hung below every handrail and floated in the air together with the active mood, which not only points out the theme of the tea house, but also removes the depressing atmosphere in the small space. The materials of different textures have realized the harmony and unification in the environment.

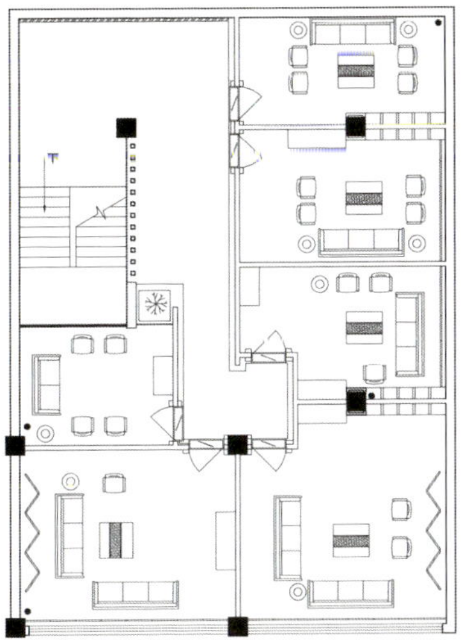

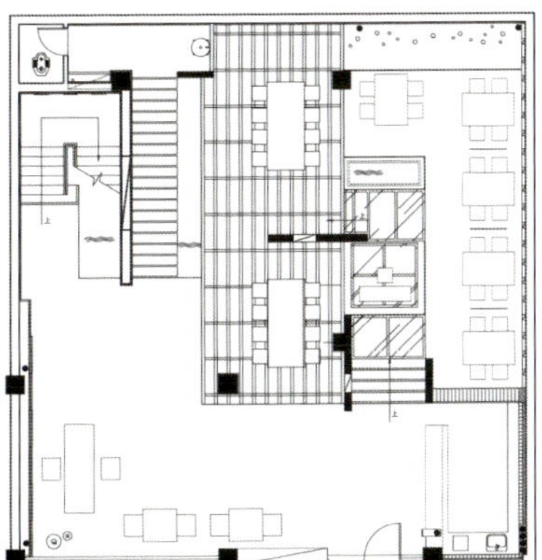

清和
清寂

设计单位:凯达环球建筑设计咨询（北京）有限公司 / 设计:冯志业 / 前期规划:HMA建筑设计事务所

无限度广场改造
Modification of Infinity Square

面积:65000m^2 / 坐落地点:上海市 / 主要材料:玻璃幕墙、钢化玻璃、铝板、人造石 / 摄影:莫尚勤

无限度广场前身为上海广场，建于1993年，后进行了二次调整。吸纳现代西方的设计理念，并融入传统室内围合构成手段，传承了新旧的文化是设计的核心。对于原本就有丰富的空间，室内装饰需要做的就是尽量简化装饰语言，用材料色彩把空间的各界面、各元素进行区分归类，使其更具层次，也就是用建筑结构界面本身进行装饰，无限度广场商业空间的成功改造则诠释了设计的精髓。

设计师慎重地解决奢华与传统文化品位这一矛盾，体现出星级商业空间特有的贵气和现代时尚的风范。通过一个通透的玻璃盒子，而演变出不同的事件发生。随着变化万千的光影、声音变化，及不同的展示、娱乐活动，整个建筑犹如一个生命体，每分每秒都在转变，为顾客营造新鲜感及无限的想象空间。再者，体宜因借，合理地借取环境意向引入室外的自然景色。

其内部空间强调了逻辑性、秩序性、整体性。在保持商场功能不变情况下，其店铺及公共空间全新划分。地下商场局部改为自行车车库及车道。地上首层由原四个入口改为在淮海路设三个入口，店铺划分由几个大空间改为小商店形式。个别扶梯移位并新增扶梯，使整个商场流向更分明合理。经过空间的动态再组合，空荡的中部将转变为演示中厅，这里是每一个游客体验的第一站。另外，公共区域地面采用人造石地台，楼面主通道两侧的玻璃隔断采用钢化玻璃，店面间隔墙采用轻钢龙骨石膏板，墙体饰面人造石进行装饰，四层为开放式空间，特别吊顶采用石膏板和铝板，无论从材质到细节都充分表现了空间的现代律动。

Infinity Square is the former Shanghai Square, built in 1993, which has been adjusted twice later. The design core is to absorb the modern western design conceptions and integrate the constituting technique of interior enclosure as well as continue the new and old cultures. As for the originally rich space, what the interior design shall do is to try to simplify the decorative languages and distinguish and classify each space interface and element by the material colors to make it more hierarchic. That is to use the structural interface of the building itself to make the decoration. The successful modification of the business space in Infinity Square has interpreted the essence of the design.

The designer has cautiously solved the conflict between the luxurious and traditional cultural tastes to demonstrate the unique nobility and modern fashionable style of the star-grade business space. The occurrences of different events are evolved through a transparent glass box. With the variations of the lights and shadows, the changes of the sounds and different displays as well as recreational activities, the whole building, just like a living body, is changing every second and minute to create a fresh and infinite imaginary space for the customers. Furthermore, it has sufficiently borrowed the natural scenes to the interior according to the actual conditions to set up a sense of green environmental protection.

Its interior space has emphasized on the logic, orderliness and integrality. In the condition that the functions of the store remain unchanged, its shops and public spaces are re-divided. The part of the underground store is modified into a bicycle garage and driveway while the first floor above the ground has been changed with the original four entrances to three entrances at Huaihai Road. The shops have been divided into small shops from several big spaces. Some individual elevators have been relocated and some additional elevators have been added so that the trends in the whole store become clearer and more reasonable. Through the dynamic reorganization of the space, the empty middle part will be changed to a middle lobby for demonstration, which is the first stop every visitor will experience. Furthermore, the ground in the public area is applied with the artificial stones while the both sides of the main path at the floor are separated by hardened glass. The partition walls between the shops are applied with the lightgage steel joist gypsum boards while the veneers on the wall are decorated by the artificial stones. The fourth floor is an open space. Especially the ceiling is applied with the gypsum boards and aluminum plates, which has sufficiently demonstrated the modern rhythms of the space either from materials or details.

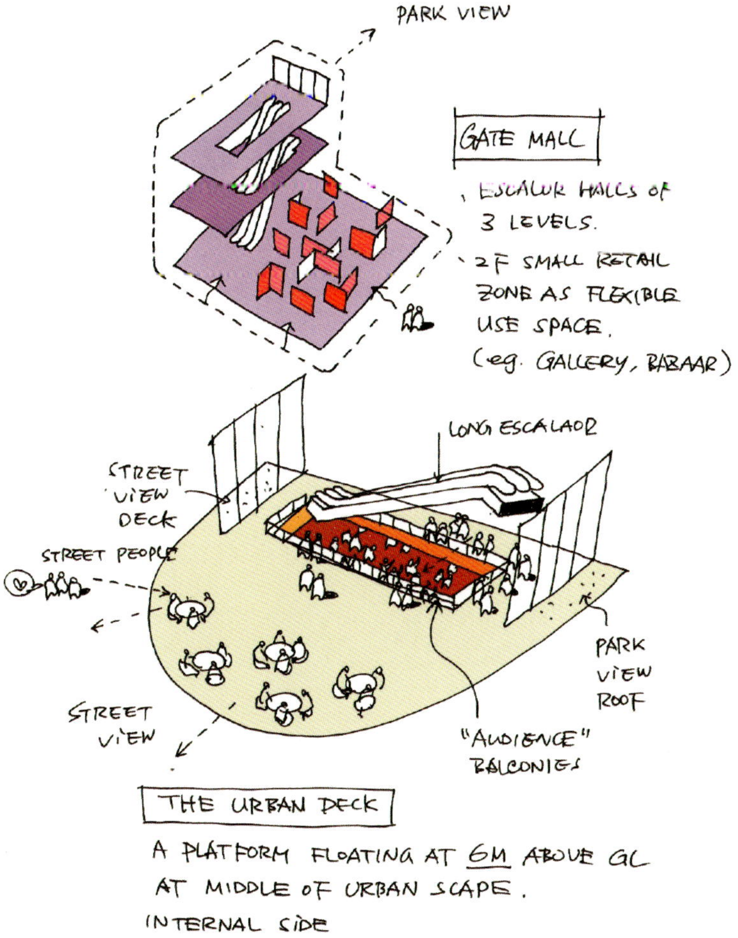
PARK VIEW
GATE MALL
. ESCALOR HALLS OF
3 LEVELS.
. 2F SMALL RETAIL
ZONE AS FLEXIBLE
USE SPACE.
(eg. GALLERY, BABAAR)
LONG ESCALAOR
STREET
VIEW
DECK
STREET PEOPLE
PARK
VIEW
ROOF
STREET
VIEW
"AUDIENCE"
BALCONIES
THE URBAN DECK
A PLATFORM FLOATING AT 6M ABOVE GL
AT MIDDLE OF URBAN SCAPE.
INTERNAL SIDE

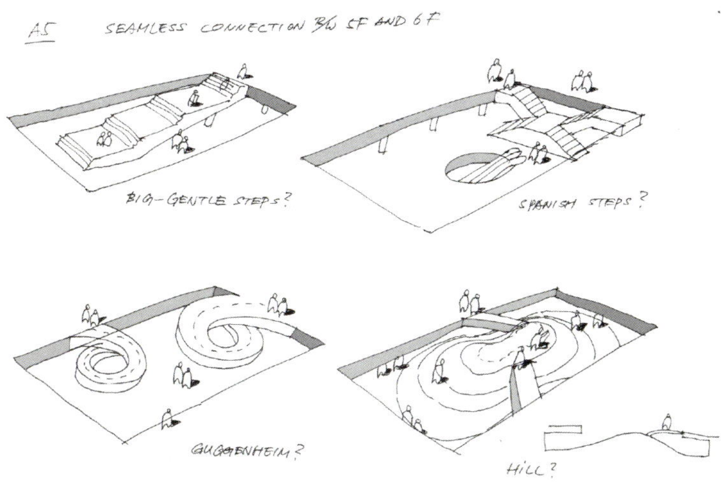
A5 SEAMLESS CONNECTION B/W 5F AND 6F
BIG-GENTLE STEPS?
SPANISH STEPS?
GUGGENHEIM?
HILL?

California
California
GUESS

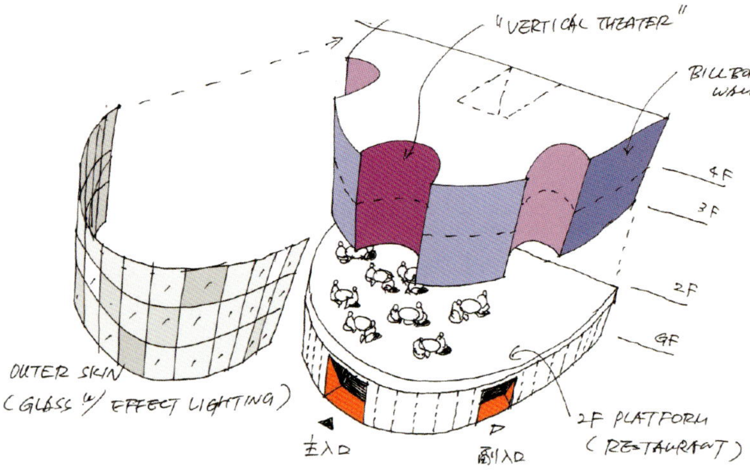
"VERTICAL THEATER"
BILLBOARD WALL
4F
3F
2F
GF
OUTER SKIN (GLASS W/ EFFECT LIGHTING)
2F PLATFORM (RESTAURANT)
主入口
副入口

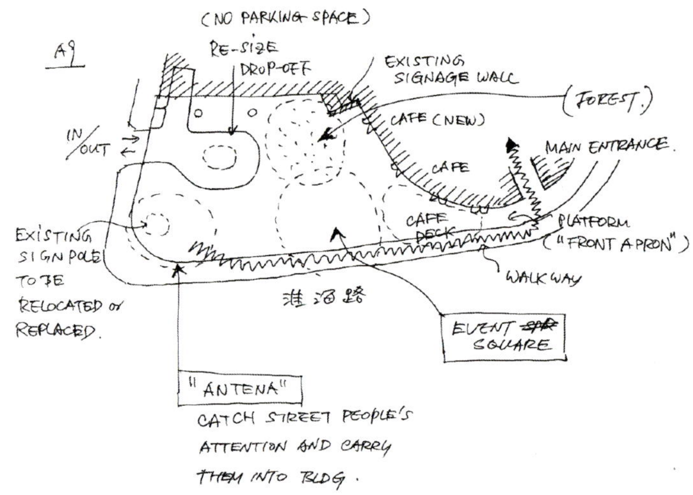
(NO PARKING SPACE)
RE-SIZE DROP-OFF
EXISTING SIGNAGE WALL
(FOREST)
CAFE (NEW)
MAIN ENTRANCE
IN/OUT
CAFE
CAFE DECK
PLATFORM ("FRONT APRON")
EXISTING SIGN POLE TO BE RELOCATED or REPLACED.
WALKWAY
EVENT SQUARE
"ANTENA"
CATCH STREET PEOPLE'S ATTENTION AND CARRY THEM INTO BLDG.

The Coffee Bean

DREAM FACTORY

10月7日正式开幕
无限度广场
ROAD
CLOSED
DO NOT
ENTER

大于设计的想象
Imagination Larger than Design

高蓓Gao bei
建筑师
美国UN+(优联加)建筑事务所 设计总监 总经理

CLUB

耿治国、俞贵云 老外滩五号公寓会所
Keiichiro SAKO、Yoshimasa TSUTSUMI 长春中日交流之窗
黄书恒 远雄建设富都接待中心
罗灵杰、龙慧祺 金地梅陇镇会所
龙域设计 绿地公馆
孙彦清 金鼎湾国际
刘延斌 东湖宾馆梅岭一号
麟研部 设计工作室 Racks桌球吧
张波 壹号公馆夜总会
吴林孟、张丰毅 新南亚宝丽金会所
吕邵苍 苏州英皇会所
孙鸣春 天香新唱KTV夜总会

会所的定义似乎一向比较模糊，从此篇中辑录的四个项目的内容就能看得出来，绿地公馆是一组小型旧建筑改造成酒店的案例；远雄建设富都接待中心其实是台北的一个售楼处；长春中日交流之窗是一个小型的访客接待和咨询公开的场所；金地梅陇镇会所比较复杂些，容纳了多样的服务和娱乐功能，为一个住区提供日常的聚会和活动空间。将它们都定义为会所，想来是因为它们都有几个比较明显和普遍的特征：一是为相对特定的人群提供活动和交流的场所（这可能是会所亦可被唤作“俱乐部”（CLUB）的原因）；二是由于提供的服务大多较为特别或优越，会所建设的投入也就相对高昂，可供“运作”的空间较为宽敞，会所的设计也更为自由。

被唤为“会所”的场所亦可分为两类，一类是像远雄建设富都接待中心这样的，或多或少带有一定的营业性，但又是某些机构和组织的附属，气派和格局都是挺惊人的，让人怀疑到底能不能收支平衡下去。现在临山近湖，市中心好地都不乏这样的出手，忙时可住可茗可饮可娱，闲时宅院深寂亦不大敞门庭，其机构、企业和组织的主人们并不借此营生，而借会所为窗口，获一个尊贵特殊的形象。从这个意义上讲，会所成为高端营销的手段，成为日趋庞大的广告体系的一部分。

作为日本旅游咨讯宣传的长春中日交流之窗和代表社区形象和定位的金地梅陇镇会所当然也是非常成功的广告和营销策略，在绿地公馆中，30个房间的酒店管理也应该是退居其次的，空间和设施特殊的表现力和个性显得更为关键。

成功的会所设计宛如一张动人的广告海报，可能没有任何内容（有时也不需要），但却让人惊心动魄，也许内容的缺失反而成就了设计的单纯质地（如同长春中日交流之窗）。这时，宛如营销手段一般，寻找主题成为会所设计中至关重要的关键。和场馆设计、酒店设计等的专项设计不同，会所设计功能类型界定上的缺失使得功能研究和流线布置等问题在设计中的重要性不再是压倒性的，合适并且强烈的气氛和场景营造需要设计语言的敏感选择和精心组织，语言感成为考验设计师的主要命题。在此篇的四个案例中，我们都能感受到设计者们单纯而强烈的运用主题语言的熟稔，尤其是在金地梅陇镇会所和长春中日交流之窗的设计中，主题性语言的排列和重复因为良好的技巧控制没有令人感官疲乏，精心的考量和尺度研究也没有带来荒诞和刻意的不适。

和电视广告一样，会所的设计也是充满指向性的，获得进入的使用者“自然而然”将个人意识和会所环境联系在一起，这种联系比在酒店空间中产生的更为紧密。会所的拥有者和设计者欣然于这种想象，一个美好的会所环境会令使用者的身份感发生改变，从而走进会所拥有者的意识——这里（包括组织和产品）是尊贵稀有充满愉悦的。于是，为了明显或深邃的商业目的，大多数会所的拥有者和设计者都非常重视“身份”的象征和“身份”的营造。在“身份”这个词带来误解之前，我倒更愿意将它描述成“对个体经验的极大尊重”，这给设计者带来的挑战不再是平面的堂皇，而是一种行为的设计与启迪，从这里开始，和“广告艺术”一样，会所本身可以超越日常，成就“精神性”的升华。

第二类会所是指具有一定营业功能的集会娱乐服务场所，它可能是酒店，也可能是餐厅，也可能是桑拿SPA中心，而为什么它们愿意被冠以“会所”的命名，除了它们确实可能具备前文所提到的两个特性之外，更重要的是，它们希望获得这种特殊的“身份感”。可惜的是，大多数的这种场所的追求“身份感”的设计中，仍然停留在“身份=尊贵”的理解模式上，而出现很多似曾相识的“会所脸”，湮没了个性和精神。这时，此篇的四个案例应该是不错的提醒和借鉴。

The definition of a club seems quite indistinct, which can be seen from the contents of the four projects compiled in this chapter: Green Town Residence is a case that a group of small old buildings are modified into a hotel; Farglory Rich City in fact is a sales pavilion in Taipei; Changchun China-Japan Communication Center is a small place for receiving visitors and public consultation; Gemdale Mellon Town club is a little bit complicated as it has contained the various service and recreation functions and is a place where to provide daily gatherings and activities for a residential quarters. They are defined as clubs because they all have several obvious and common characteristics: first, it is to a place to provide the activities and communications for the relatively specific groups (this may be the reason why a club can be called “club”); second, as the most of available services are quite special or superior and the investment on building a club is comparatively high. The space available for “operation” is quite spacious and the club can be designed more freely.

The places that can be called “club” also can be divided into two categories: one category is like Farglory Rich City; they more or less have certain business operation, but are also the annexes to some institutions and organizations. Their style and arrangement are so surprising that people feel doubted if they can make both ends meet. Now, you can find many such kinds of works beside the mountains, close to the lakes and in the good places in downtown area. You can stay, enjoy teas, drink and have fun here when they are busy while the courtyard is quiet but the doors are not widely open when they are not busy. The owners of their institutions, enterprises and organizations do not live on them. The clubs are only a window to obtain a noble and special image and leave certain interest of home banquets. Speaking in this sense, the clubs have become a technique for high-end sales and part of the advertisement system that tends larger and larger every day.

Changchun China-Japan Communication center as the consultation and propaganda for Japanese tourism and Gemdale Mellon Town club representing the image of the community and orientation of course is a very successful advertisement and sales strategy. In Green Town Residence, the hotel management of 30 rooms also shall retire to the second position while the expressive forces and individualities of the space and facilities look more critical.

A successful club design is just like a charming advertisement poster. It may not have any content (does not need sometimes), but it is exciting. Maybe, the lack of contents instead has accomplished the pure character of the design (just like Changchun China-Japan Communication Center). At this moment, just like the sales techniques, finding the theme has become the crucial key point in the design of clubs. Different from the special designs such as the design of the venues and hotels, the missing of the function category for club design in definition has made the importance of the problems such as study on functions and arrangement on streamlines in the design no more overwhelming. The creation of proper and strong atmospheres and scenes requires the sensitive selection and delicate organization of the design languages. The language senses have become the important assigned topic that tests the designers. In the four cases in this chapter, we all can feel that the designers are familiar with simple and strong application of the thematic languages. Especially in the design of Gemdale Mellon Town club and Changchun China-Japan Communication Center, the sensory organs do not feel tired of arrangement and repeat of the thematic languages due to their proper and tactful control while the delicate examination and scale study also do not cause any absurd and deliberate discomfort.

Same as TV advertisements, the design of clubs is also filled with directions. The users obtaining access right will “naturally” combine the individual awareness with the environment in the club. This connection is more intimate than that generated in the hotel. The proprietor and designer of the club are satisfied with this imagination. A beautiful club environment can change the user’s sense of identity to enter the consciousness of club owner - it (including organizations and products) is noble, rare and full of happiness. Therefore, for the obvious or profound business purposes, the proprietors and designers of most clubs have paid special emphasis on the symbolization and creation of “identity”. Before the word “identity” has brought with misunderstanding, I prefer to describe it as “great respect on individual experience” so that the challenges brought to the designers are no longer the plan magnificence, but a design and enlightenment of a behavior. Since now, same as “advertisement art”, the club itself can surpass the ordinariness and realize the “spiritual” sublimation.

The second category of clubs refers to the place for recreational services of certain business functions. It can be a hotel, a restaurant or a SPA center. But why they are willingly called “club” - an unspecific name. In addition that they may indeed have the above-mentioned two characteristics, more importantly, they hope to obtain this kind of special “sense of identity”. Regrettably, in the design of most of such places that pursues “sense of identity”, they still stay in the understanding mode of “identity=nobility” so that many familiar “club faces” have appeared and the individualities and spirits are neglected. Here, the four cases in this chapter shall be a good reminding and reference.

设计单位：飞形设计事业有限公司 / 设计：耿治国、俞贵云 / 参与设计：OFA / Hotel 设计部

老外滩五号公寓会所
No. 5 Apartment of Old Bund

坐落地点：宁波市老外滩 / 主要材料：莱姆石、铁刀木、烤漆玻璃、烤漆钢材、亚克力、不锈钢、石材、涂料、橡木、茶镜、织品、玻璃 / 摄影：周宇贤

这是一座精装修精品酒店式公寓，坐落于甬江、余姚江、奉化江交汇处的老外滩码头区。公寓内部规划以“第二居所”为概念，为房客经营一座如家般的居巢。基地是一座9层楼的复合机能大厦，一楼至三楼为对外营业的精品商场，四楼至九楼是精品公寓。

不同于传统酒店规划花园偏好张扬热闹的设计手法，四楼露台诉求开阔视野以及与城市对话等特质，规划水池、草坪，架高的木平台作为赏景的座区，静静眺望街道上多国文化碰撞的典雅建筑，透过简约低调的姿态与周围环境融合，细心聆听城市脉动。

因应本案垂直错落的特殊房型，设计师在公共廊道以米色石材以及齐整林立的深色房门营塑纯净端庄的氛围，当人们推开门片踏进一座座格局各异的住房时，也触动了探索内在风景的趣味。

It is a well-decorated fine hotel-style apartment, located in the dock area of the old bund where Yong River, Yuyao River and Fenghua River are joined. The interior planning of the apartments, based on the conception of “the second home”, is operating a homely living place for the residents. It is a nine-storey comprehensive functional building. It is an opening boutique shopping mall from the first level to the third level while the boutique apartments are located from the fourth level to the ninth level.

Different from the planned gardens of the traditional hotels that are in favor of the aggressive and lively design technique, the open terrace at the fourth floor strives for the features of the wide view and the dialogue with the city etc. the water pool, lawn and raised wood platform are planned as the sitting area to enjoy the sceneries, overlooking the elegant buildings with the confliction of multi-cultures in the streets quietly and carefully listing to the city pulsation through the simple and low-profiled poses and the integration with the surrounding environments.

In correspondence with the special vertical and random room styles, the designer has applied the cream-colored stone materials and the well ordered serried dark room doors at the public corridors to create a pure and grand atmosphere. When the visitors push the door and step on the rooms with different styles, their exploration on the interior scenes is also actuated.

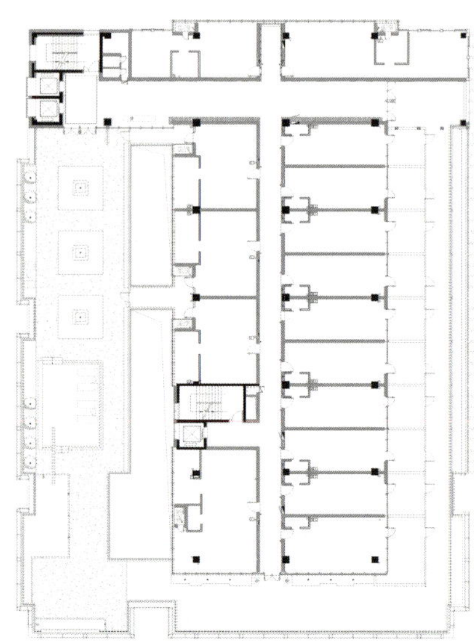

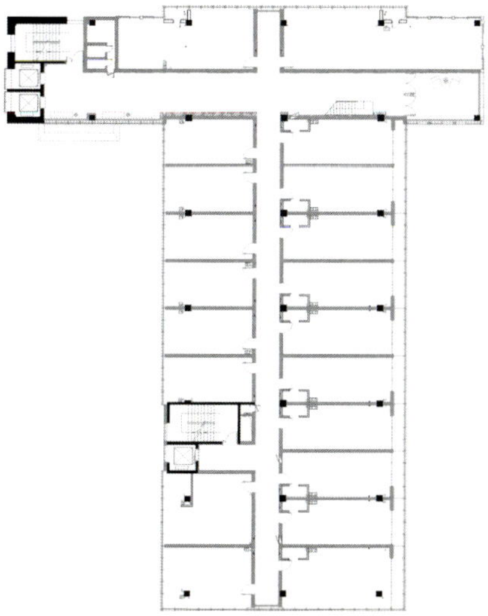

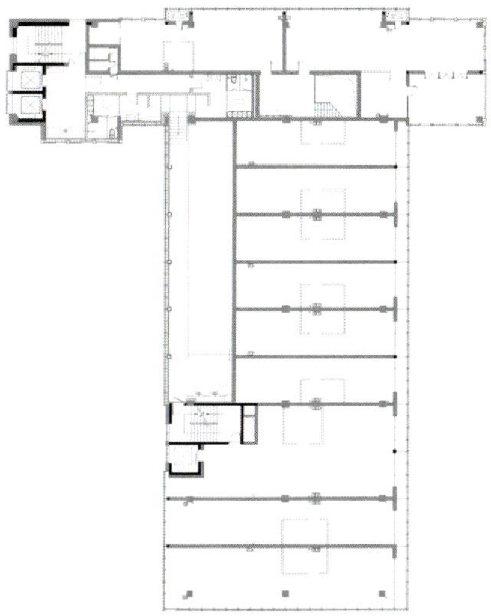

404

设计:Keiichiro SAKO, Yoshimasa TSUTSUMI

长春中日交流之窗
BRANCH in Changchun

面积:280m^2 / 坐落地点:长春 / 摄影:Misae Hiromatsu (BEIJING NDC STUDIO, INC.)

这是一处向中国介绍日本文化的交流设施，旨在创造出既适合一大群人聚集在一起，也适合一个人放松的地方。根据目的和氛围的不同这里都会让人心情愉快并喜爱，这是靠人们自己能够发现的多样的空间。器物的脉络通过宛如树木一样进行形态上的复杂变幻，使其用途变得多种多样。这个不断重复着分支与合成的生动的空间，也是对两国之间的交流与成长的一种隐喻。

It is a communication facility to introduce Japanese cultures to China with an aim to create a place that is suitable not only for the gathering of a large number of people but also for the people to relax themselves. People will feel happy here and like it according to the different aims and atmospheres. It is a place where people can find the varieties by themselves. The sequence of the objects has the complicated changes in their forms, just like the tree leaves so that their functions become various. This vivid space with constant repetitive branches and convergences is also a metaphor to the communication between the two countries and their growths.

设计单位：玄武设计 / 设计：黄书恒 / 参与设计：欧阳毅、陈新强、王建益、胡春惠、胡春梅

远雄建设富都接待中心
Farglory Fortuna

面积：1488m² / 坐落地点：台北市中山北路 / 主要材料：硅酸钙FA板、半反射玻璃、瓷砖漆、水泥压花、抿石子 / 摄影：王基守

一条无法复制的百年官拜大道、一栋见证台北城历史风华的经典旅店，形塑了远雄富都承袭百年荣耀的光荣使命。“再现当下”成为一个重要的课题，也代表了富都接待中心中介于过去旧饭店记忆与未来建筑量体被建构前鲜明属性。

本案坐落于中山北路林荫树海之一方，优越的地理位置在蓬勃发展的房产市场中可说是兵家必争之地。如何在有限的基地条件下强化建案特色、突显当地文化、塑造独有风格，考验着设计者的学养。由于两侧皆为邻屋，因此前后立面成为主要迎接来客之面向。进入接待中心，以大器刻划的家徽图腾镂空板材首先映入眼帘，除了成为主要立面的构成语汇之外，亦为象征富都于时间轴中三阶段的一种重要符号与记忆载体，建立欲盖弥彰，大隐于市之气势。刻意缩小与迂回建筑的开口，必须经过刻意引导或探寻才能发现，进而绕经清澈池水边之柱廊，方能真正进入建筑内，动线上予人以高潮迭起，柳暗花明又一村之空间印象。

进入室内后，挑高8m的壮阔大厅，让造访者立即感受到气势不凡的视觉飨宴。而层层延伸的长廊、渐次连接的门拱、简练现代的建筑柱廊及杜拜式的水晶吊灯，也在隐约中透露连接古典及现代、过去与未来的设计巧思。行脚至此，翻转而上则另有一番洞天。连接上、下楼层的楼梯则犹如圣经故事上“雅各的天梯”从天而降。拾级而上，楼梯两侧的壁面设计，镶嵌罗列着天使群的注目礼。清透明亮的天光洒落，仿佛置身幸福国度，也正呼应了远雄富都汇集巨擘领袖的辉煌全记录及大宅传世的终极向往。

A centennial official avenue that can not be reproduced and a classic hotel that has witnessed the historic primes of Taipei city have vividly created the Farglory Fortuna’s glorious task to inherit the hundred years of glories. The “reappear instantly “ becomes an important subject. It also represents the Farglory Fortuna reception center that is a clear-cut attribute between memories of the old restaurant and the building to be planned.

This case is located among the avenue of Zhongshan North Road. The advantageous geographic location can be called a place over which the competition is unavoidable in this prosperously developing real estate market. How to strengthen the characteristics of the building project, show the local culture obviously and create a unique style in the limited base conditions has been testing the knowledge and accomplishments of the designer. As both sides are the houses, the front and back facades have become the main faces to greet the visitors. Once entering the reception center, you will be greeted by a hollowed-out board grandly carved with the armorial totem. In addition that it becomes the composing language of the main facades, it is also an important symbol and memory carrier that symbolizes the three stages of Fortuna in the time axis to delivery a momentum of being exposed more while it tries to hide as well as ensconcing itself in a busy street. The intentionally shrunk openings on the detoured building can only be discovered under the guidance or search, then you must detour some colonnades by a clear water pond before entering the house. The moving line for guidance can provide to us a space impression, namely, one surge after another and a new scene.

After stepping in, the visitors immediately feel an extraordinary visual banquet from an 8m high spacious hall. The long corridor that extends step by step, the door arch that are connected gradually, the simple modern colonnade of the building and the crystal pendant lamps in Dubai style have also indistinctly been revealed with a clever idea of connecting the classics with the modern. Once walking till here, you will find another fascinating world when you turn upwards. The staircase connecting the upstairs and downstairs appears very unexpectedly just like “Jacob’s Stairways to Heaven” in the biblical stories. Along the steps upwards, the wall surfaces on both sides of the staircase are designed with the group of angels. The clear and bright day lights spray over as if you are staying in a happy county, which is also echoing with the Farglory Fortuna’s complete glorious record on gathering the giants and leaders as well as the ultimate hope of handing the house down.

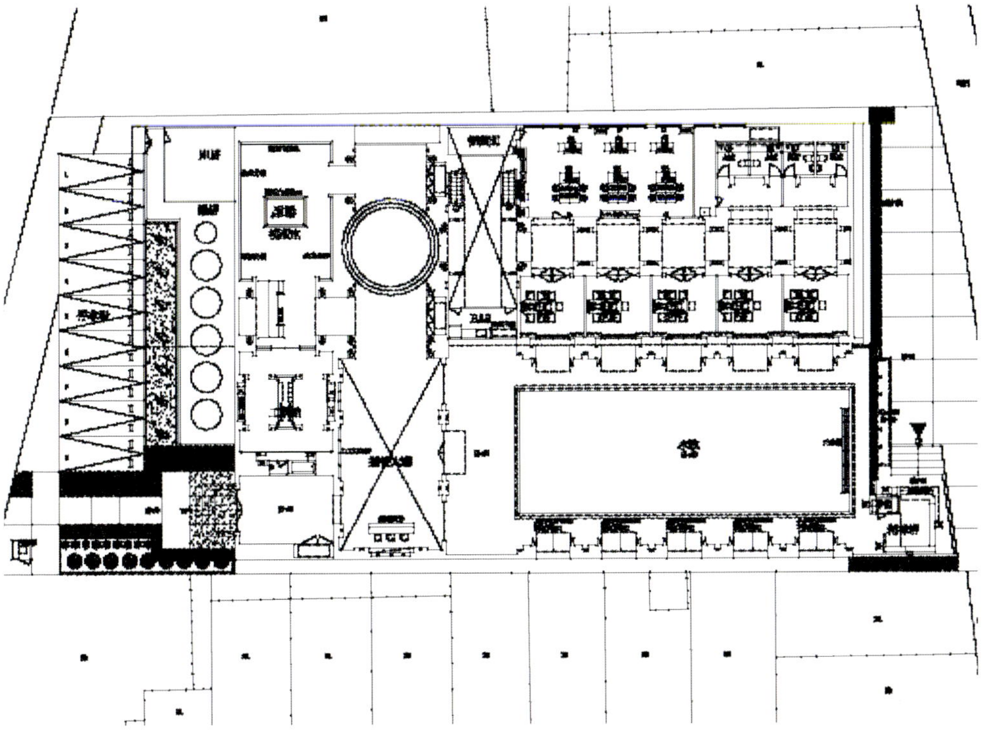

THE LEGEND of Architecture Master
創世紀大師演繹
國家級 室內空間大師

设计单位:壹正企划有很公司 / 设计:罗灵杰、龙慧祺

金地梅陇镇会所
Mellon Town Now Club

面积:2000m² / 坐落地点:深圳 / 主要材料:三星Staron仿石、黑色地砖、黑色地毯、白色乳胶漆、铁刀木地板、黑镜钢、不锈钢、灰镜 / 摄影:龙慧祺、罗灵杰

梅陇镇会所的出现，打破了人们对住宅会所的惯性期望。“设计风格”不再是风格，取而代之的是设计者、规划者突出的个人品位的汇集。

会所原本建筑空间是一个单层面积约为400m²，净高约为12米的金黄色玻璃盒子，室内设计方案一开始就摈弃了常规，以搭积木的空间想象力，贯穿整个会所的功能布局设计，同时融入数码化及科幻主题，使得它浑身上下都充满了电影感。高达三层的巨型玻璃窗，在放下巨型窗帘时，总让我想起欧式大会堂里的帷幕，都一样的宏丽。帷幕烘托着整个会所，空间里弥漫着的是最为永恒的白与黑，白色的大体量造型夸张地缠绕着黑灰色的几何线性，横切、竖挺、斜叉、穿行着，无时无刻不充满着艺术感的建筑语言。

这次空间的功能布局处理将常规的平面纬度，创意地设计为立体纬度，巧妙地将离会所入口最远端的空间，往高度方向搭积木似的堆砌出休闲咖啡区、健身房、击剑室、乒乓球室四层空间，并在平面方向交错着阅览室和一二层卫生间、更衣间，最出乎想象力的是从二层楼板贯穿至顶层天花的斜面灰黑色攀岩墙，将所有的空间有机地联系起来。这所有的功能场所融合在一起就如一张巨大的立体地图，诠释着富有建筑语言的独特空间设计。

We started this project from a bare shell. We carved out volumes within this 12-meter high box and created different shape of floor plate in different levels that promote visual interaction. Visitors no longer need a floor guide for this club house because architecturally this club house is already a three dimensional map. Right when visitors arrive the club, they can see where the activities are. We conceived a dynamic club that use architecture to attract visitors, make them feel the fun right when they step into the club but they can only have a glimpse through the slots. All these inviting elements lead the visitor to explore the club. The various slab height, slots opening, dynamic walls and white form create different scenario at different point at the club.

The dynamic interlacing space initiated from the gymnasium. The gymnasium is interlacing with the pure white blocky fencing room. This floating elongated fencing room is right across the middle of the gymnasium and formed the ceiling of the gymnasium. The fencing room protruded out from the full height glass overlooking the lobby area wanting every visitor to notice its presence. Inside the gymnasium, through a slanted opening, there is a dramatic rock climbing area with 9.1 meter ceiling height. The result is a continuous sequence of surprises. Passing a short staircase, the reading area is embracing the table tennis room and has a strong interplay of enclosed and enclosure. Slots on the wall of the gymnasium, the fencing room, the table tennis room and the yoga room encourage visual interaction and strengthen proximity.

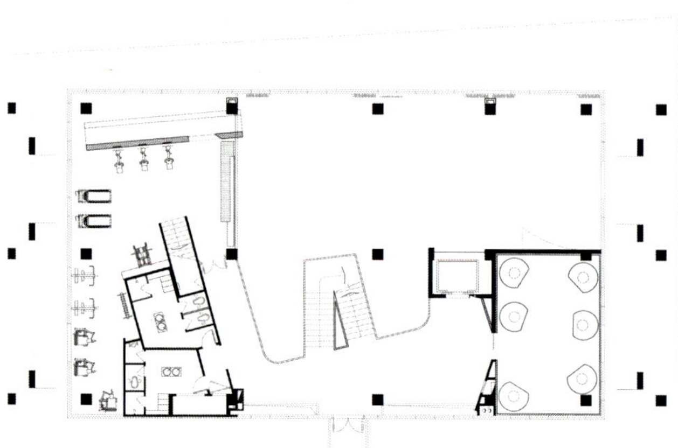

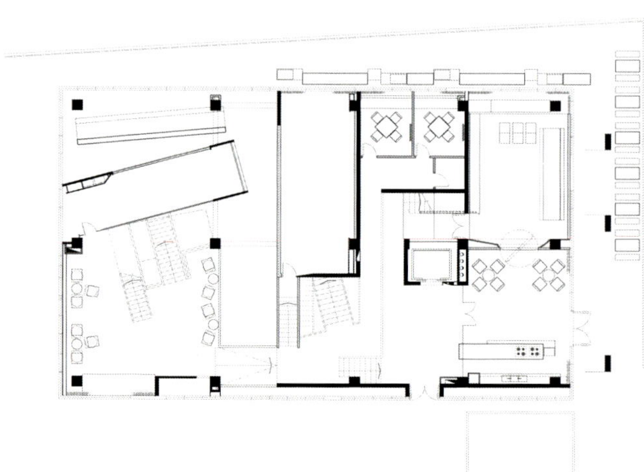

DIGIZONE

DIGI ZONE

设计单位：龙域设计

绿地公馆
Green Garden Hotel

坐落地点：上海 / 主要材料：玻璃、地毯、地砖

要把7栋上世纪90年代初建成的别墅改建成一座小型高档酒店，是一个不小的挑战。因为这7栋别墅是中国改革开放后房地产发展初期的产物，无论从外观还是内部使用来说都存在很多不足，要把一个富有特色的酒店功能植入进来，无疑是对设计智慧的考验。设计师把酒店公共功能安排在新建建筑中，通过新建建筑将7栋别墅串起来。张弛有度、辗转灵动的空间是成功的关键，内部空间充分地借用室外景观，步移景异，内外交融，原来别墅不规整的平面反而造就了风格和大小各异的30间客房。材料的使用并不奢华怪异，仅仅是黑色木地板、黑色石材的独到使用营造出黑、白、灰的调子，配以静静的水面，就完成了对于本土和传统的“宁静致远”的追求，和对于国际化价值观念的兼顾。

It is quite a big challenge to modify seven villas built in the early 1990s into a small but top-grade hotel. Because these seven villas are the production at the early stage of the development of the real estate after China' s reform and opening to the outside. There are many shortages as viewed from their visual appearances or interior functions. It is absolutely a test on the design intelligence to transplant the functions of a featured hotel here. The designer has arranged the public functions of the hotel in the new building and connected the seven villas in series through the new building. A space that is exaggerated and controlled properly full of nimbuses is the key to success. The interior space has fully borrowed the exterior scenes. The scene varies along with the movement and the interior and exterior are integrated together. The originally irregular planes of the villas instead have been produced with 30 guestrooms of different styles and sizes. The applied materials are not luxurious and weird, but the special application of the black wood floor boards and black stones materials has created a melody of black, white and gray. Together with the peaceful water surface, a pursuit to the local and traditional "indifferent to fame" is be accomplished as well as and the seeking of internationalized values are also realized.

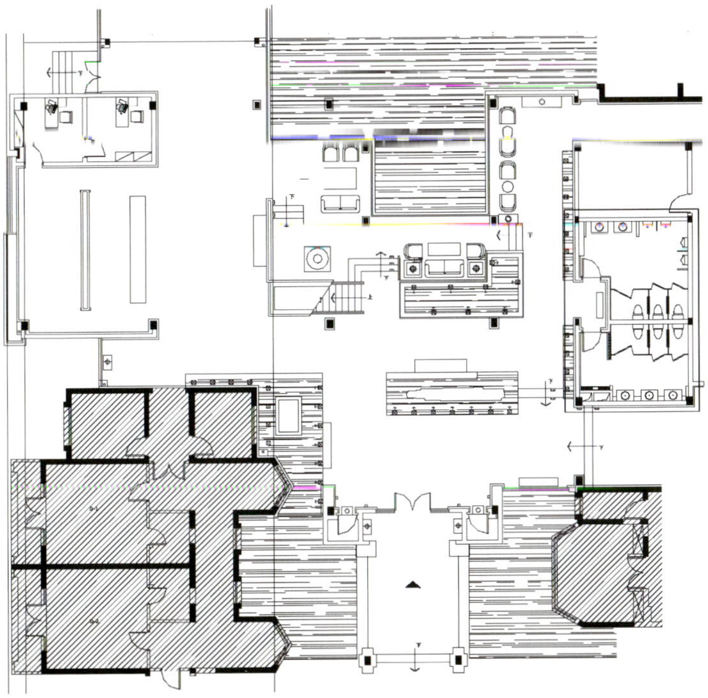

设计单位：苏州金螳螂建筑装饰股份有限公司 设计七分院三所 / 设计：孙彦清

金鼎湾国际
Global Golden Tripod Bay

面积：400m² / 坐落地点：南京 / 主要材料：米色石材、镜面不锈钢、皮革、木饰面 / 摄影：文宗博

在设计之初，业主就希望营造一处格调近乎酒店的住宅会所，以贴合来此消费的人群的品位。故设计师在创意最初就希望通过一些材质的运用以及细部的刻画来达到一种低调而又尊贵的室内环境。因此不管是桌上的鲜花配饰，还是墙边的荷花雕塑，抑或是墙面上错综复杂、纷繁交错的铁艺布景，都是经过精益求精的施工打造而来。整个空间宛如一幅静逸流淌的水墨画，耐人寻味。

At the beginning of the design, the Owner hopes to create a housing club in a style similar to the hotel in order to meet the tastes of the groups who come for consumption. Therefore, at the preliminary conception, the designer hopes to reach a kind of low-profiled but noble interior environment by the application of some materials and emphasis on the details. So regardless of the ornaments of fresh flowers on the table, the sculpture of water lilies beside the wall, or the anfractuous and complicated iron art scenes on the wall, they all are created by the constantly improved construction techniques. The whole space looks like a quietly flowing ink painting, very meaningful.

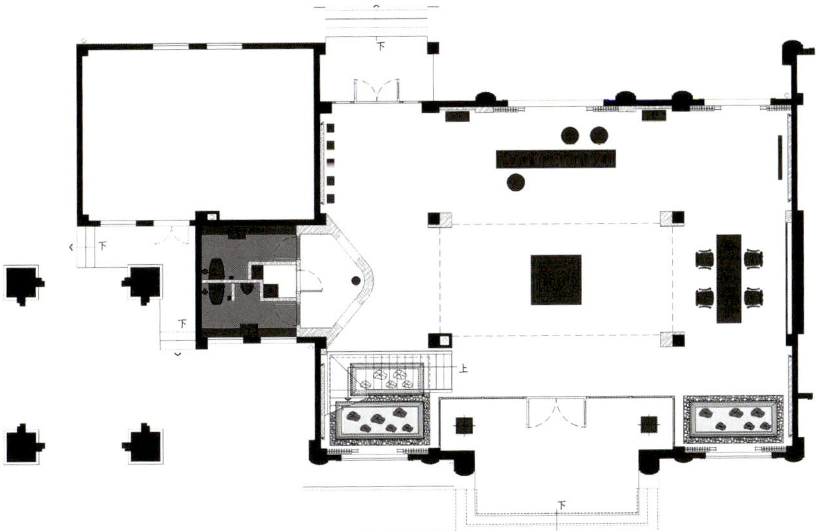

VISION

设计单位：南京测建装饰设计顾问有限公司 / 设计：刘延斌

东湖宾馆梅岭一号
Meiling No. 1 of East Lake Hotel

面积：5500㎡ / 坐落地点：武汉 / 主要材料：毛石、木纹石、桃花芯饰面板

坐落于武汉东湖宾馆内的梅岭一号是一栋极具历史价值的庭院式建筑，在绿树掩映中显得格外清新与质朴。它建于20世纪60年代，曾经是毛泽东六七十年代去湖北视察时数十次下榻的地方，这里的建筑与一草一木已深深烙下了历史的印记。

东湖宾馆是湖北省委省政府接待宾馆，由于毛泽东对其的特殊偏爱曾有“湖北中南海”之誉。改造后的梅岭一号将主要用于最高级别的政治接待任务。

业主的期待是：不事张扬，无需太多的繁琐粉饰，低调沉稳中透出不凡的品位。这也正是设计之所想。推开梅岭一号的大门，呈现在眼前的是深红棕色的球形桃花芯饰面板、意大利木纹石地面与米色基调的软包布所营造出的恬静、典雅、沉稳的意境。现代中式家具配合中国的水墨画，厅内流露出文化滋养下的底蕴。

会客厅两片毛石墙面是原建筑中的重要元素，这次设计中刻意将其完整地保留下来，自然、质朴充满了怀旧的气息。扇形过厅的连接处是宴请餐厅，满墙的梅花图与墙面的毛泽东书法字相得益彰。书房内的背景屏风上是毛泽东曾于东湖宾馆内书写的“水调歌头·游泳”，它与满墙的书架、宽大的皮质沙发、厚重的书桌这些陈设共同营造出浓厚的书卷气息。主卧室宽大而舒适与总体风格一脉相承，精致的家具、灯具与精巧的中式陈设融为一体。

Meiling No. 1, located in Wuhai East Lake Hotel, is a courtyard-type building with high historical values and looks extremely fresh and simple in the setoff of green trees. It was built in 1960s and used to be the place where Chairman Mao had been stayed in tens of times when he inspected Hubei. The buildings as well as every grass and tree here have been deeply left with historical marks.

East Lake Hotel is the reception hotel of the provincial government of Hubei Province. It was reputed as “Zhongnanhai in Hubei” because of the special preference to it by Chairman Mao. The renovated Meiling No. 1 will be mainly used for the highest political reception tasks.

What the owner expects is: not aggressive, without excessive complicated ornaments and revealed with the extraordinary tastes from low profile and calmness, which is also just what the design expects. Pushing the front door of Meiling No. 1, you will be confronted with a peaceful, elegant and calm atmosphere created by the dark sorrel ball-shaped Maghony veneers, Italian wood-grain ground and soft off-white package cloth. The modern Chinese furniture to go with the Chinese ink paintings has revealed the deposits in the hall nourished by the cultures.

The two rubble walls in the reception hall are the important elements in the original building and are intentionally retained intact in this design, natural, simple and full of reminiscent feelings. The connection part of the fan-shape hallway is the feting restaurant. The club paintings full of the wall and the calligraphic characters on the wall set off each other. On the background screen in the study is “A Prelude to Water Melody · Swimming” ever written by Chairman Mao in East Lake Hotel, which has jointly created a strong cultural atmosphere together with the displays such as the full-wall book shelf, wide leather sofa and heavy desk. The spaciousness and comfort in the main bedroom have come down in one continuous line with the overall style. The exquisite furniture, lighting fixtures and delicate Chinese display have been merged into one whole.

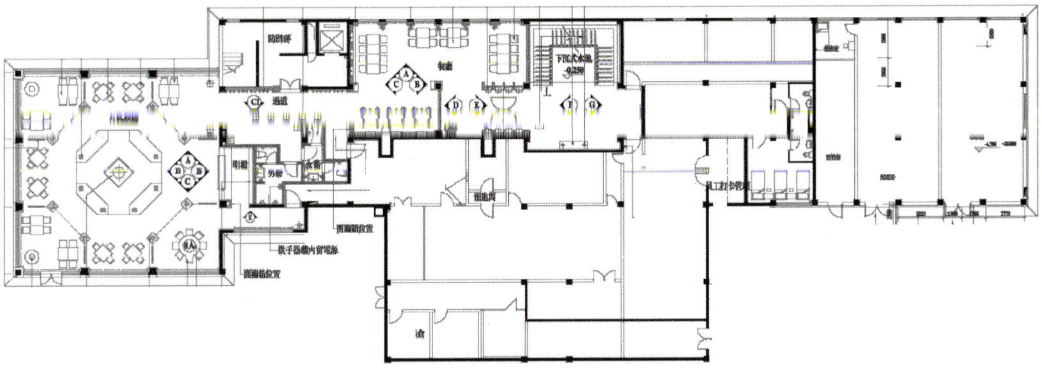

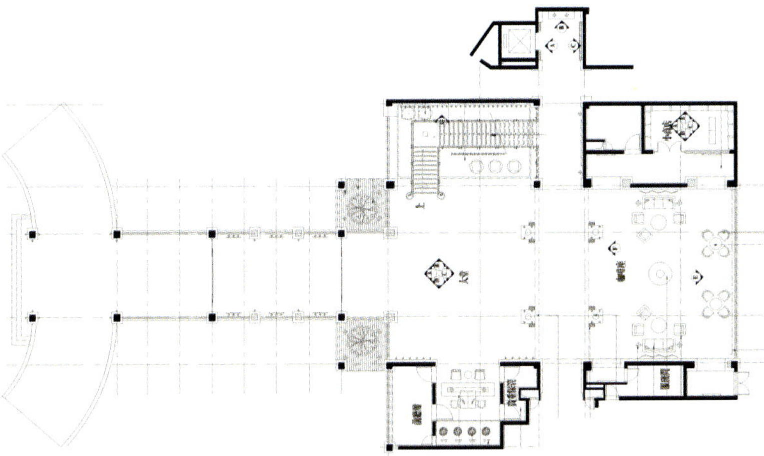

设计单位:麟研部 设计工作室

Racks桌球吧
Racks Snooker Bar

面积:1300m² / 坐落地点:上海 / 主要材料:绣花钢板、不锈钢、玻璃 / 摄影:莫尚勤

桌球吧的主要设计目标是符合建筑的多功能性及时间的阶段性需求，坐落在时尚流量高的地段上海新天地。

主题概念源自于美式古典主义。在设计中，建筑和室内在同一设计概念被完整统一的考量并结合的非常紧密。设计师在黑与白之间的无色系中玩味现条的交错与变化，有点像斑马纹的那种趣味。从平面、立面、转换到家具、饰品和画，再加入材质间的粗细对比衔接，逐渐演变成一种三度空间的思维。以黑白为两个空间分开，建筑的内部被大面积的玻璃幕墙包裹，入口处Racks的标志游走在墙面上，LED灯很好地控制着整个空间。吧台透过折射玻璃所透进的光线，呈现有趣的光影线条，与处在室内的人流动线交织成一个完整的建筑巡礼。利用光线变化、空间的结合和穿透，材料不再是给与颜色的主角，一切回到灯光原理的本质——万物的颜色都是光线所给予的。结构空间本身的线条利落单纯，甚至可以用更口语一点的表达方式来形容——挑逗。

跟酒吧台结合在一起的空间到走道甚至包房，没有一丝多余的装饰赘线。而包房的设计别有韵味，一朵玫瑰花的VIP房在不同角度看呈现不同的结果，有红色的玫瑰，有白色的镂空隔断。如此这般的概念，重复在此区域，主题完整而明确，凸显出设计者的用心。包房在保持基本的极度的舒适度之外，其他倒是处处透露着古怪却又充满了惊奇。

The snooker bar is mainly designed to meet the multi-functionalities of the building and the periodic requirements of the time and is located in an area of high fashion and stream of people - Xintiandi, Shanghai.

The thematic conception is derived from American classicism. In the design, the architecture and interior in the same design conception are fully and uniformly considered and integrated intimately. The designer has ruminated on the intercross and change of the lines in a colorless series of black and white, which is something like zebra lines. It gradually evolves to a tri-dimensional thought from the plane, facade to furniture, ornaments and paintings together with the contrast and connection of the fineness and roughness among the materials. It is divided into black and white spaces. The interior building is covered by large glass curtains while the symbol of Racks at the entrance is floating over the wall while LED lamps have controlled the whole space properly. The lights permeated in through the refractive glass in the bar counter have delivered the interesting light-and-shadow lines, which are interwoven into a compete patrol on the building together with the streams of the people in the room. The materials are no longer a principal role to provide the colors with the application of the light changes, space combination and permeation. Everything now returns to the nature of the light principle —— the color of every thing on the earth is given by the lights. The line of the structural space itself is brisk and simple and can even be described in a more oral expression —— flirting.

There is not any excessive decorative lines from the space combined with the bar counter to the corridor and even VIP rooms. The design of VIP rooms has a special charm; a VIP room with one rose has been demonstrated with the different results from the different angles. There are red roses and white through-carved partitions. Such a conception is repeated in this area. The theme is complete and clear with an obvious display of the designer ' s intentions. In addition to that VIP rooms has maintained the basically extreme comfort, it is revealed with weirdness and filled with surprise other else.

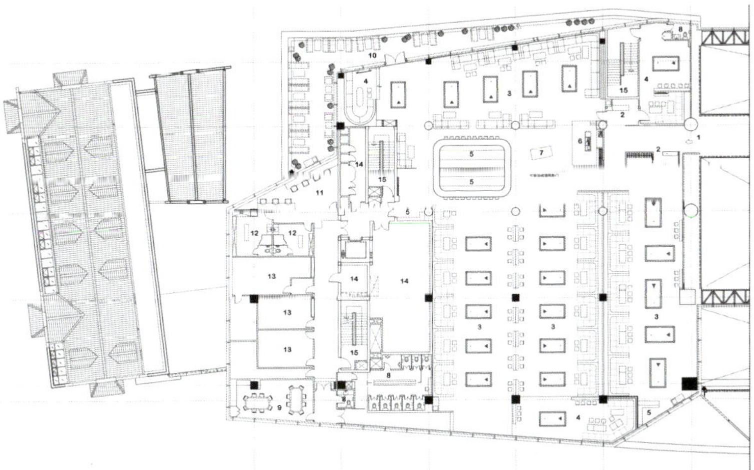

设计单位：张波设计工作室 / 设计：张波 / 参与设计：刘继烨、吴川文

壹号公馆夜总会
No.1 Residence Night Club

面积：6000m² / 坐落地点：合肥市阜南路 / 主要材料：土耳其玫瑰大理石、啡网大理石、紫罗兰大理石、珊瑚红大理石、黑白根大理石、进口墙纸、仿古浑水裂纹漆、金箔、订制金色木线条、订制木雕

走进壹号公馆俱乐部，进入的是一场装饰主义的盛宴，一个古典装饰元素与现代工艺交融的异度空间，一段极尽奢华而又充满无数惊喜的奇幻梦境。

二楼是中包房、红酒屋、洋酒屋及休息区。淳厚深邃的墨绿色墙面配上繁复华丽的金色木线条，品质上乘的大理石，错落在走廊、包房中的古典油画，加以点缀上精心挑选的摆件、布幔、水晶灯……双眼被这一处处的匠心独具弄得迷离、迷失。仿佛这已不是俱乐部，而是一座欧洲贵族的官邸。连接二三楼的是气派非凡的宽大通道。巨大的红色水晶吊灯，直到天花的巨幅壁画，通向的又是另一个怎样的奇妙空间？

三楼的特色是它层次繁复的墙群。运用浑水裂纹漆的独特做法，营造出充满古典主义装饰味道的氛围，仿佛存在的年代已经久远。甚至以独特的做法，运用油漆制作出石材的效果。装饰主义的精髓被把玩得如此恰到好处而又惊喜连连。

Once entering No.1 residence of VIP night club, you will enter a decorative feast: another space integrated with classic decorative elements and modern arts and crafts and an extremely luxurious fanciful dreamland full of numerous surprises.

The second floor includes medium-size VIP rooms, red wine room, liquor room and resting area. The pure and profound blackish green wall in arrangement with the complicated and magnificent golden wood lines as well as the superior marbles, the classic oil paintings scattered in the corridors and VIP rooms, the decoration of elaborately selected ornaments, cloth curtains, crystal lamps…… both eyes become blurred and lost by one after another unique arrangement. It seems to be the official residence of a European aristocrat instead of a club. The spacious passage connecting the second and third floors is extraordinary and grand. From the huge red crystal pendant lamp to the huge mural, what kind of another wonderful space it will lead to?

The feature of the third floor lies in its wall group with complicated arrangement. The unique method of applying the muddy crackle paint is used to create an atmosphere filled with a classic decorative taste, as if its existing ages have been long time ago. Even the paint is made in a unique way to make the effects of stone materials. The soul of decoration is demonstrated not only so appropriate and but also with constant pleasant surprises.

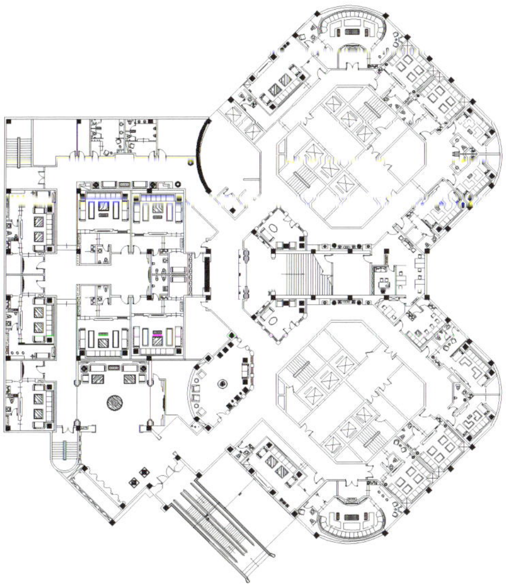

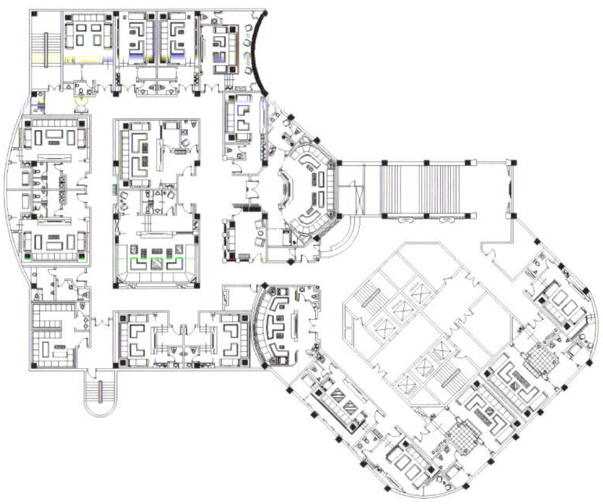

Heineken

设计：吴林孟、张丰毅

新南亚宝丽金会所
New Nanya PolyGram Club

面积：6000㎡ / 坐落地点：温州 / 主要材料：玫瑰金不锈钢、皮革软包、微晶石、金镜

本案设计时尚、浪漫、奢华，大都会的主题设计风格贯穿始终，置身于如此清新畅快的时尚派对之中，必定会令您度过轻松开怀的一夜。本案设计师力求在视觉、音乐、文化之交融下，营造一种另类的休闲文化。

This design is fashionable, romantic and luxurious with the metropolitan subjective design style throughout it. Staying among such a fresh and lightsome fashion party, it will definitely make you spend a relaxing and happy night. The designer for this case tries to create an unusual creational culture in the integration of vision, music and culture.

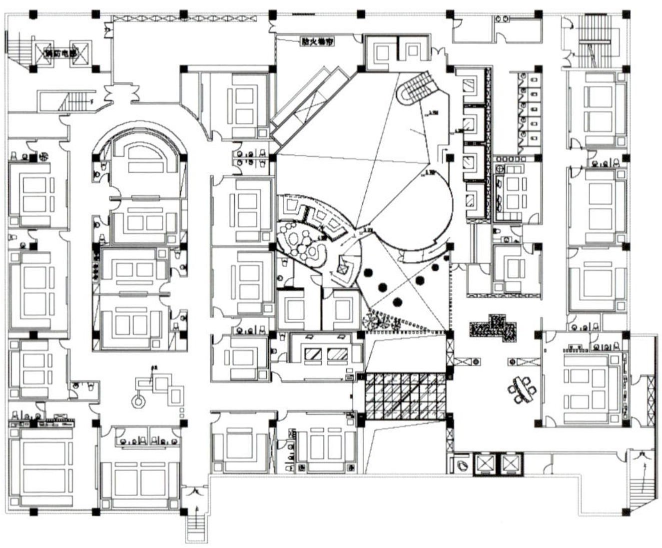

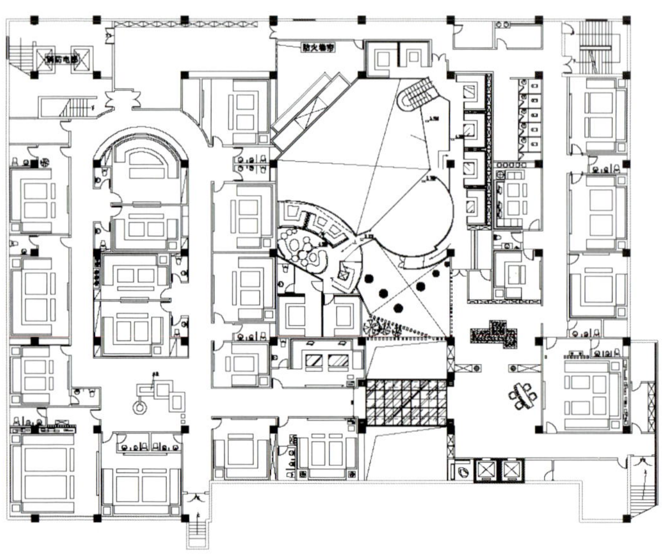

设计单位：无锡观点设计 / 设计：吕邵苍 / 设计执行：单昱俊

苏州英皇会所
Emperor Club Suzhou

面积：3000m^2 / 坐落地点：苏州锦帆路

低调奢华后新古典的娱乐设计——英皇会所设计说明娱乐设计的时尚性与商业性在目前的中国是设计师必需考虑的两个问题。我是把夜总会设计分代的：第一代为众所周知的欧式；第二代为中式新古典，以符号和软装陈设为主要设计手段；第三代则以镜面、不锈钢、LED为主要设计手段。这就回答了为什么从南到北的夜总会设计模式几乎类同的原因。因为投资的是甲方不是设计师，本案的设计就基于上述原则做了一次尝试性的整合，所以就有我们这个后新古典的英皇会所娱乐设计。

英皇会所是倡导精英消费的高级娱乐场所。本案设计我们充分利用LED光源对氛围的营造，镜面、不锈钢对梦幻与迷离的映衬，新古典的欧式灯具突显奢华。皮革硬包软包，具装置感的软陈设，洗去了浮华的喧嚣，一种精致的低调奢华扑面而来。

Low-profiled, luxurious and post-neoclassic, recreational design —— the design of Emperor Cub has shown that the fashion and business characteristics in the recreational design are the two questions the designer must take into consideration in present China. I have divided the night clubs into different eras. The first era, as it is well-know, is the continental style; the second era is neoclassic in Chinese style to take the symbols and soft decorations as the main design technique; the third era is to take mirrors, stainless steels and LED as the main design technique. These have answered to the question why the night clubs throughout the country look similar in their design styles. As the investment party is the owner instead of the designer, based on the above-mentioned principles, the design of this case has been made with a tentative reorganization. Thus, this post-neoclassic recreational design of Emperor Club comes into being.

Emperor Club is a superior recreational place advocating the consumptions by the elites. In this design, we have made full use of the creation of the atmosphere by LED lighting sources. Mirrors and stainless steels have set off the reverie and loss while the neo-classic continental-style lighting fixtures have shown the luxuries prominently. The soft and hard leather packages and the soft exhibits with strong decorative sense have washed away the showy nosiness. A kind of exquisite low-profiled luxury greets you abruptly.

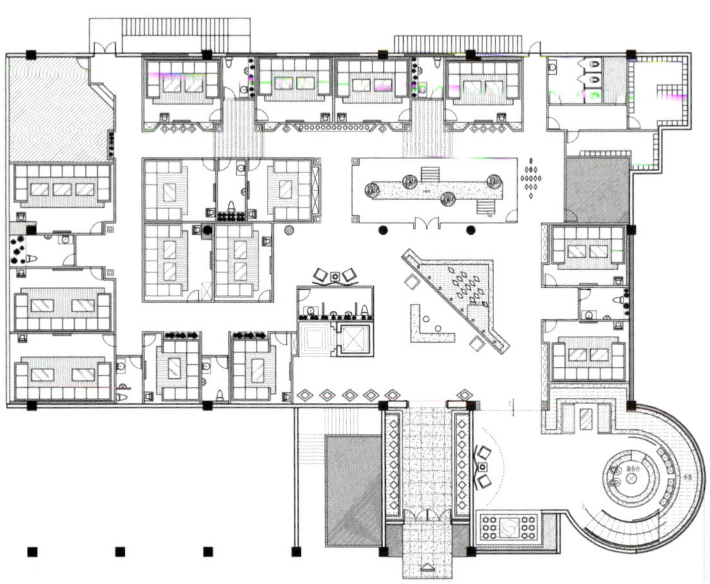

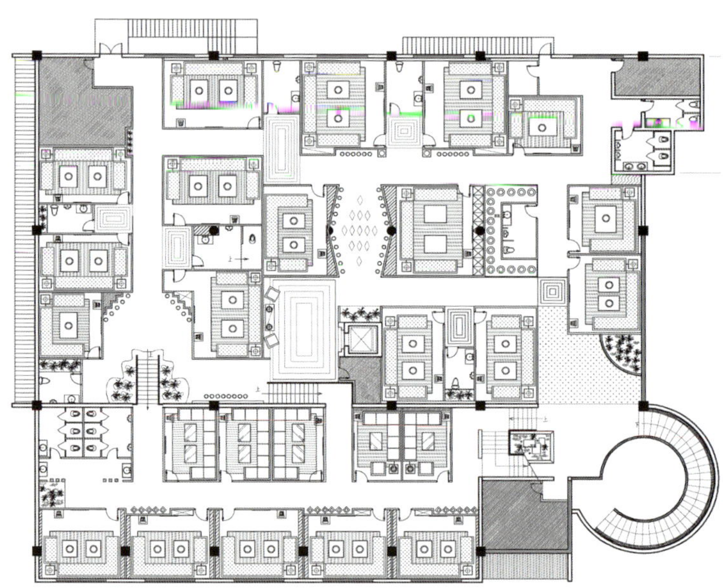

设计单位：西安汇典环境设计装饰工程有限公司 / 设计:孙鸣春

天香新唱KTV夜总会
Tianxiang Xinchang KTV Nightclub

面积：2400m² / 坐落地点：西安 / 主要材料：焗油玻璃、银镜面、纱缦、吸声板 / 摄影：孙鸣春、焦长虹

本案为西安市第一家具有独特风格设计模式的娱乐城，建筑面积约2400m²，整个室内的空间设计营造力求表达一种变换莫测、亦真亦幻的视觉体验。

在整个空间、平面的布局中，弧形通道的天空处理使整个室内空间的视觉意识加大，给整个夜总会带来神秘而活泼的气氛，在公共环境的设计中，使风格、色彩、灯光充分融入空间，同时对材料的运用大胆而有新的节奏。

以现代风格为设计表现，区域造型上不仅形体抽象、线条简洁，而且镜面玻璃与不锈钢相互呼应、反射，使整个区域视觉意识无限扩大，同时充满强烈的时代感。

This is the first amusement city of the special style and design mode in Xi' an with a floor area of about 2400m2. The space design of the whole interior space is attempted to create a visual experience, changeable and unpredictable as well as real and illusory.

In the arrangement of the whole space and plane, the sky is made in an arced passage so that the visual consciousness of the whole interior space is enlarged to bring the whole nightclub a mysterious and vivid atmosphere. In the design of the public environment, the styles, colors and lights are sufficiently integrated into the space while the materials are applied boldly with new rhythms.

The modern style is revealed in the design. As for the modeling of the areas, the form is abstract with simple lines while the mirror glass echoes and reflects with stainless steel so that the visual consciousness of the whole area is enlarged infinitely, full of strong sense of the times.

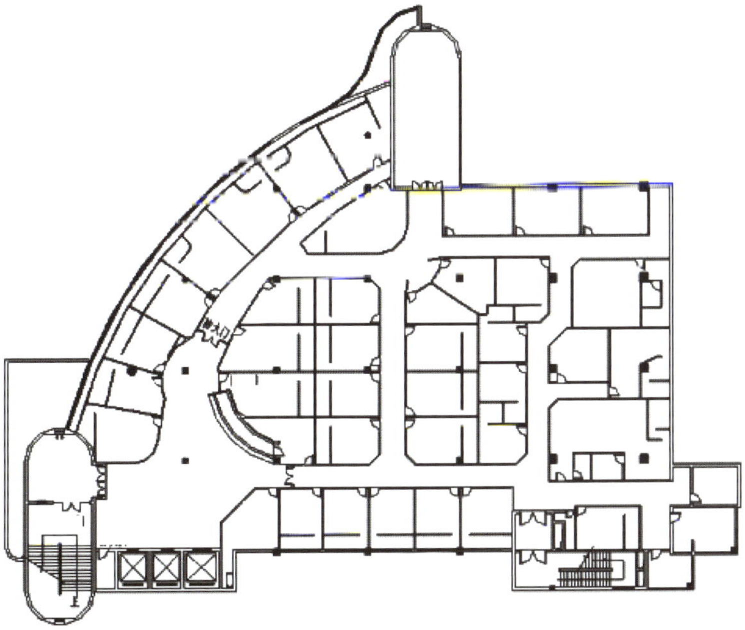

林道恩

毕业于英国诺丁汉大学，英国建筑联盟建筑学硕士，英国皇家建筑师。1997年起历任马来西亚TR HAMZAH AND YEANG建筑事务所建筑师，英国纵向建筑事务所建筑师，上海同济大学建筑设计院项目设计师，2004年至今任上海万恩佑设计有限公司董事。

蔡旭民

从事室内设计20年，于2002年创立香港天艺空间设计制作有限公司、深圳天艺空间装饰设计有限公司任设计总监。致力于餐饮及酒店设计工作，完成近200个餐饮酒店设计项目。

查波

2003年毕业于湖北美院，2003-2006年于宁波市富华建筑装饰有限公司任设计师，2007年任宁波大学教师。

陈骏

1993年毕业于汕头大学美术设计系，本科，学士学位。从事室内设计14年，1994年创建蓝鲸室内设计有限公司，担任经理兼创意总监。

陈旭东

1987年毕业于吉林艺术学院美术系，1995—2002年任吉林省大东室内设计有限公司总经理兼总设计师，2003年至今任吉林艺术学院设计学院教授。设计作品多次获奖："凝固的意象"获得2005年第二届海峡两岸室内设计大赛办公类一等奖，"别墅设计"获得吉林省第五届建筑画展一等奖，"咖啡小镇"获得2006年华耐杯中国室内设计大奖赛二等奖等。

陈耀光

1987年毕业于中国美术学院首届环艺系，就职于浙江省建筑设计院，1995年创立杭州典尚建筑装饰设计有限公司，任创意总监兼总经理。CIID副会长，中国建筑装饰协会设计委员会副会长，浙江省建筑装饰行业协会设计委员会会长。1996—2005年荣获中国室内设计大赛七次一等奖。被中国建筑装饰协会授予"全国有成就的资深室内建筑师"称号。

仇益国

先后在国家级设计机构工作，并担任总建筑师。2003成立中派设计有限公司,并与The Arcop Group 合作成立了Arcop (China) Design Consultants Limited。室内设计是在十几年的设计经历中的重要组成。其主要作品有：北京西钓鱼台主题酒店、深圳凤凰卫视大厦、深圳大剧院、乌鲁木齐机场等。

大帅

毕业于辽宁省工艺美术学校，现任孙志刚空间设计工作室主案设计师。

董静

西安交通大学艺术设计专业硕士，现任西安财经学院讲师，中国建筑学会室内设计分会会员。

范江

1999年创立高得设计公司,现任设计总监兼负责人。作品常见于《室内设计与装修》、《现代装饰》、《美国室内》中文版等专业杂志。2006年，"高丽使馆贵宾厅"获亚太室内设计双年大奖赛金奖；2007年获《现代装饰》（国际）室内设计传媒奖、年度精英设计师。2003年出版《当代建筑与室内设计工作室实录》，2007年出版《新中式主义》。

王闳

北京包达铭室内装饰设计有限公司负责人。主要作品有：安盛咨询，德勤会计师行，施耐德(中国)大东电报局，中信银行，中国建设银行，德累斯顿银行等。

冯志业

1999年加入凯达环球,现任高级主任建筑设计师，专长于办公及商业设计。其最为瞩目的项目是香港渣打大厦商业部分设计及香港历山大厦商业裙楼改造工程。2002年开始专注中国市场,重点项目有城隍庙广场商业改建项目、上海广场改建项目、北京市东方广场新天地美食街改造等。

富元

1995年毕业于鲁迅美术学院，2000年深造于新加坡，2003年创建海南富元设计公司。2004年荣获海南十大杰出设计师，2006年荣获亚洲十大设计师，擅长休闲度假风格设计及异域风情设计、园林景观设计。

高超一

毕业于北京中央工艺美术学院室内设计专业，美国室内设计师协会会员，CIID会员。现任美国CY设计事务所主要合伙人兼总设计师，上海金螳螂环境设计研究有限公司总经理兼总设计师，苏州金螳螂建筑装饰股份有限公司酒店设计事务所总设计师。主要作品有上海世和中心、上海飞机厂办公楼、世贸新城会所、苏州国际青庭会所及样板房等。

高立平

CIID会员，IFI会员，高级室内建筑师，2004-2006年全国百名优秀室内建筑师。毕业于吉林艺术学院油画系，现任北京优恩空间环境设计有限公司董事长/设计总监。2006年获IFI大奖赛一等奖，2006年中国室内设计师十大年度人物，2008年获上海国际建筑及室内设计金外滩大奖。

高轶

1999年取得平面设计专业学士学位，1999年-2003年从事平面设计工作，专业设计企业标志和品牌视觉形象，2004年取得室内设计专业学士学位，2004年创办南京目达室内设计有限公司，任设计总监。CIID会员。

耿治国 Gustaf Kan

出生于台湾，东海大学建筑系毕业，1998年创立台北飞形设计，2001年创立上海飞形设计。主要设计作品有：台中中友百货系列主题商品馆、台北Bando 8扮桌会时尚宴会厅、上海898 Cabinet、宁波老外滩五号公寓、北京福鹿名肴会等。

谷腾

中国室内高级设计师。IFI会员，CIID会员。荣获第六届中国室内设计双年展室内设计大赛银奖，14届香港亚太设计大奖办公类金奖，2007《现代装饰》国际传媒奖室内设计大奖，2007中国室内设计大赛最具创意奖等。

郭轲蔚

金螳螂浙江设计研究院院长、金螳螂建筑装饰股份有限公司副总设计师、浙江省设计委常务理事、高级室内建筑师、全国百名优秀室内建筑师和IAID最具影响力中青年设计师。主要获奖：镇江移动电信大厦获2004年中国建筑工程鲁班奖，2005年获中国最佳酒店设计奖，2007年获室内设计艺术成就奖等。

郭锡恩、胡如珊

如恩设计研究室负责人。

韩冠恒

室内建筑师，CIID会员,哈尔滨市唯美源装饰设计有限公司经营经理。获奖作品：2001年获中国室内设计大奖赛“佳作奖”，2006年获中国室内设计大奖赛商业工程类“佳作奖”，2007年作品获中国室内设计大奖赛商业工程类“优秀奖”。

何潇宁

毕业于中央工艺美术学院，1995年获得日本东京艺术大学硕士学位，2000年至今任鼎贺环境设计（深圳）有限公司董事长兼设计总监，深圳市长城家具装饰工程有限公司副院长。2004年获中国建筑装饰协会全国杰出中青年室内建筑师称号，2006年获中国社科院中国地标建筑最具影响力室内设计师称号，2007年CIID授予全国百名优秀室内建筑师称号。

黄斌

毕业于上海工业大学建筑与室内设计专业，高级室内建筑师，先后从事及参与多项大型室内设计工程项目及上海近代优秀保护建筑的改建与装饰设计，包括交通银行上海分行、国家电力调度中心、武汉美术馆改建方案、上海老饭店改建等。

黄书恒

伦敦大学建筑硕士，黄书恒建筑师事务所主持人，台湾民间文化基金会董事，玄武设计有限公司设计总监。

姜峰

建筑学硕士、国务院特殊津贴专家、教授级高级建筑师、高级室内建筑师。现任姜峰室内设计有限公司总经理兼总设计师、CIID副会长、中国建筑装饰协会设计委副主任。先后荣获美国室内设计杂志中文版年度十大封面人物、深圳市十大杰出青年、中国室内设计20年设计功勋奖、中国国际设计艺术观摩展终身艺术设计成就奖等荣誉。

姜湘岳、吴海燕

姜湘岳，毕业于南京艺术学院工艺系装饰专业，同年任职于江苏省建筑科学研究院，1997年任江苏省建科院装饰设计所所长，2001年至今任江苏省海岳建筑装饰工程设计有限公司总设计师和南京金陵国际装饰设计工程实业有限公司设计总监。致力于研究酒店的设计及发展。CIID会员，高级室内建筑师。
吴海燕，CIID会员，室内建筑师。

李益中

1993年毕业于大连理工大学建筑学专业，1998年创立深圳市派尚环境艺术设计有限公司，2007年出版著作《样板生活》，现为CIID理事及第三专业委员会副主任。近年来主要致力于地产商的样板房、售楼处、会所等的专业室内设计，并在此领域积累了大量的成功案例和经典作品。

琚宾

就读于中央美术学院设计学院,主修室内设计。毕业后致力于研究中国文化在建筑空间里的运用和创新，并与国内国际多位知名建筑师合作设计项目，以个性化及独特的视觉语言来表达设计理念，以全新的视觉传达来解读中国文化元素。

方钦正

建筑师　台湾人。毕业于英格兰的一所学校，在中国有7年工作经验。负责亚洲地区酒店、零售及活动等有关的各种项目。在施工方面具有丰富的经验。

Margaux Lhermitte

室内设计师 法国人。毕业于伦敦的一所学校并具有5年工作经验。派遣到上海开办工作室之前在巴黎纳索工作，拥有酒店、餐饮及零售设计经验。

赵牧桓

1995年毕业于纽约室内设计学院，1997年成立牧筑空间暨造型研究室（台湾），2002年成立上海牧桓+灯光设计研究室。

李越

毕业于上海同济大学建筑城规学院建筑学专业，高级工程师，上海同济室内设计工程有限公司设计部主任，国家一级注册建筑师，高级室内设计师。

梁毅锋

高级室内建筑师，IFDA注册国际室内设计师，中国建筑装饰协会会员，现任毅锋设计（香港）策划顾问有限公司CEO，宁波简艺艺术毅锋室内设计有限公司设计顾问。主要设计项目：宁波象山春夏秋冬休闲中心、浙江上虞三源大酒店、江西高安假日大酒店娱乐城、宁波甬港饭店KTV、杭州临平SOS酒吧、骆驼光辉岁月娱乐城、浙江外国语学校生态园等。

林志锋

毕业于广州美术学院，高级室内建筑师，CIID会员，IFI会员，森瀚室内设计顾问有限公司设计总监。主要作品：广州天河汽车客运站大楼、顺德君兰高尔夫俱乐部、梦幻巴黎俱乐部室内外设计，广州市银鹏机电公司办公大楼室内设计、番禺映月翠庭样板间等。

林志宁

毕业于广州大学艺术系室内设计专业，清华大学建筑系建筑与室内设计专业研究生，CIID第14专业委员会副秘书长，广东省环境艺术设计行业协会理事，RYB国际设计机构董事，森瀚室内设计顾问有限公司首席设计师。荣获IDIA中国最具影响力的中青年设计师、全国杰出中青年室内建筑师等称号。

刘卫军

美国IARI国际注册高级设计师，CIID理事兼深圳专业委员会常务副会长，高级室内建筑师，中国百名优秀室内建筑师，PINKI（品伊）创意机构创始人兼首席设计师。2005 年在深圳关山月美术馆成功举办“十人・十年联展”，2004年6月于深圳科技馆举办刘卫军个人专题学术报告，2005年获邀参加国际双年展及美国室内设计展。

刘延斌

毕业于南京艺术学院,CIID会员。1996年赴香港接受专业室内设计培训和工作，2001年组建南京测建装饰设计顾问有限公司任设计总监至今。

罗灵杰、龙慧祺

罗灵杰，香港理工大学室内设计学士，香港李惠利工业学院室内设计专业高级文凭，壹正企划设计总监。拥有多年室内及平面设计的资历及管理经验，荣获多项室内设计大奖。
龙慧祺，美国俄勒冈州立大学室内建筑学士，香港李惠利工业学院室内设计高级文凭，壹正企划设计总监。曾任职多间大型国际设计师楼并参与跨国设计项目及发展工程。

卢文伟

高级室内建筑师、中国建筑装饰协会会员、浙江省装饰设计协会会员、浙江省优秀室内建筑师、2006年度中国杰出中青年室内建筑师。1999年创办杭州历程装饰设计有限公司。

卢迅

毕业于广州美术学院，CIID理事，广东省环境艺术设计行业协会专家组专家。2003年获中国建筑艺术奖及综合商业建筑奖、2003年获年度最具影响力十大地产项目案例金奖、2005年获建设部中国最佳商业环境设计奖、2008年荣获第三批全国杰出中青年室内建筑师贡献奖。

陆文星

南京陆文星室内设计事务所设计总监，CIID会员。荣获奖项有2003年度南京十佳首席设计师、2006年度雷士杯全国设计师大奖赛二等奖、2006年度中国（深圳）国际品牌与设计大奖赛二等奖等。

陆希杰

出生于台北，毕业于东海大学建筑系，英国AA建筑联盟硕士学位，在英国期间曾于Raoul Bunschoten事务所担任设计师，而后回国成立CJ STUDIO，从事建筑及室内设计、家具设计、产品设计等相关研究开发。曾任教于实践设计学院空间设计系、铭传大学空间设计系、东海大学建筑系、国立交通大学建筑研究所，著有《锻造视差》。

吕邵苍

中国百名优秀室内建筑师，金螳螂设计八分院院长，CIID理事、无锡委员会副主任，国际ICIAD室内设计学会理事，IFI会员。获奖：2008年获中国上海“金外滩”最佳配色运用奖，2007年获中国室内设计大奖赛优秀奖，2006年获中国室内设计大奖赛优秀奖等。

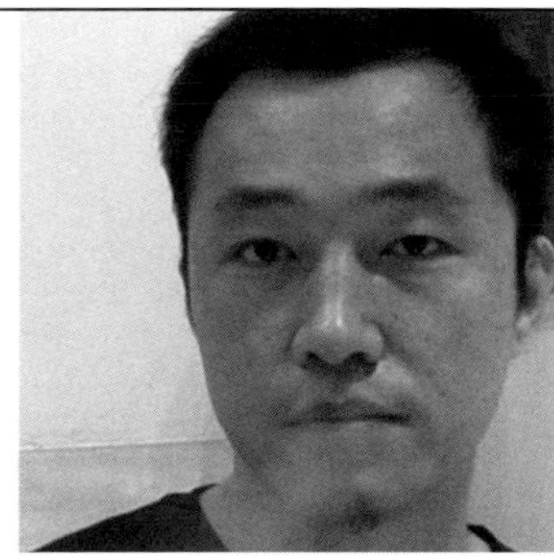

吕永中

毕业于同济大学建筑系，现任教于同济大学建筑与城市规划学院艺术设计系，上海市工业设计协会家具设计专业委员会副主任，唯品设计董事长及设计总监，半木原创家具品牌创始人及主持设计师。长期从事空间设计及家具研究。

麦德斌

高级室内建筑师、CIID理事、东莞专业委员会主任、东莞创达维森设计机构执行总监。荣获多项殊荣：2006年中国室内设计双年展金奖和银奖、2007年金外滩室内设计大赛最佳照明设计奖、2007年国际室内设计传媒奖年度设计大奖等。

朱永春

毕业于南京航空航天大学，组建了朱永春设计有限公司，任董事长兼设计总监。擅长雅致及新锐风格的设计，代表作有南通大饭店、味彩阁餐厅、艾邸俱乐部、立文办公室、锦绣苑私宅、良渚文化村样板房等。曾先后获得中国室内设计大奖赛一等奖、二等奖、最佳设计原创奖以及入编《IFI国际室内设计年鉴》、参加全球华人设计成就展等荣誉。

潘开富

1994年室内设计专业毕业,同年进入江苏省建筑装饰设计研究院工作，2000年组建南京智点设计顾问有限公司任设计总监。中国建筑学会室内设计分会会员、室内建筑师、南京市室内设计学会理事。

潘礼旺

美国南加州建筑研究所建筑硕士，台湾东海大学商业管理学士。现任七观国际有限公司总经理，七域国际创意有限公司总经理，七艺室内装修有限公司总经理，七能建筑师事务所总经理，宅修生活国际股份有限公司总经理，台湾民间文化基金会董事，铭传大学建筑系专任讲师。

陈石林

1997年毕业于中国美术学院环境艺术系，2000年合伙创立杭州正午装饰设计公司，2005年赴法国游学。从事高档娱乐、餐饮、会所、酒店、宾馆等大型公共场所的设计。主要作品有纽约LINDEN PLACE娱乐会所、纽约SHANG HAI LEE餐厅、上海顺风大酒店、上海金碧辉煌娱乐城、杭州世贸西湖四季酒店、萧山凯撒皇宫娱乐城等。

王国元

CIID会员，室内建筑师，以酒店 、餐饮、娱乐等商业设计见长。2007年度主持设计的作品：宁波新金星宾馆、华东饭店、八点半酒吧、青莲酒家、帕玛美业、北仑太阳雨宾馆等。

迫庆一郎

日本东京工业大学研究生毕业，1996-2004年就职于山本理显设计工场，2004年成立SAKO建筑设计工社（中国北京），2004-2005年赴哥伦比亚大学担任客座研究员，日本文化厅外派艺术家驻外研修员。主要作品：广尾T（珠宝首饰加工店）、北京芬理希梦、北京蒲蒲兰绘本馆、天津万花景、巴塞罗那·IMAGINARIUM等。

钱 宇

毕业于沈阳大学艺术工程学院环艺系，现任孙志刚空间设计工作室主案设计师。

秦岳明

1999年组建朗联设计顾问公司任设计总监、高级室内建筑师、CIID理事和第三专业委员会委员、IFI及亚洲室内设计学会联合会中国地区委员会深圳代表处专业委员。从业十余年来，同时涉足建筑和室内设计，设计作品在全国和亚太区曾多次获得大奖，并在专业杂志及书籍上发表。

盛燕

毕业于城市学院室内设计与环境艺术专业，并取得清华大学美术学院文学学士学位，高级建筑师。1993年开始于中国建筑设计研究院工作至今，现在环境艺术设计研究院室内设计研究所任主任建筑师。2004年获“全国百名优秀室内建筑师”及“全国杰出中青年室内建筑师”称号。

史永杰

高级室内建筑师、CIID会员、深圳委员会委员、深圳室内设计协会理事、深圳市专家工作联合会建筑业专家工作委员会委员、深圳市鼎维室内设计有限公司执行董事。

宋国梁

全国百名优秀室内建筑师、ICIAD理事会宁波区理事、IFI会员、CIID会员、宁波[V]室内设计策划顾问有限公司总设计师、宋国梁设计事务所有限公司总设计师、上海维特锐配饰设计有限公司艺术总监。

孙鸣春

毕业于西安美术学院，在中国建筑西北设计研究院工作6年，后调入西安美术学院设计系工作。负责了许多重大工程项目的设计工作，并获得了多项荣誉奖项，在专业刊物上发表著作、作品。西安美术学院副教授，全国百名优秀室内建筑师。9年以来连续为黄帝陵祭奠活动承担总体策划、设计、制作，任总设计师之一。

孙天文

高级室内建筑师，中国建筑学会室内设计分会会员、理事。

孙彦清

1995年毕业于郑州轻工业学院室内设计专业，曾在首届清华大学建筑工程与设计高研班学习，中国建筑学会室内设计分会会员，南京市室内设计学会会员。现为金螳螂第七设计研究院主案设计师。

孙志刚

毕业于中央美术学院美术史系，现任孙志刚空间设计工作室设计总监。

谈星火

1986年考入天津职工工艺美术学院环境艺术系学习并于1988年7月毕业回到建设部建筑设计院工作，现任中国建筑设计研究院环境艺术专业设计院室内所设计室主任建筑师。获IAID最具影响力中青年设计师奖，全国百名优秀室内建筑师奖。

童明

东南大学建筑学专业硕士，同济大学建筑与城市规划学院城市规划理论与设计专业博士，1999年至今担任同济大学城市规划系副教授。一直从事于较为广泛的理论研究与实践工作，已发表多篇论文并出版若干专著和译著。主要作品有苏州大学文正学院教学楼、南京高新区国际俱乐部、十院宅、南宁武鸣西江八大处园林、上海嘉定艺术家工作室等。

童武民

高级室内建筑师。全国百名优秀室内建筑师，江西首位获得大型房地产（千户建筑群）样板房设计资格设计师，设计作品发表于全国性专业杂志。IFI会员，CIID会员、第27（江西）专业委员会副秘书长。

万宏伟

1991年毕业于四川美术学院，全国百名优秀设计师。宁波市汉文设计工作室总经理、设计总监，IFI会员，CIID会员，宁波专业委员会理事，第三届中国国际设计艺术观摩展室内设计十大新锐人物。作品曾收入在《中国室内设计年鉴》、《样板房设计经典》、《中国室内设计年刊》、《亚太室内设计年鉴》、《样板房设计经典》、《第十四届亚太室内设计大奖作品选》等。

王善祥

毕业于华东师大艺术系国画专业,从事现代艺术创作及建筑、室内设计。努力使纯艺术及建筑设计相互融合，曾参加多项艺术展览。主要项目有:瑞典ABB公司上海办公室改造、上海来福楼火锅餐厅、浙江杭州拜占廷画廊、北京菩提谷山庄会馆、上海中中家缘家具商场、云南红河州体育馆办公楼及宿舍楼等。

王祎华

中国建筑装饰协会及中国饭店协会2006年度中国十佳酒店设计师，室内建筑师/工艺美术师，CIID会员，太湖新天地餐饮设计获中国建筑装饰协会及中国饭店协会2006年最佳餐饮包厢奖，西太原晋祠国宾馆六号楼获第二届IFI国际室内设计大赛暨2006年华耐杯中国室内设计大奖赛优秀奖。

王仲文

毕业于哈尔滨市轻工学院室内设计系，室内建筑师、CIID会员、哈尔滨市唯美源装饰设计有限公司创意室主任。主要设计作品有：哈尔滨医院办公楼、南京月牙湖小区样板房、哈尔滨农业银行接待中心、哈尔滨新空间酒吧、牡丹江海燕美容生活馆、哈尔滨采珍集火锅等。

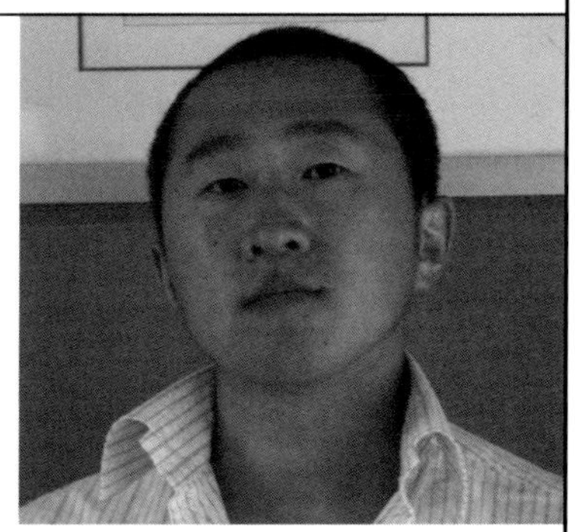

温明

2002年毕业于哈尔滨学院环艺专业，现任黑龙江国光建筑装饰设计院一所主任设计师。主要作品：一品制造坊、黑龙江省广播电视中心、哈尔滨市工人文化宫、吉林省宾馆。

朱岚

1999年以来先后任职于苏州金豹实业有限公司设计师、苏州金螳螂建筑装饰有限公司助理设计师、上海金螳螂环境设计研究有限公司项目经理、上海那格装饰设计有限公司总经理兼陈设策划。主要项目有：扬州迎宾馆总统楼、广州南美大酒店、上海green spa、南京招商局依云溪一期别墅样板房及会所整体等。

吴君华

2000年毕业于浙江工商职业技术学院，2003年创立翰禹·印加空间设计机构任总经理兼设计总监，擅长办公空间的设计。设计作品有：舟山华文周渔大酒店、宁波大红鹰职业技术学院、丽水庆元国际大酒店、镇海国家石油储备基地、余姚科学馆、杭州湾跨海大桥工程指挥部办公楼等。

吴立成

从事室内设计15年，有丰富的酒店和餐饮设计经验。

吴政道

出生于台湾，1984年复兴美术工艺科毕业。1998-2000年开展中国大陆地区业务，2000年定居上海。2001年与王镇华、谭精忠合伙于上海成立大隐室内设计公司。

吴宗勋

1989年毕业于台湾东海大学建筑系，曾先后任职台湾姚仁喜建筑师事务所设计师、浙江台亚建筑工程有限公司副总經理、昆山得利裝飾工程有限公司总经理，现任宁波市富华建筑装饰有限公司董事兼设计顾问。

伍昱

2002年-2004年，在东易日盛室内设计公司期间，专业设计高档家装和别墅。2004年至今，创办南京目达室内设计有限公司，任设计总监。CIID会员。

叶铮

CIID理事，IFI会员，中国饭店协会设计委员会常务理事，中国建筑装饰协会设计委员会会委员，美国室内设计学会（ASID）国际会员，上海应用技术学院副教授，上海泓叶室内装饰有限公司总设计师，首批入选美国INTERIOR DESIGN China《名人堂》。至今已获国内外专业设计奖项及荣誉28项。

谢智明

CIID佛山专业委员会副秘书长。个人获奖：2005年中国室内设计大赛酒店类优秀奖、2005年中国室内设计大赛商业类佳作奖、2006年中国(上海)国际建筑及室内设计节设计师走廊“金外滩”奖中荣获最佳材料运用奖。

徐 慧

毕业于辽宁省工艺美术学校，现任孙志刚空间设计工作室主案设计师。

许国维

高级室内设计师，2000年从事室内设计工作，曾在北京皇府井装饰公司担任设计师。以酒店、办公、娱乐场所等商业设计见长。2008年主要设计项目：上海朗地投资公司办公楼、南京泰州别墅、宁波江北教堂、宁波丽人KTV等。

许志国

南京师范大学美术学院毕业，南京光线装饰设计工作室设计师，中国建筑学会室内设计分会会员。

杨颜江

1992年毕业于上海轻工业高等专科学校，高级室内建筑师，杰出中青年室内建筑师，中国建筑学会室内设计分会会员，南京市室内设计学会会员。现任职苏州金螳螂第七设计研究院三所所长。

殷艳明

深圳市创域设计有限公司总经理兼设计总监、深圳长城家具装饰工程有限公司长城设计院副总设计师、 IFI深圳委员会委员、亚洲室内设计学会联合会深圳委员会委员、香港室内设计协会中国深圳代表处委员、CIID第三专业委员会委员、高级室内建筑师、深圳市室内设计师协会常务理事。

于强

先后毕业于吉林师范学院美术系和中央工艺美院环艺系。1999年组建于强室内设计师事务所。2001年在亚太室内设计大奖赛上获奖，2002年在深圳何香凝美术馆成功举办个人室内设计作品展，2005年在深圳关山月美术馆参加以“深圳-设计师-我们”为主题的深圳设计师十人十年联展。

余平

西安电子科技大学副教授，中国建筑学会室内设计分会理事。

曾建龙

毕业于福州大学工艺美术学院和美国国际联盟大学艺术设计学院，CIID会员、室内建筑师、国际室内建筑师与设计师理事会理事。荣获奖项有2005年海峡两岸三地别墅空间设计三等奖、2007年桑格拉杯首届中国建筑学会室内设计分会大奖赛工程类一等奖及方案类二等奖、金羊奖，2007年度中国十大设计师奖等。

张波

广州美术学院设计专业毕业后从事电影美术指导工作，90年代移居美国，于加州斯坦福大学深造设计专业，90年代末回国发展，成立张波设计工作室。作品散见于国内外多种技术及时尚杂志：《装点》、《室内设计与装修》、《建筑设计及技术》、《城市住宅》、《时尚》等。作品《金钻潮庭》曾获得中国酒店饭店业协会2005年度优秀设计奖。

上瑞元筑设计制作有限公司成立于2001年,由原作设计工作室与三逸潜流环境艺术设计有限公司组合而成。发展至今,已建立起由5位主案设计师领导下的30余人规模的专业工作团队,服务内容涵盖了以星级酒店、大型餐饮及时尚娱乐业为主的商业空间。2006年江南大学设计学院环境艺术系选定本公司为实习基地。2006年6月成立江苏省建筑装饰设计研究院无锡分院。

张成喆

国际建筑协会（ICU）中国区常务理事，1993年成立个人设计事务所，1997年旅居巴黎，1999年定居上海，成立IADC国际涞澳设计公司，2007年出版个人作品集《喆思空间》。多次荣获国内外重要设计奖项：2006年度亚太精英设计师奖、2008年度设计师网络传媒奖、2008年度金外滩最佳居住空间奖等。

张奇永

1999年就读于黑龙江建筑职业技术学院，室内建筑师。现任哈尔滨唯美源装饰设计有限公司设计师兼摄影师。主要作品：现代粗粮（大连）、大连湾海鲜酒店（锦州）、现代食尚（大连）、人和美容美体会馆（哈尔滨）、女王美容SPA馆（哈尔滨汉水路店）等。

张星

香港东仓设计策划顾问有限公司董事兼首席设计师、中国杰出中青年室内建筑师、中国最具影响力室内建筑师、2007金羊奖中国十大设计师。荣获奖项有：2001年 Singapore ommercial Space-The Second Place、首届海峡两岸三地室内设计大赛特等奖、2007年现代装饰国际传媒奖年度杰出设计师等。

张晔

毕业于重庆建筑大学建筑学室内设计专业，清华大学建筑学院工程硕士，任中国建筑设计研究院环境艺术设计研究院副总建筑师。曾获中国建筑装饰协会评选的"杰出青年室内建筑师" 称号,"全国百名优秀室内设计师"称号,2004年"INTERIOR DESIGN CHINA"杂志十大封面人物。

东仓"D5"组

东仓"D5"组是香港东仓设计策划顾问有限公司应市场需求而成立的精英组，该组成员由公司拔尖设计师、创意总监余霖、创作助理李杰智、执行助理杨锋组成。

莫意波

毕业于浙江工商职业技术学院环境艺术系，现任职宁波市富华建筑装饰有限公司设计师。

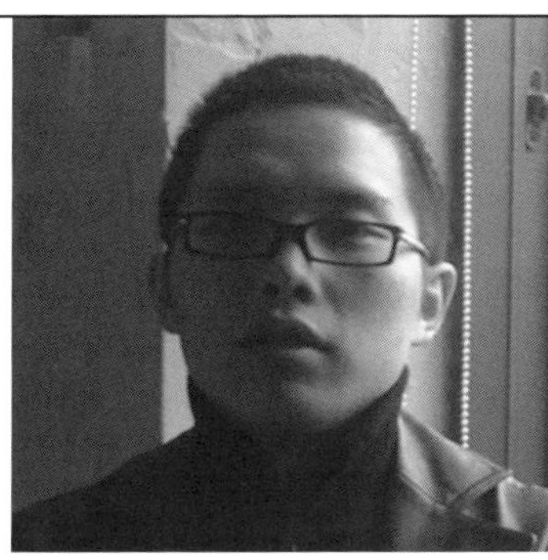

竺春光

2002年毕业于浙江工商职业技术学院环境艺术系，2002年就职于昆山得利装潢有限公司设计师，现任宁波市富华建筑装饰有限公司董事兼設計总监。

仲松

1999年毕业于中央美术学院雕塑系。作品"东方之光"，获上海浦东开放开发十年精品项目雕塑金奖、"艺术与科学"，获国际艺术大展评委特别奖，出版物有《中国艺术的进程-70年代后出生的艺术家》、《中国当代雕塑全集》、《中国当代艺术年鉴》。演讲过《艺术家的公众意识》、《公共艺术在新时代的转变》。

编委（排名不分先后）
陈耀光、陈南、高蓓、胡文杰、黄玉枝、满登、张乘风
孙天文、沈雷、宋桥林、王琼、王兆明、吴海燕、叶铮

主编
陈卫新

国书在版编目（CIP）数据

2008中国室内设计年鉴. 2 /《中国室内设计年鉴》编委会编. —沈阳：辽宁科学技术出版社，2008.11
ISBN 978-7-5381-5585-3

Ⅰ.2… Ⅱ.中… Ⅲ.室内设计—中国—2008—年鉴
Ⅳ.TU238-54

中国版本图书馆CIP数据核字（2008）第151005号

出版发行：辽宁科学技术出版社
（沈阳市和平区十一纬路29号 邮编：110003）
印 刷 者：上海美雅延中印刷有限公司
经 销 者：各地新华书店
幅面尺寸：230mm × 300mm
印　　张：87
字　　数：300千字
插　　页：8
印　　数：1~2000
出版时间：2008年11月第1版
印刷时间：2008年11月第1次印刷
责任编辑：陈慈良
封面设计：赵宝伟
版式设计：赵宝伟
责任校对：周　文

书　　号：ISBN 978-7-5381-5585-3
定　　价：498.00元 (1) (2) 册

联系电话：024-23284360
邮购热线：024-23284502
E-mail:lkzzb@mail.lnpgc.com.cn
http://www.lnkj.com.cn